# Strategies to Overcome Superbug Invasions:

## Emerging Research and Opportunities

Dimple Sethi Chopra
*Punjabi University, India*

Ankur Kaul
*Institute of Nuclear Medicine and Allied Sciences, India*

A volume in the Advances in Medical Diagnosis, Treatment, and Care (AMDTC) Book Series

IGI Global
PUBLISHER of TIMELY KNOWLEDGE

Published in the United States of America by
IGI Global
Medical Information Science Reference (an imprint of IGI Global)
701 E. Chocolate Avenue
Hershey PA, USA 17033
Tel: 717-533-8845
Fax: 717-533-8661
E-mail: cust@igi-global.com
Web site: http://www.igi-global.com



Library of Congress Cataloging-in-Publication Data

Names: Chopra, Dimple Sethi, 1973- editor | Kaul, Ankur, 1979- editor
Title: Strategies to overcome superbug invasions : emerging research opportunities / Dimple Sethi Chopra and Ankur Kaul, editors.
Description: Hershey, PA : Medical Information Science Reference, [2020] | Includes bibliographical references. | Summary: "This book examines current research and potential strategies to overcome the emergence and re-emergence of drug resistant pathogenic microbial strains"--Provided by publisher.
Identifiers: LCCN 2019022166 | ISBN 9781799803072 (hardcover) | ISBN 9781799803096 (ebook)
Subjects: MESH: Drug Resistance, Microbial | Drug Resistance, Multiple | Complementary Therapies
Classification: LCC QR46 | NLM WB 330 | DDC 616.9/041--dc23
LC record available at https://lccn.loc.gov/2019022166

This book is published in the IGI Global book series Advances in Medical Diagnosis, Treatment, and Care (AMDTC) (ISSN: 2475-6628; eISSN: 2475-6636)

British Cataloguing in Publication Data
A Cataloguing in Publication record for this book is available from the British Library.

# Advances in Medical Diagnosis, Treatment, and Care (AMDTC) Book Series

ISSN:2475-6628
EISSN:2475-6636

**MISSION**

Advancements in medicine have prolonged the life expectancy of individuals all over the world. Once life-threatening conditions have become significantly easier to treat and even cure in many cases. Continued research in the medical field will further improve the quality of life, longevity, and wellbeing of individuals.

The **Advances in Medical Diagnosis, Treatment, and Care (AMDTC)** book series seeks to highlight publications on innovative treatment methodologies, diagnosis tools and techniques, and best practices for patient care. Comprised of comprehensive resources aimed to assist professionals in the medical field apply the latest innovations in the identification and management of medical conditions as well as patient care and interaction, the books within the AMDTC series are relevant to the research and practical needs of medical practitioners, researchers, students, and hospital administrators.

**COVERAGE**

- Disease Management
- Diagnostic Medicine
- Chronic Conditions
- Critical Care
- Internal Medicine
- Experimental Medicine
- Emergency Medicine
- Patient-Centered Care
- Medical Testing
- Patient Interaction

IGI Global is currently accepting manuscripts for publication within this series. To submit a proposal for a volume in this series, please contact our Acquisition Editors at Acquisitions@igi-global.com or visit: http://www.igi-global.com/publish/.

The Advances in Medical Diagnosis, Treatment, and Care (AMDTC) Book Series (ISSN 2475-6628) is published by IGI Global, 701 E. Chocolate Avenue, Hershey, PA 17033-1240, USA, www.igi-global.com. This series is composed of titles available for purchase individually; each title is edited to be contextually exclusive from any other title within the series. For pricing and ordering information please visit http://www.igi-global.com/book-series/advances-medical-diagnosis-treatment-care/129618. Postmaster: Send all address changes to above address. 

## Titles in this Series

*For a list of additional titles in this series, please visit: http://www.igi-global.com/book-series/advances-medical-diagnosis-treatment-care/129618*

***Cases on Applied and Therapeutic Humor***
Michael K. Cundall Jr. (North Carolina A&T State University, USA) and Stephanie Kelly (North Carolina A&T State University USA)
Medical Information Science Reference • © 2021 • 267pp • H/C (ISBN: 9781799845287) • US $255.00

***Artificial Intelligence for Accurate Analysis and Detection of Autism Spectrum Disorder***
Sandeep Kautish (Lord Buddha Education Foundation, Nepal & Asia Pacific University, Malaysia) and Gaurav Dhiman (Government Bikram College of Commerce, India)
Medical Information Science Reference • © 2021 • 305pp • H/C (ISBN: 9781799874607) • US $295.00

***Diagnostic and Treatment Methods for Ulcerative Colitis and Colitis-Associated Cancer***
Ashok Kumar Pandurangan (B. S. Abdur Rahman Crescent Institute of Science and Technology, India)
Medical Information Science Reference • © 2021 • 309pp • H/C (ISBN: 9781799835806) • US $295.00

***Enhancing the Therapeutic Efficacy of Herbal Formulations***
Rajesh Kumar Kesharwani (Nehru Gram Bharati (Deemed), Prayagraj, India) Raj K. Keservani (Faculty of B. Pharmacy, CSM Group of Institutions, Prayagraj, India) and Anil K. Sharma (GD Goenka University, India)
Medical Information Science Reference • © 2021 • 367pp • H/C (ISBN: 9781799844532) • US $265.00

***Exploring the Efficacy and Beneficence of Complementary and Alternative Medicinal Products in Disease Therapy***
Etetor Roland Eshiet (Sustainable Energy Environmental and Educational Development (SEEED), USA)
Medical Information Science Reference • © 2021 • 315pp • H/C (ISBN: 9781799841203) • US $295.00

IGI Global
PUBLISHER of TIMELY KNOWLEDGE

701 East Chocolate Avenue, Hershey, PA 17033, USA
Tel: 717-533-8845 x100 • Fax: 717-533-8661
E-Mail: cust@igi-global.com • www.igi-global.com

# Editorial Advisory Board



# List of Reviewers

Rahul Mehta, *Max Planck Research Laboratory, Poland*
Ashu Mittal, *KIET School of Pharmacy, Ghaziabad, India*
Sahil Moza, *National Center for Biological Sciences, India*
Sanjay Patil, *Pune University, India*
Sujata P. Sawarkar, *Mumbai University, India*
Shelley Sehgal, *Central University of Jammu, India*
Pranav Shah, *Uka Tarsadia University, India*
Dilpreet Singh, *Maharaja Ranjit Singh State Technical University, India*
Herratdeep Singh, *Indian Council of Medical Research, India*
Sachin K. Singh, *Lovely Professional University, India*
Richa Sood, *Dayanand Sagar University, India*

# Table of Contents

# Foreword

It is a privilege for me to write the foreword for the *Strategies to Overcome Superbug Invasions: Emerging Research and Opportunities*. In today's world, antibiotic resistance in bacteria is an emerging public health problem and actually a crisis. Some bacteria have evolved to survive the most effective antibiotics in the clinical setting leading to death of increasing number of people. These bacteria are mentioned as superbug metaphorically and they can invade living tissues, systems and also our environment. The loss of effective antibiotics nowadays weakens our ability to fight infections caused by these superbugs. Recent studies analyzing the antibiotic pipeline indicates there aren't enough antibiotics in development to meet current needs. Therefore, we need novel strategies to fight with antibiotic resistant bacteria. In this book, the researchers have evaluated the current literature to probe into the antibiotic resistance problem and to review the promising approaches to overcome this public health problem. I hope, this book will contribute to this aim and will help the researchers in this field.

*Mümtaz Güran*
*Eastern Mediterranean University, Turkey*
*1st June 2021*

# Preface

*"Declare the past, diagnose the present, foretell the future." – Hippocrates*

Technology in this 21st century has hugely impacted our lives in almost everything we do today. It has assisted us as humans to make 'smart' advances; our ancestors could only dream about. As we utilize intelligent devices and tools to improve our quality of life, bacteria have also evolved rapidly to become quite 'smart.' These superbugs can cause antimicrobial resistance (AMR), which is one of the reasons for failed antimicrobial chemotherapy. In fact, the superbugs have outpaced have our ability to get newer antibiotics to the market. Antibiotics had always been considered as one of the most relevant discoveries in the last 100 years. Unfortunately, the rising occurrence of AMR has caused the down sliding of our most existing antibiotics.

The phenomenon of AMR develops when bacteria acclimatize and grow in the presence of antibiotics. When many antibiotics belong to a common class of anti-bacterial agents, become resistance to a specific antibiotic drug can ultimately lead to resistance to an entire class of the bacteria. Once resistance develops in one class of organism, it can unpredictably and rapidly spread to another creature/species. Hence, it becomes increasingly challenging to treat commonly seen bacterial infections in clinical settings. Globally, it is reported that there is loss of 7 lakh lives to AMR annually, and this figure could escalate to 10 million by 2050. Undoubtedly, AMR was considered one of the most severe current threats to the global health after HIV/AIDS, Ebola and non-communicable diseases in the pre COVID-19 era.

Moreover, now, the recent COVID-19 experience has been fairly dreadful to our global public health systems that were quite ill prepared to tackle a new unpredictable viral strain. The irrational use of antimicrobial agents during COVID-19 pandemic times will further be a source for AMR snowballing.

Hence, it becomes imperative to study about development of AMR, the challenges to overcome these superbugs. This book has been structured into eleven chapters. It includes the development of novel approaches in searching for new antimicrobials, designing more effective preventative measures, and, essentially, enhanced understanding of bacterial ecosystems and antibiotic resistance. The development of conceptual frameworks created on current expansions in the field of diagnosis of systemic bacterial disorders, and application of in-silico methodologies to pace up the speed of drug discovery for COVID-19 are also discussed herein.

We wish to thank all the contributors for their engagement and collaboration which played a monumental role in bringing this book to fruition. To conclude, we are also grateful to the publisher, IGI Global for sincere co-operation throughout the project.

We hope that this book will be enlightening source for, clinicians, and other healthcare professionals; microbiologists and students in the life sciences; and scientists and decision makers in medical research.

*Dimple Sethi Chopra*
*India*

*Ankur Kaul*
*India*
*20th May 2021*

# Acknowledgment

We express our sincere gratitude towards the authors and their respective collaborators for their scientific contributions to this book entitled *Strategies to Overcome Superbug Invasions: Emerging Research and Opportunities*.

We are also highly grateful to the Editorial Advisory Board members for their commendable suggestions and inputs for improving the scientific content of this book. In fact, the reviewers had worked tirelessly in improving the quality of this publication.

The patience and co-operation of the entire IGI Global publishing team, who were directly or indirectly involved in the fruition of this book, is highly appreciated.

It goes without saying that we are sincerely indebted to our families for their unconditional support, either morally or physically, throughout the successful journey of this book.

Last but not least, we bow down to the power of Almighty for His blessings.

# Section 1

# Challenges and Antimicrobial Resistance (AMR): A Diagnostic Perspective

*This section consists of chapters that contain an introduction to AMR and current trends of diagnosis of bacterial infection with special emphasis on mycobacterium tuberculosis.*

# Chapter 1
# Resistance Phenotypes and Surveillance

**Shahrukh Khan**
*Center for Disease Dynamics, Economics, and Policy, India*

**Jyoti Joshi**
*Center for Disease Dynamics, Economics, and Policy, India & Amity Institute of Public Health, Amity University, India*

## ABSTRACT

*The emergence of drug resistance complicates surveillance and treatment of antimicrobial phenotypes. For example, the rise of Methicillin-resistant Staphylococcus aureus and carbapenem-resistant Enterobacteriaceae influence delivery of care. Moreover, a lack of surveillance programs in most of the developing world exacerbates the problem of MDR. Existing studies in humans are mostly retrospective single-center surveillance-based studies that look at the molecular makeup and prevalence of phenotypic resistance for several pathogens. Very few studies examined infection prevention measures or antimicrobial stewardship activities, and of those that did, none of them were multicenter. The aim of this chapter is to explore prevalent phenotypes in clinical settings and antimicrobial resistance (AMR) surveillance programs throughout the world.*

## INTRODUCTION

Antimicrobial resistance (AMR) threatens to upend decades of progress on managing infections. Antimicrobial drugs have saved life on earth from once-fatal diseases and risky procedures like surgeries and transplants, but their systematic misuse and overuse have transformed the bacterial population and limited the effectiveness

DOI: 10.4018/978-1-7998-0307-2.ch001

of these miracle drugs. Moreover, the interconnectedness of biomes has made the exchange of bacteria in the environment, animals and humans much easier. The challenge of AMR has led many experts to reconsider the effectiveness of current antibiotic therapies over the next 100 years with a dwindling pipeline of new drug research and development (R&D). Antibiotic usage outside of human medicine, like in agriculture for growth promotion creates another entry for antibiotics to enter our biome through our environment (Diallo, 2020). The broad scope of drug resistance requires effective collaboration among disciplines and countries to build robust surveillance program health outcomes. Similar plans to educate stakeholders and surveil infections to prevent antimicrobial resistance through hygienic infection control practices would sustainably optimize use of antimicrobial medicines in humans and animals. monitor antimicrobial resistant phenotypes to inform interventions. A "one health" approach, like the one proposed by the World Health Assembly in 2015, outlines an action plan that encourages collaboration among actors in human and veterinary medicine, finance, agriculture, environment and consumers. By creating programs, policies and research through multiple sector collaboration, a "one health" approach can achieve better public well being (GLASS WHO Report, 2017).

*Figure 1. The One Health triad*

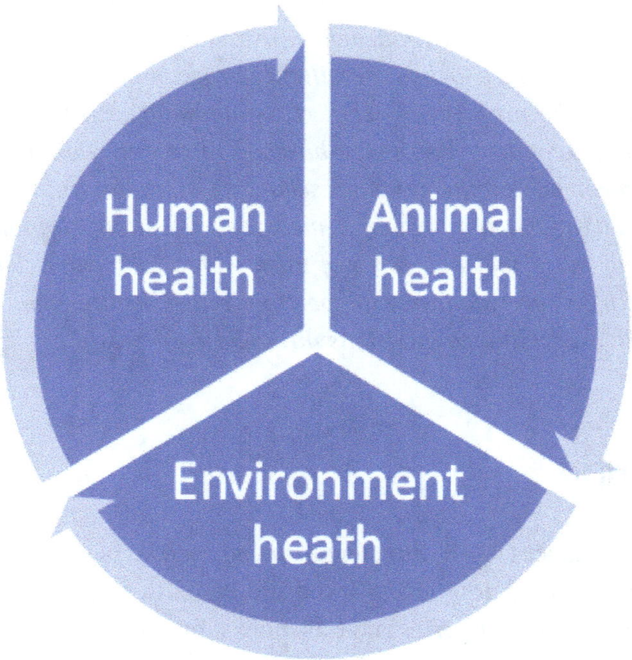

*Figure 2. AMR within the one health ecosystem, where One Health*
*Photo credit: Dr Sabiha Essack, UKZN South Africa and Paula Cray, NC state Veterinary medicine*

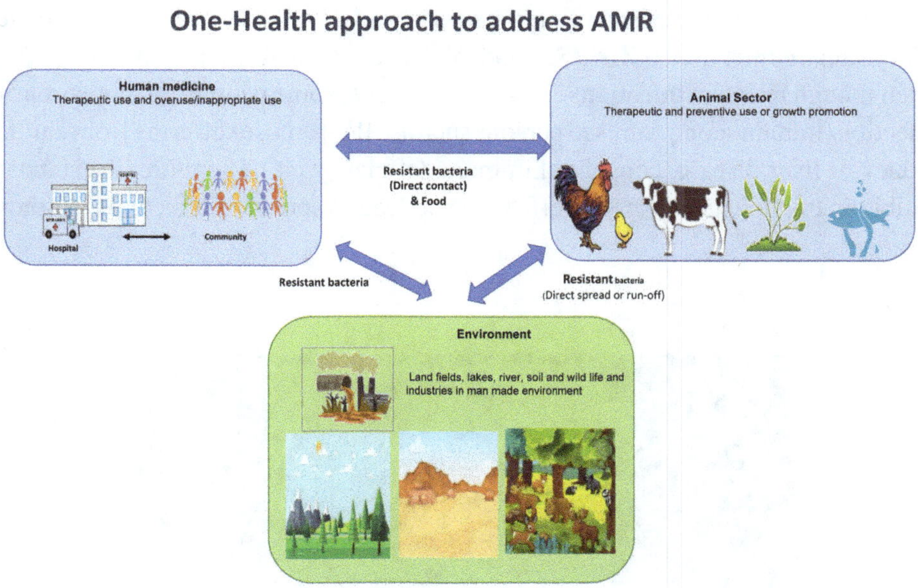


The emergence of drug resistance complicates surveillance and treatment of antimicrobial phenotypes. For example, the rise of *Methicillin-resistant Staphylococcus aureus* (MRSA) and *carbapenem-resistant Enterobacteriaceae* (CRE) exacerbate delivery of care. Most at risk to these infections are those living in developing countries, which often lack the surveillance programs necessary to track and address these infections. Additionally, very few research studies in these countries examine infection prevention measures or antimicrobial stewardship activities, and of those that do, none are multicenter (Gandra, 2017). The aim of this chapter is to explore prevalent phenotypes in clinical settings and antimicrobial resistance (AMR) surveillance programs throughout the world.

## What is AMR Resistance and how Does it Develop?

Antimicrobial resistance, or AMR, refers to a microorganism's (like bacteria, viruses, and some parasites) ability to prevent an antimicrobial (like antibiotics, antivirals, and antimalarials) from being effective. Antibiotic resistant infections are associated with higher rates of medical costs and hospital stays. Accordingly, Healthcare acquired infections (HAIs) or nosocomial infections are a concerning type infection that patients get while receiving treatment for medical or surgical conditions at hospitals and long-term facilities like nursing homes. These are infections that are unrelated to the original illness and appear within 30 days after patient discharge (Revelas, 2012). Among the most prevalent types of HAI's include infections from MRSA, Carbapenem- resistant Enterobacteriaceae (CRE), Central Line Associated Bloodstream Infection (CLABSI) and Ventilator-associated pneumonia (VAP). Even though resistant infections can impact anyone, some people are more prone to infection (immunocompromised persons such as HIV+, those suffering from chronic kidney or liver disease, congenital immunodeficiency etc).Therefore, the standard antibiotic treatments may not be effective and lead to spread of infections to others.

*Figure 3.*

*HOW TRANSMISSION WORKS:*

*THE TRANSMISSION OF ANTIMICROBIAL RESISTANCE IS A TWO-STEP PROCESS: (LANCET PETER)*

- *IT CAN START WITH EITHER DE-NOVO MUTATIONS OR AN ACQUISITION OF RESISTANCE GENES (OR CAN BE BOTH) VIA MOBILE GENETIC ELEMENTS LIKE PLASMIDS THAT CAN MEDIATE AN ANTIMICROBIAL-RESISTANCE PHENOTYPE.*
- *THERE IS A SPREAD OF RESISTANT STRAINS, OR CONTAGION, THAT CAN HAPPEN THROUGH MANY VECTORS (EXAMPLES: HUMANS, BIRDS, WATER, INSECTS, AND AGRICULTURE)*

## What is Phenotypic Resistance?

When it comes to antibiotic resistance, the conversation typically involves infections caused by a genetic alteration through the acquisition of genetic genes or mutations. However, there are situations where antibiotic resistance can come about without any genetic change. This phenomenon of phenotypic resistance, or a non-inherited form of resistance, can be seen via growth in biofilms, not in classical clinical susceptibility tests at microbiology laboratories Corona, 2013). Recent research has highlighted the importance of bacterial metabolism on susceptibility of antibiotics (Corona, 2013). This illustrates the important role antibiotics play as inter-microbial signaling molecules, regulators of gene expression and mediators of host immune response. A better understanding of these processes is essential to develop novel therapeutic solutions based on bacterial susceptibility.

## History of AMR

To understand the rise of phenotypic resistance, it is imperative to revisit the origins of antibiotics. Famously, Sir Alexander Fleming discovered penicillin through a failed experiment 1928 (Ban, 2006). His discovery was a useful treatment for all kinds of bacterial infections and eventually opened the gate for fourteen other classes of antibiotics for clinical and veterinary use (Yap, 2013). Together, these innovations became the miracle drugs of the twentieth century and treated life-threatening infections for the first time. However, extensive use of antibiotics over the last few decades have dramatically reduced the efficacy of these drugs and contributed to the rise of antibiotic resistance.

Bacteria commonly acquire resistance to an antibiotic through gene mutations or by transferring genetic materials with other bacteria. They also possess other characteristics that limit portals of entry for antibiotics. These include: using enzymes to break down antibiotics; "efflux pumps" that reduce antibiotic concentrations to ineffective levels; changes to the bacterial membrane that prevent antibiotics from entering the cell, thereby reducing a core mechanism by which antibiotics like Penicillin's work (Rosenblatt-Farrell, 2009).

Bacterium that exhibit more than one of the functions above are likely resistant to more than one type of antibiotic and are known to confer multi-drug resistance. Yet, the implications of antibiotic misuse were predicated long before the first case of penicillin resistant *Staphylococcus aerus* by the very man that had discovered it. Upon winning the Nobel Prize for his discovery in 1945, Fleming said, "The time may come when penicillin can be brought by anyone in the shops. Then there is the danger that the ignorant man may easily underdose himself and by exposing his microbes to non-lethal quantities of the drug make them resistant (Yap, 2013). Not surprisingly, within 10 years of widescale production of penicillin, penicillin resistance to *Staphylococcus aureus* began to emerge (Rosenblatt-Farrell, 2009).

## Phenotypic Resistance

Similarly, other types of antibiotic resistance began to show after the development of new antibiotics and spread quickly throughout the world. For example, infections from Methicillin- resistant *Staphlyoccoccus aureus* (MRSA) is one of the most pressing healthcare-acquired infection today. MRSA resistance began in the United Kingdom on 1961, two years after the discovery of methicillin (Molton, 2013). Over the next few decades, MRSA spread first to North America and then to Europe and

Asia (Molton, 2013). Today, rates of MRSA differ around the world. The percentage of MRSA isolates in Europe range from 0.9% in the Netherlands to 30.7% in Greece (Curtis, 2015). Likewise, a study that looked at 15 tertiary hospitals within India found that rates of MRSA varied; western India presented a 25% MRSA rate compared to 50% in South India (Joshi, 2013). The spread of MRSA is similar to the transmission of other infections in the 20th century like *Vancomycin-resistant enterococci* and *Klebsiella pneumoniae carbapenemase–producing K. pneumoniae*.

Fortunately, rates of MRSA infections are falling in Europe and the United States primarily due to improved infection control, screening and restrictions on some antibiotics (Molton, 2013). On the other hand, multi-drug resistant MRSA has become endemic in many Asian countries. Approximately 28% of clinical S. aureus isolates in Hong Kong and Indonesia and 70% contained MRSA, while two studies in India found MRSA rates to be 41% to 45% (Chen, 2014).

Another concerning bacteria is *Carbapenem-resistant Enterobacteriaceae*, which is resistant to the last resort Carbapenem class of drugs. Its rates vary across the world; from 6% in United States to 14% in China and 46% in India (resistance map). Similarly, *Acinetobacter baumannii* presents its own share of challenges with a carbapenem resistance rate of 30% in the US, 82% in China and 73% in India. On the other hand, *Pseudomonas aeruginosa* is a multi-drug resistant infection with carbapenem resistance rates of 10% in US, 25% in China and 43% in India.

## Factors Contributing to AMR

There are many factors that create favorable environments for the exchange of bacteria. Many of these resistant organisms develop through inappropriate use of antibiotics and are often spread through the hands of healthcare workers (Multidrug Resistant Organisms MDROs What Are They). Objects like medication carts, bedside tables and IV poles are other ways in which these resistant organisms can be transmitted, in addition to direct contact like sores (Multidrug Resistant Organisms MDROs What Are They). Hand hygiene is the simplest way to reduce the spread of infections, but there is a lack of a compliance among healthcare providers throughout the world (Pittet, 2009). Consequently, poor hand hygiene is the leading cause of spread of multi- resistant organisms (Pittet, 2009). Several studies over the past three decades have found that implementation of a hand washing program notably reduced the percentage of patients colonized with *Klebsiella pneumoniae*, hospital acquired MRSA, urinary tract infections and other hospital acquired infections (Pittet, 2009). For example, a study done in a Russian neonatal unit estimated that one health care-associated infection would cost US$1100, which could cover 3265 patient- days of hand antiseptic (Pittet, 2009). Most importantly, hand washing programs are the most cost-effective and efficient way of limiting the spread of germs and bacteria.

Similarly, rising incomes, poor access to quality primary health care, and easy over-the-counter access to antibiotics has further led to increasing rates of resistance (Singh, 2017). Between 2000-2015, global consumption of antibiotics rose by 65% and doubled in low and middle- income countries (Klein, 2018). Among the most at risk of drug resistance are key populations like newborns suffering from sepsis and patients requiring antibiotics to prevent infection during surgery, organ transplant care or cancer treatment (Saha, 2018). This creates a dilemma where more people still die because of lack of access to antibiotics than from resistant infections.

Unrestricted and inappropriate use of antibiotics results in more infections that are tougher and more expensive to treat (Laxminarayan, 2016). With increased antibiotic use comes a greater risk to acquire a multi-drug resistant pathogen which directly correlates with prolonged hospital stay, exposure to infected or colonized patients, and receiving broad-spectrum antimicrobial agents, especially third-generation cephalosporin. It is important to comprehensively manage these risk factors to limit AMR.

Identifying risk factors for MDR is also important to diagnose the underlying factors of antimicrobial infection. Patients with preexisting conditions like diabetes, skin lesions, and previous colonization with a multi-drug resistant infection are at a higher risk of MDR. Previous prolonged use of antibiotics or hospital stays are other risk factors that can cause a variety of infections from skin to urinary tract. Once infected, multi-drug resistant infections are difficult to treat because they are unresponsive to many common and powerful antibiotics. Treatment with an inappropriate antibiotic can also make the infection difficult to cure, which underscores the importance of patient and prescriber education on antibiotic use (Multidrug Resistant Organisms MDROs What Are They).

Socioeconomic inequality is another major cause of low health among the poor. This type of inequality amplifies the perception that some people are worth more than others, fueling a sense of superiority among the rich and subordination amongst the poor. In the same way, antimicrobial resistance is linked to poverty. A country's quality of governance, public spending on health per GDP and education are known to affect health outcomes and are likely to cause antibiotic resistance as well (Collignon, 2018). For example, in Europe, poor governance and corruption are as closely associated with variances in antimicrobial resistance levels between countries as are antibiotic consumption patterns (Collignon, 2018). It is estimated that antimicrobial resistance could push 28 million people into extreme poverty by 2050 without effective measures to control AMR (Adeyi et al., 2017).

## Surveillance

Surveillance data on drug resistance is extremely important. It can help assess disease burden, improve diagnostic and treatment and monitor the effectiveness and design of control programs, which in turn helps create a sufficient public health response (Zignol, 2018). However, many low- and middle-income countries (LMIC) do not have the capacity for routine testing for patients' resistant to drugs and instead rely on surveys to get estimates on drug resistance. These gaps in surveillance contribute to greater burden of AMR because healthcare professionals lack the data to make effective treatment decisions and preserve antibiotics. This has led the World Health Organization (WHO) to establish the Global Antimicrobial Resistance Surveillance System (GLASS) to monitor commonly reported resistant bacteria in high, middle and low- income countries. The purpose of GLASS was to collect AMR data from 40+ partner countries in order to report health outcomes and surveillance capacities across countries. The first report (WHO GLASS report, 2017) described high rates of resistance across the world for select bacteria like Acinetobacter spp., Escherichia coli, Klebsiella pneumoniae, Neisseria gonorrhoeae, Salmonella spp., Shigella spp., Staphylococcus aureus, and Streptococcus pneumoniae. This functional and comprehensive global surveillance system provides reliable data across socio-political and economic contexts (Iskandar et al., 2021).

To make a greater, coordinated effort to provide practical guidance on antibiotics across countries, the United Nations established the Interagency Coordination Group on Antimicrobial esistance (IACG) was created in consultation with the World Health Organization (WHO), Food and Agriculture Organization (FAO) and World Organization for Animal Health (OIE). In a recent discussion paper, the agency highlighted the importance of an effective surveillance program that covers human, animal, food and plant populations. Such a system would provide easily comparable data that can be exchanged across nations. Many low-income countries, in comparison, lack the capability to establish and maintain surveillance systems to collect data on AMR, which means they are unable to use data to study trends that can inform effective policies and programs.

*Figure 4. Antibiotic Resistance of Escherichia Coli (Resistance Mapping Charts)*

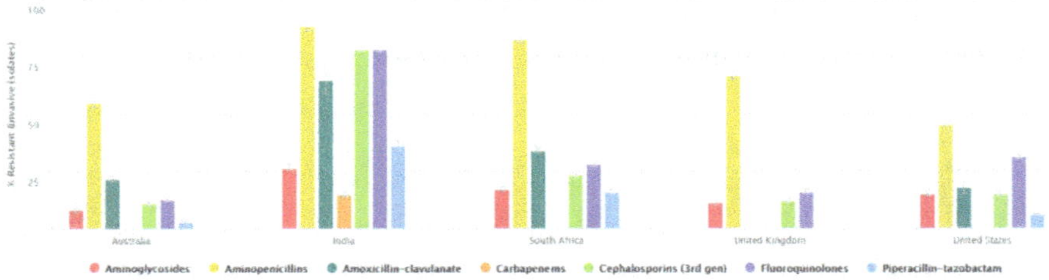

*Figure 5. Antibiotic Resistance of Acinetobacter baumannii*

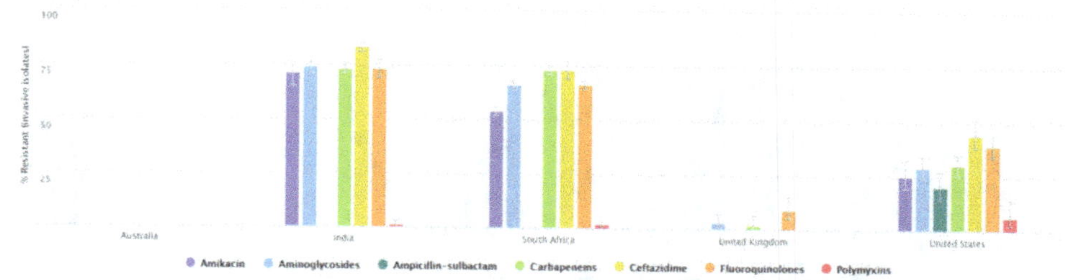


*Figure 6. Antibiotic Resistance of Klebsiella pneumoniae*

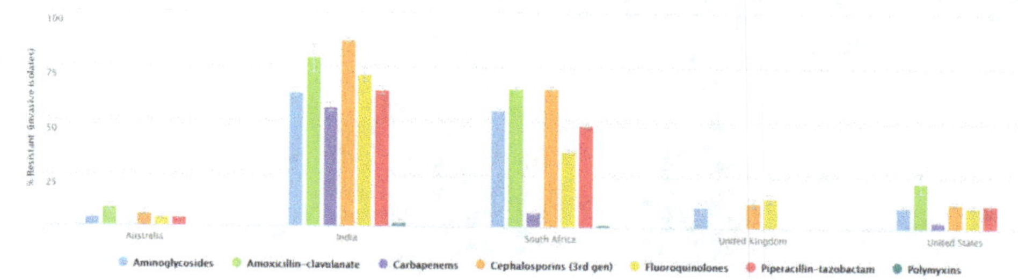


*Figure 7. Antibiotic Resistance of Salmonella paratyphi*

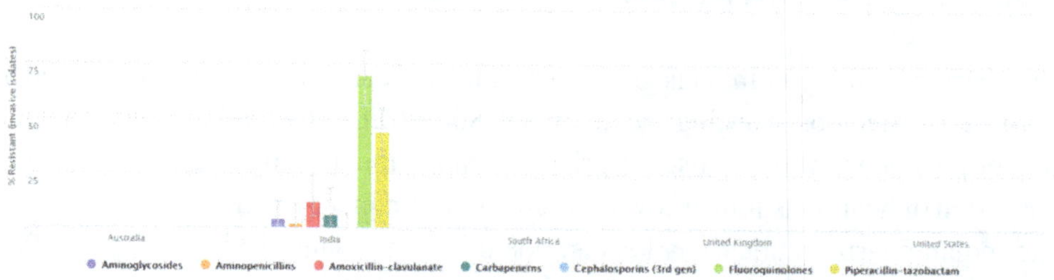


*Figure 8. Antibiotic Resistance of Salmonella typhi*

*Figure 9. Antibiotic Resistance of Staphylococcus aureus*

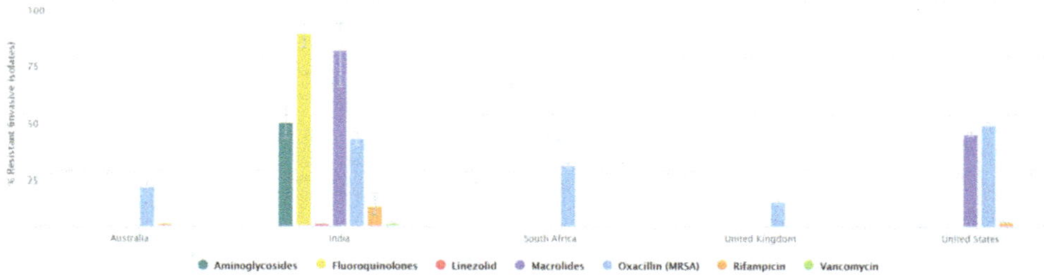


*Figure 10. Antibiotic Resistance of Streptococcus pneumonia*
*All charts courtesy: resistance map from www.cddep.org*

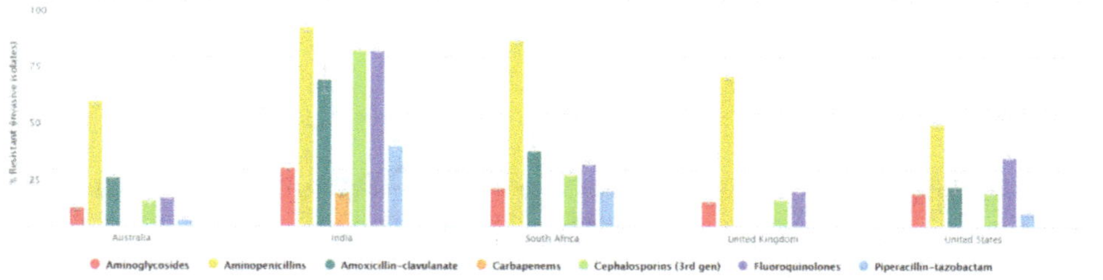


## ROLE OF PHARMA AND ANTIBIOTICS

The pharma industry in India is growing rapidly and can deliver some of the needed innovation in diagnostic testing and treatment. Some factors for market growth include greater prevalence chronic diseases, rising income, modern medical infrastructure, expansion of health insurance coverage, and more frequent patent product launches. The Indian pharmaceutical market can potentially grow up to $55 billion by 2020 because of greater affordability and market access (McKinsey, 2020). Local players are especially well-positioned to lead because of their dominance in formulation development capabilities and early investments (McKinsey, 2020).

Yet, bottlenecks in drug discovery have added to antimicrobial challenges. Pharmaceutical companies have been reluctant to invest in antibiotic research and development because of low returns, not to mention the how difficult it is to predict when resistance can rise to an antibiotic. Therefore, the government and the pharmaceutical industry must do more to develop accessible diagnostic tools for healthcare providers to use if drug resistance is to be prevented. One solution could be for the government to provide grants or tax incentives to companies that develop better technological innovations to calculate the prevalence of and prevent antimicrobial infections (Plackett, 2020).

## CURRENT PROGRESS AND UPDATES

In this section, standard practices throughout the healthcare industry will be covered, along with an overview of gaps and other potential solutions. For example, hospitals and other healthcare facilities have rules in place to prevent MDR infections. As mentioned earlier, one of the most effective rules is hand washing. Healthcare workers are encouraged to wash their hands with soap and warm water or alcohol-based hand sanitizer when handling protective clothing, after touching surfaces, and before and after every patient (Multidrug Resistant Organisms MDROs What Are They). Antibiotic control and private rooms for patients are other effective precautions.

Greater innovations in health care can improve AMR surveillance. For example, genomic sequencing is a valuable tool for drug resistance surveillance. It completes drug susceptibility testing in two rounds; it starts with first-line drugs and then moves to second-line drugs (Zignol, 2018). Genomic testing runs around US$130 on average per sample. The emergence of rapid molecular tests like GeneXpert MTB/RIF are improving access to susceptibility tests (Zignol, 2018). Greater affordability for these tests can expand testing and revolutionize public health.

For surveillance efforts to work, robust foundations are necessary. Therefore, it is essential for DNA extraction methods and recording of nomenclature and data to be standardized for stronger surveillance (Zignol, 2018). Likewise, targeted gene or whole-genome sequencing on sputum samples directly can overcome the need for culturing through a standardized approach (Zignol, 2018). When genomic sequencing is standardized and economically feasible, it could be a powerful tool in countries with low lab and referral capacity to better monitor antimicrobial resistance.

## FUTURE TRENDS AND CONCLUSION

An important measure of success for antibiotic control would be how cost-effective genetic First sequencing, which in turn would improve testing for antimicrobial resistance by reducing lab work load through cost savings. Already, the price of genetic sequencing is lower than some and second line phenotypic testing of drugs and grouping specimens to sequence several isolates in one single run offers additional savings (Zignol, 2018). To expand genomic sequencing in low-income countries, molecular biology and bioinformatic skills need to be developed along with mentoring from supranational reference labs. Other models of antibiotic stewardship should also be adopted. For example, the United States Center for Disease Control approaches antibiotic resistant through a surveillance-based database and educating prescribers on the importance of antibiotic control. This has led to clarity on the prevalence of infections like *Clostridium Difficile* and that patients were given powerful antibiotics more often than other drugs (Antibiotic Use in the United States). Similar programs in India and other LMIC can transform healthcare by improving treatment and public health programs.

It isn't enough to just limit antimicrobial consumption to prevent rise of phenotypic resistance. A combination of better access to quality care, robust health policies, community engagement, and intersectoral action can go a long way in changing the prevalence of AMR. For example, by improving supply chain management, inappropriate use of antimicrobials due to drug shortages can be prevented (Ganguly, 2011).

Resistant phenotypes and surveillance are difficult to monitor but the future is bright. Investments in global health particularly by LMIC would make emerging technology more accessible, improve health outcomes and control the factors that lead to poverty. Similarly, health programs would be an important tool to educate and curb multi-drug resistance rates. When these investments are made, healthcare delivery and public health will improve immensely.

## REFERENCES

Adeyi, O. O., Baris, E., Jonas, O. B., Irwin, A., Berthe, F. C. J., Le Gall, F. G., Marquez, P. V., Nikolic, I. A., Plante, C. A., Schneidman, M., Shriber, D. E., & Thiebaud, A. (2017). *Drug-resistant infections: A threat to our economic future*. World Bank.

Antibiotic Use in the United States. (n.d.). *CDC*. www.cdc.gov/antibiotic-use/stewardship- report/pdf/stewardship-report.pdf

Ban, T. (2006, September). *The role of serendipity in drug discovery*. Retrieved from https://www.ncbi.nlm.nih.gov/pmc/articles/PMC3181823/

Chen, C. J., & Huang, Y. C. (2014). *New Epidemiology of Staphylococcus Aureus Infection in Asia*. Advances in Pediatrics, U.S. National Library of Medicine. www.ncbi.nlm.nih.gov/pubmed/24888414

Collignon, P. (2018). *Anthropological and Socioeconomic Factors Contributing to Global Antimicrobial Resistance: a Univariate and Multivariable Analysis*. www.thelancet.com/journals/lanplh/article/PIIS2542-5196(18)30186-4/fulltext#

Corona, F., & Martinez, J. L. (2013, April 18). *Phenotypic Resistance to Antibiotics*. Retrieved from https://www.mdpi.com/2079-6382/2/2/237/htm

Curtis. (2016). MRSA Infections among Patients in the Emergency Department: a European Multicentre Study. | *Journal of Antimicrobial Chemotherapy*. academic.oup.com/jac/article/72/2/372/2290949

Diallo, O. O., Baron, S. A., Abat, C., Colson, P., Chaudet, H., & Jean-Marc Rolain, J. M. (2020). Antibiotic resistance surveillance systems: A review. *Journal of Global Antimicrobial Resistance*, *23*, 430–438. doi:10.1016/j.jgar.2020.10.009 PMID:33176216

Gandra, S. (2017). *Scoping Report on Antimicrobial Resistance in India*. www.dbtindia.nic.in/wp-content/uploads/ScopingreportonAntimicrobialresistanceinIndia.pdf

Ganguly, N. (2011, March). *Situation Analysis Antibiotic Use and Resistance in India Global Antibiotic Resistance Partnership-India National Working Group*. Retrieved from https://cddep.org/wp-content/uploads/2017/06/india-report-web_8.pdf

Iskandar, K., Molinier, L., Hallit, S., Sartelli, M., Timothy Craig Hardcastle, T. C., & Haque, M. (2021). Surveillance of antimicrobial resistance in low- and middle-income countries: A scattered picture. *Antimicrobial Resistance and Infection Control*, *10*(1), 63–82. doi:10.118613756-021-00931-w PMID:33789754

Joshi, S., Ray, P., Manchanda, V., Bajaj, J., Chitnis, S., & Gautam, V. (2013, February). *Methicillin resistant Staphloccocus aureus (MRSA) in India: Prevalence & susceptibility pattern*. Retrieved from https://www.ncbi.nlm.nih.gov/pmc/articles/PMC3657861/#ref2

Klein, E. Y., Van Boeckel, T. P., Martinez, E. M., Pant, S., Gandra, S., Levin, S. A., Goossens, H., & Laxminarayan, R. (2018). Global increase and geographic convergence in antibiotic consumption between 2000 and 2015. *Proceedings of the National Academy of Sciences of the United States of America*, *115*(15), E3463–E3470. doi:10.1073/pnas.1717295115 PMID:29581252

Laxminarayan, R., Matsoso, P., Pant, S., Brower, C., Røttingen, J. A., Klugman, K., & Davies, S. (2016). Access to effective antimicrobials: A worldwide challenge. *Lancet*, *387*(10014), 168–175. doi:10.1016/S0140-6736(15)00474-2 PMID:26603918

McKinsey and Company India. (n.d.). *India Pharma 2020 Propelling Access and Acceptance, Realising True Potential*. www.mckinsey.com/~/media/mckinsey/dotcom/client_service/Pharma and Medical Products/PMP NEW/PDFs/778886_India_Pharma_2020_Propelling_Access_and_Acceptance_Realising_True_Potential.ashx.

Molton & Ang. (2013, January 18). *Global Spread of Healthcare-Associated Multidrug-Resistant Bacteria: A Perspective From Asia.* Retrieved from https://academic.oup.com/cid/article/56/9/1310/293355

Multidrug Resistant Organisms MDROs What Are They. (n.d.). *CT.gov - Connecticut's Official State Website.*, portal.ct.gov/DPH/Infectious-Diseases/HAI/MultidrugResistant-Organisms-MDROs-What-Are-They

Pittet. (2009). *WHO Guidelines on Hand Hygiene in Health Care: a Summary.* WHO. www.who.int/gpsc/5may/tools/who_guidelines-handhygiene_summary.pdf

Plackett, B. (2020). Why big pharma has abandoned antibiotics. *Nature*, *586*(7830), S50–S52. doi:10.1038/d41586-020-02884-3

Resistance mapping: Antibiotic Resistance. (n.d.). Retrieved from https://resistancemap.cddep.org/AntibioticResistance.php

Revelas, A. (2012, June). *Healthcare – associated infections: A public health problem.* Retrieved from https://www.ncbi.nlm.nih.gov/pmc/articles/PMC3530249/

Rosenblatt-Farrell. (2009). *The Landscape of Antibiotic Resistance.* www.ncbi.nlm.nih.gov/pmc/articles/PMC2702430/

Saha, S. K., Schrag, S. J., El Arifeen, S., Mullany, L. C., Shahidul Islam, M., Shang, N., Qazi, S. A., Zaidi, A. K. M., Bhutta, Z. A., Bose, A., Panigrahi, P., Soofi, S. B., Connor, N. E., Mitra, D. K., Isaac, R., Winchell, J. M., Arvay, M. L., Islam, M., Shafiq, Y., ... Baqui, A. H. (2018). Causes and incidence of communityacquired serious infections among young children in south Asia (ANISA): An observational cohort study. *Lancet*, *392*(10142), 145–159. doi:10.1016/S0140-6736(18)31127-9 PMID:30025808

Singh, P. K. (2017). *A universal good: how increased health coverage can help beat back antimicrobial resistance.* World Health Organization, Regional Office for South-East Asia. (http://www.searo.who.int/mediacentre/features/2017/howincreased-health-coverage-can-help- beat-back-amr-resistance/en/.

WHO Glass Report. (2017). Retrieved from https://apps.who.int/iris/bitstream/handle/10665/279656/9789241515061-eng.pdf?ua=1

Yap, M. (2013, July). *The Double Life of Antibiotics*. Retrieved from https://www.ncbi.nlm.nih.gov/pmc/articles/PMC3856192/

Zignol, M. (2018). *Genetic Sequencing for Surveillance of Drug Resistance in Tuberculosis in Highly Endemic Countries: a Multi-Country Population-Based Surveillance Study*. www.thelancet.com/action/showPdf?pii=S1473-3099(18)30073-2

# Chapter 2
# Emerging Strategies for Sensing of Blood-Stream Bacterial Diseases

**Ankur Kaul**
https://orcid.org/0000-0002-1935-6133
*Institute of Nuclear Medicine and Allied Sciences, India*

**Dimple Chopra**
*Punjabi University, India*

**Abhishek Thakur**
*Lovely Professional University, India*

**Pranav Shah**
*Uka Tarsadia University, India*

**A. K. Mishra**
*Institute of Nuclear Medicine and Allied Sciences, India*

## ABSTRACT

*Early prognosis of infection has been a major concern for clinicians worldwide. The diagnostic investigation in clinical practice focuses on resolving the complex mysteries of the deep-seated systemic infections, which are often difficult to decipher. Amongst the infectious diseases, the bacterial infections are the most ubiquitously found infections. The clinics have moved to application of molecular imaging techniques for early detection of systemic bacterial infections which even allow the follow-up during antimicrobial therapy to assess the efficacy of a particular course. Further, new age diagnostic methodology has seen a paradigm shift to the detection of biomarkers in the blood samples of patients.*

DOI: 10.4018/978-1-7998-0307-2.ch002

## 1.1 INTRODUCTION TO INFECTION: HISTORICAL AND CURRENT PERSPECTIVE

In the past, primitive human populations most probably have suffered for long time from infectious diseases similar to diseases of other wild primate populations.

The chronicle of infection control began nearly 30 years before the discovery of *Vibrio cholera* by Robert Koch, a renowned microbiologist. It was the time when John Snow, a physician had found the epidemic spread of London cholera was due to faecal contamination of drinking water facility in the golden square locality. And in the case of AIDS(Acquired Immuno Deficiency Syndrome (AIDS), where causative organism retrovirus type of Human Immunodeficiency Virus (HIV) was identified in year 1984 long after its occurrence since 1981 (Moorhead, 2002).

Thus, an important activity in the field of medical research is discovery of pathogens which cause diseases in humans, and many bacteria, fungi, viruses, helminthes, protozoa and prions are recognized as potential pathogens.

The various types of systemic infectious diseases can be classified based on the causative organism and the description along with disease associated is mentioned in Table 1.

Despite advances in the comprehension of pathophysiology of infection at the molecular level, it remains one of the major causes of morbidity and mortality world-over, particularly in developing countries (Lopez, Mathers *et al.*, 2006; Kok, Pechère *et al.,* 2004).The infections of bacterial etiology are the most common in clinical practice. As per the World Health Organisation, (WHO) report (2008), billions of money is spent on healthcare in mitigation of infectious diseases globally. Of these infections which manifest the presence of viable bacterial load in the blood pool are categorized as Blood Stream infections (BSIs). Although, currently focus is on ongoing pandemic viral infection, significant global burden still remains bacterial pathogens. (WHO Fact Sheet, 2019).

There are two main reasons to this changing global scenario with rising cases of the communicable diseases. As humans are encroaching into the biomes, at the interface, there are 'hot-spots' for human transmission of new pathogens from the wild. This phenomenon has been shown by the most recent pandemics and epidemics like SAR-CoV-2 (Severe Acute Respiratory Syndrome Coronavirus 2 or COVID-19), SARS-CoV1, Ebola and Zika viral outbreaks, MERS (Middle East respiratory syndrome), chikungunya infections (CDC Report,2019). In addition to this, the looming secondary bacterial infections are result of low immune system following a viral illness(Langford BJ et al. 2020). Moreover, the irrational use of antimicrobial agents is leading us to the ever increasing number of drug resistant pathogenic bacteria.The rising rates of Antimicrobial resistance (AMR) is another area of public health threat as well as a significant cost burden to the global economy.

*Table 1. Etiology Of Infectious Diseases In Humans*

| Infection causing microorganism | Characteristic(s) | Size range of microorganism | Disease(s) |
|---|---|---|---|
| Bacterial | Prokaryotic unicellular | 0.2 - 100 µm | TB, anthrax, diphtheria, toxic shock syndrome,Meningitis, |
| Fungal | Eukaryotic unicellular,<br>Eukaryotic multicellular | 5.0- 10 µm<br>2.0-10.0 µm by several mm | Systemic mycosis, Aspergillosis |
| Viral | Small non-cellular parasite | 0.015-0.2 µm | AIDS, dengue fever, SARS, |
| Protozoa | Eukaryotic unicellular | 2.0- 200 µm | Chagas disease, African trypanosomiasis, Malaria |
| Helminthes | Parasitic worms | 1.0 mm to several metres | Filariasis, Schistosomiasis, Hookworm infection |
| Prion | Misfolding of protein | – | Creutzfeldt–Jakobdisease |

## 1.2 BACTERIAL INFECTIONS AND THEIR CLASSIFICATION

The bacterium is the simplest organism and most abundant of all the organisms on the earth. Life on earth cannot survive without it and these omnipresent bacterial microorganisms are mostly to blame for infections (Pfaller, Diekema et al., 2007).

Broadly classified as Gram variable and Gram Indeterminate bacteria depending upon the type of staining they take. Since, the first step in the microbiological detection of bacteria is almost always done by Gram staining, hence the classification (Pelczar M J, 1993).

Further both the types of bacteria can be classified based on their physical properties. Besides these, there are also some bacteria with unusual properties but it is not discussed here to avoid complexity. The rationale for the Gram staining is due to basic anatomical differences in the cell wall of these two groups of bacteria. The Gram indeterminate bacteria are stained by acid fast staining method.

The typical examples of the different genera of bacteria are:

**Gram Positive**: *Staphylococcus, Streptococcus*
**Gram Negative:** *Escherichia, Neisseria*
**Acid Fast:** *Mycobacteria, Nocardia*

In numerous clinical cases, it is very essential for therapeutic decision making and planning not only to establish whether or not there is an inflammatory reaction, but also to differentiate between infection and sterile cases of inflammation e.g. in case of wound healing and post surgical infections(Palestro, Love *et al.*, 2007; Das, Hall *et al.*,2002).

## 1.3 SYSTEMIC BACTERIAL INFECTIONS: DIAGNOSIS METHODOLOGY

The superficial infectious spots of bacterial origin can be simply accessed and diagnosed easily. The blood-stream infections pose a grim challenge to the clinical practice and result in uncertain findings through use of inappropriate procedures.

Accurate and prompt diagnosis which includes rapid and sensitive detection techniques of bacterial pathogens of the deep-seated infections is crucial for initiating appropriate therapy and patient management.

The correct evaluation by clinicians for prognosis of the Bacterial Infections is currently is in demand because of following reasons:

- Increase in number of immune-compromised individuals who are quite susceptible to even non-pathogenic microbes as a consequence of cancer chemotherapy, transplant case or case of opportunistic infections like HIV, TB etc
- The global population is aging with ever increase in demand of orthopedic grafts and cardiovascular implants and other prosthetic devices that further become the reason for microbial colonization post surgical intervention (Baldoni, 2009)
- The spectrum of the causative microorganism is changing, with the presence of antibiotic resistant bacterial strains

The various predisposing factors for the occurrence of Infectious Diseases are elaborated in Figure 1.

The early prognosis and competence to discriminate between bacterial and sterile inflammation is vital to treat patients efficiently and prevent the pathological complications (IAEA Report 2000).

*Figure 1. Predisposing factors for occurrences of Infections in Humans*

| PATIENT FACTORS | HOSPITAL SET-UP | SURGICAL AND MEDICAL INVOLVEMENT |
|---|---|---|
| • Age, Debility | • Cross infections | • Protheses |
| • Immunocompromised status | • Understaffing | • Intravascular stents |
| • Trauma | • Overcrowding | • Urinary catherers |
| • Potential pathogen: Normal Flora | • Hospital pathogens | • Antibiotic overuse/ Misuse |
| | • Staff as transient carrier | • Surgical Incisions |
| | | • Anesthesia and ventilation |

Presently, the diagnosis of most of the infections can be regarded as a two step procedure. The first step is based on classical assessment which includes physical examination, clinical history, and laboratory tests. The second step is based on the use of imaging techniques like Ultrasound, Computed Tomography (CT), and Magnetic Resonance Imaging (MRI) which are currently employed for pinpointing of infectious lesions.

### 1.3.1 Traditional Laboratory Investigations

The microbial infection on the body may be indicated by clinical manifestations such as high fever, pain, lack of appetite, general illness and abnormal laboratory reports. The routine techniques of diagnosis depend on physical examination of microbes from infectious lesions. There is a plethora of laboratory investigations available which are based on white blood cell-count, acute-phase proteins, cytokines, markers of the inflammatory response, but none of them can specifically discriminate between infection and inflammation. These techniques are time-consuming and conclusive findings often are quite late to lead a correct decision-making in clinics (Wareham, Michael *et al.*, 2005).

## 1.4 DIAGNOSTIC INFECTION IMAGING

The scope of diagnostic imaging modalities in terms of electromagnetic spectrum is from ultrasonic to gamma ray and X-ray frequencies. Commonly, these techniques are *in vivo* procedures, with the exception like optical projection tomography (OPT) of optical imaging technique, where the sample is a 'fixed' *ex vivo* preparation. Commonly, these techniques are *in vivo* procedures (Wickline, Lanza *et al.*, 2002).

With the recent resurgence of significance in molecular imaging strategies, the design of target-specific molecular imaging agents based on enhanced understanding of the biochemistry of the Infectious lesions may ultimately lead to the development of an ideal agent for infection imaging.

### 1.4.1 Introduction to Imaging Modalities For Infection

Diagnostic Infection imaging studies can be further classified into functional and anatomic imaging technique based on the type of information they provide about the infection.

Diagnostic Infection imaging studies can be classified into **(a) Anatomic** and **(b) Functional imaging techniques** based on the type of information they provide about the infection.

The morphological imaging modalities viz., X-Ray radiograph, Ultrasound, Computed Tomography, and Magnetic Resonance Imaging reveal anatomical/ structural alteration in tissues or organs as a result of microbial invasion with the inflammatory response of the host to the invasion (Brant, Helms *et al.*2006). CT and MRI have higher sensitivity and resolution and the detection of anatomical changes can help predict the precise site of the pathological condition (Scatliff, Morris *et al.*2014). However, in early phases especially in absence of any anatomical alteration infection specificity cannot be predicted. Moreover, as a result of distortion in normal anatomy due to post surgical scarring or presence of vascular grafts, the role of these diagnostic techniques is limited(Gemmel, Dumarey *et al.* 2009).Some of key characteristics and main issues related to application of these techniques in infection are listed below:

#### 1.4.1.1 Planar X- Ray and Computed Tomography

- It is based on the application of X- Rays where the planar X-Ray is a pure 2D-form of visualization procedure while the Computed Tomography produces a 3D imaging technique
- It is relatively inexpensive, easily available for clinical practice, and their clinical findings are highly reproducible and reliable with excellent spatial resolution of 50-200 μ
- These X-ray based modalities can map only anatomical aspect. So they have a limited utility in infection imaging (Palestro, Love *et al.* 2007; Peterson, 2006)

### 1.4.1.2 Ultrasonography

- It is widely available in clinics as well as portable forms, quick to perform and with excellent spatial resolution attaining below 1mm
- It is versatile diagnostic tool based on use of ultrasound waves
- It is usually employed as a first line investigation by medical practitioners suspecting some internal infections around abdominal region and can be utilized to obtain functional information to a very limited extent (Gotthardt, Bleeker-Rovers *et al.*, 2010)
- The limitations of ultrasonographic results are extremely dependent on operator's discretion and the mapping and the visualization of internal body structures is intricate as a result of dense regions like intestinal gas and bony structures which hindered penetration of the ultrasonic waves (Cootney, 2001)

### 1.4.1.3 Magnetic Resonance Imaging

- It utilizes nuclear magnetic resonance (NMR) to image nuclei of atoms in internal organs to get detailed information
- MRI is based on the detection of molecules that contain nuclei that possess the property of nuclear spin and provides a resolution of 10 μ
- The advantage of MRI is excellent anatomic details to pinpoint the exact location, and its contraindications are scanning of patients with pacemakers, metallic implants and prostheses
- Furthermost, long duration of imaging procedures make the scans prone to movement artifacts especially for elderly and children (Olsen, 2013)

### 1.4.1.4 Endoscopic Techniques

- It requires a simple instrument called Endoscope which is used to scan the interior of a hollow cavity of the organ
- The main complications associated with this technique are perforation in the lining of stomach and esophagus, which may bleeding especially at the site of a biopsy removal
- The endoscopic techniques have limited utility only for gastro-intestinal tract infection (Neumann *et al.*, 2013)

#### 1.4.1.5 Optical Tomography

- It is an another emerging modality based on detecting the transmission of photons through biological systems and a specific tissue contrast is created by different absorption or scattering patterns of photon at various wavelengths
- Its use is limited on soft tissues, like breast and brain tissue and on the organ under investigation being partly transparent or translucent (Haisch 2012; Sharpe 2004)

### 1.4.2 Functional Imaging Techniques

These techniques are exemplified by nuclear medicine studies make use of tracer amounts of radioactive materials (Brant, Helms *et al.,* 2012) and they assume a significant role in diagnosing the infections focus and in planning targeted antimicrobial chemotherapy. The advances in understanding pathogenesis of infectious disease combined with progress in radiopharmaceutical sciences have boosted the development of nuclear medicine (NM) techniques for the diagnoses of infection more precisely (Elgazzar, 2006). The nuclear medicine technique comprises of Single photon emission computer tomography (SPECT) and Positron emitting tomography (PET) which provide detailed localization to give crucial information in clinics. Diagnostic radiopharmaceuticals, as a component of Nuclear medicine technique, offer a methodology for the assessment of the infectious disease state and monitoring the antimicrobial chemotherapy non-invasively.

The Radionuclide Imaging techniques Like Positron Emission Tomography (PET) and Single Photon Emission Computed Tomography (SPECT) can offer considerable benefit over these less sensitive diagnostic methods for imaging specific biological targets, predominantly those present in low concentration. The nuclear medicine techniques show specificity in locating the infectious focus with alterations in functional aspects that are early clinical manifestations.

Also, SPECT can further offers the prospect to extend the time-window of patient examination (due to the longer physical half-life of single photon emitters) thus, permitting clinicians to examine biological processes *in vivo* several hours or days after administration of the labeled compound. And there are plethoras of radiotracers are currently available for infection and the list is expanding every year.

The radiotracers are being utilized to prepare these radiopharmaceuticals could be Single photon emitting radioisotopes like $^{111}In$, $^{67}Ga$, and $^{99m}Tc$ and PET radiotracers can be $^{18}F$, $^{68}Ga$.The most widely used SPECT radioisotope for this technique is $^{99m}Tc$ due to its favorable physical properties. The 140 keV energy of $\gamma$ emission with half-life of 6 h offers ideal nuclear medicine imaging properties (Saha, 2010).

SPECT/ PET scans are commonly employed imaging modality which provides more comprehensive information especially in patients with osteomyelitis and suspected of deep-seated infected joint prostheses or implants. Perhaps the least expensive and simplest technique to boost the specificity of 'Three-Phase Bone Scintigraphy' is the delayed imaging schedule. In contrast to normal bone scintigraphy, in which uptake of radiotracer decreases by about 4 h post injection, the tracer-uptake in condition like osteomyelitis builds up for several hours, resulting in a higher lesion-to-background ratio on the delayed scintigraphic scans (Kim *et al.,* 2016).

## 1.5 NEW-AGE LABORATORY BASED BLOOD TESTS

### 1.5.1 Biomarker Detection

Early diagnosis of systemic infections is considered complicated through the conventional blood-culture reports, as it can lead to unnecessary delay in treatment and enhance possibilities of adverse outcomes.

Most sensitive diagnostic approach based on host biomarker detection for bacterial infection include polymerase chain reaction (PCR) to detect nucleic acids (DNA or RNA), (Tkadlec *et al.*, 2019) and immunological tests like ELISA that are based on antigen–antibody interactions (Tsao *et al.*, 2020).

These conventional techniques show high specificity in distinguishing viable bacterial cells. But, the complicated and laborious process requirements in such diagnostic techniques renders them unsuitable for rapid diagnostics (Samuel *et al.,* 2004).

The biomarkers have always play an important role in the prognosis and diagnosis of human disease. Since, the levels of Biomarker in the blood circulation are dynamic, they can be used to assess the changes in response to different inflammatory stimulus, especially in the case of bacterial infections. The levels of C-reactive protein (CRP),an acute phase protein, procalcitonin (PCT) and presepsin shows elevation in certain conditions such as inflammation and bacterial infections. It was found that elevated CRP levels (over 120 mg/l) and decreased PCT levels (less than 1.25 ng/l) was observed in the immunocompromised hematological patients with fungal infection (Marková *et al.*, 2013). Whereas elevated PCT level and normal levels of CRP was primarily the prognosis for - blood stream infection of bacterial origin (Stoma *et al.*, 2019). Recently, TRAIL (tumour necrosis factor-related apoptosis-inducing ligand) biomarker was found to suitable for distinguishing viral infections from the bacterial infections due to higher concentration in the cases of viral dieases only (Does *et al.*, 2018).

### 1.5.2 Systemic Inflammation Response Syndrome (SIRS) Analysis

SIRS analysis represents the early signs of infection as an inflammatory response to any foreign invasion. The SIRS is manifested by several conditions such as:

1. Increase in core temperature above 38 °C or decrease in core temp below 36 °C.
2. Change in rhythm of heart beat, which may be more than 90 beats per minute (Tachycardia).
3. More than 20 breaths per minute or decrease in arterial Carbon dioxide tension below 32 mmHg.
4. Abnormal increase or decrease in white blood cell count, which may be more than 12000 per µL or less than 4000 per µL or greater than 10% abnormal white blood cells.

When two or more than two parameters are there and among statement (a) and (d) at least one is confirmed, indicates the systemic inflammatory response (Dellinger *et al.*, 2013). This method cannot be used as marker of infection alone because change in white blood cell count may occur by several other reasons including just the inflammation, but can be used as important prognostic parameter (Conti *et al.*, 2015).

### 1.5.3 T2 Magnetic Resonance

T2 magnetic resonance (T2MR) offers superior tool for the early detection of systemic infection with low limit of bacterial load detection as 1 CFU/mL of whole blood (Figure 2).

This diagnostic tool is amalgamation of MRI along with the PCR based amplification of DNA infection causing pathogens.T2 Magnetic Resonance can detect various targets such as cells and nucleic acid within blood. The first medical application of T2MR in diagnosis was T2Candida and T2Bacteria panels, found successful in detection of microbial cells within whole blood (Clancy and Nguyen, 2018).

The detection of pathogenic cells from blood is based on lyses of blood cells, concentrates cellular debris and bacterial cells, breakdown of cells by bead-beating, amplification of DNA by thermostable polymerase and detection of amplified product by amplicon-induced agglomeration and T2 measurements (Mylonakis*et al.*, 2015).

*Figure 2. Schematic representation of diagnosis based on T2 Magnetic Resonance*

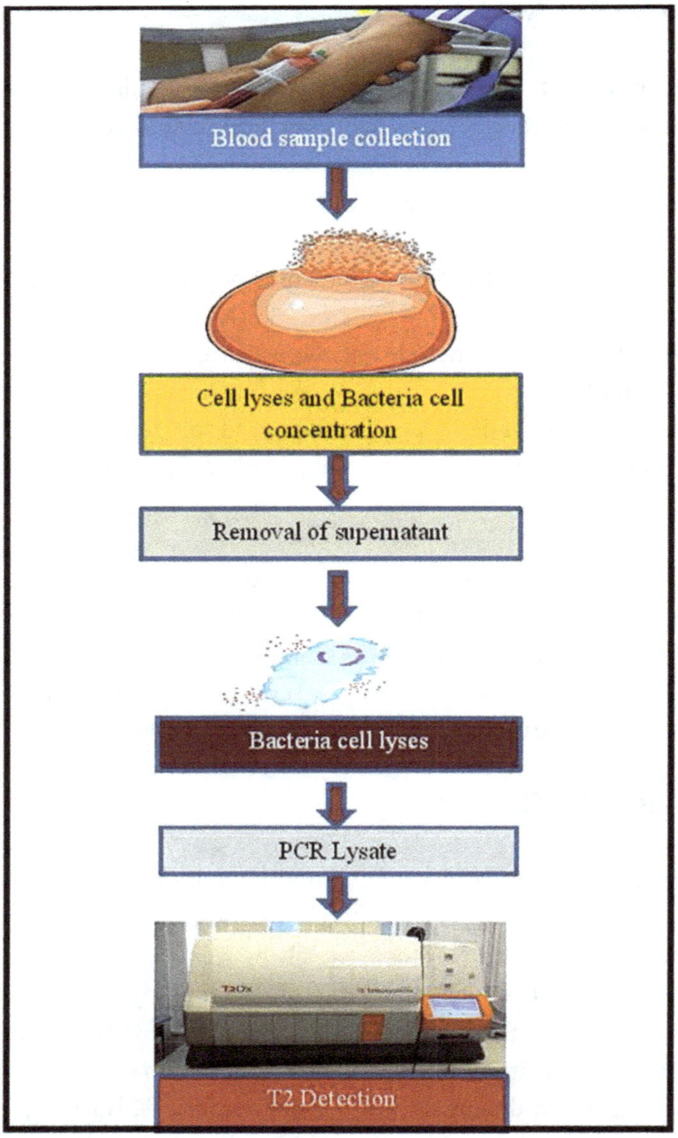


## 1.5.4 16s Metagenomics

16S Metagenomics is one of the newer molecular approach for the diagnosis of Infections, but this technique has not been commercialized yet due to optimization issues. The advantages offered by 16S Metagenomics over Blood culture techniques are given as (Rutanga *et al.*, 2018):

- High sensitivity
- Detect all bacteria in a given sample
- Culture independent
- Polymicrobial infections can be detected
- Low blood volume requirement for diagnosis
- Shorter turnover time

Beside these advantages offered certain loopholes are also there, owing to this it become difficult to implement this technique over Blood culture. Limitations of this technique are given below as:

- False positive results
- Difficulty in identification between pathogen and environmental contaminant
- Do not provide information about antibiotic susceptibility profile

## 1.6 NEW AGE BIO-SENSING DIAGNOSTIC PLATFORMS

The new generation diagnostic tests aim to tap the potential of body fluids in identifying and evaluating the human well-being. The non-invasive easy collection process and convenient storage makes saliva more clinical relevance for tracking the physiological processes. (Roi *et al.* 2019) The biosensor based diagnostic platforms have the prospective to make commercially and clinically viable diagnostic systems [Figure 3]. Such biosensor based devices will convert a biological response into quantifiable signal which may be electrochemical (Simoska *et al.*, 2019), thermal (Tlili *et al.*, 2013), chemiluminescence (Wu *et al.*, 2018),or optical (Massad N *et al.*, 2016) as an output indicator.

Current approaches in investigational tools include application of nanomaterials/ nanoparticle (NP) (Tallury *et al.,* 2010) and utilizing their unique physiochemical properties for designing a biosensoring platforms. Moreover, convergence of nanotechnology with advanced fabrication techniques has created a new dimension 'nanobiosensor' hold a great promise for future application as smart POC device. (Zarei, 2018)

These 'smart' diagnostics in collaboration with AI tools (Smith *et al.*, 2020) can further augment the more accurate diagnosis and these investigative measures can be more accessible to the remote settings.

*Figure 3. Representative image of new-age bio-sensing platform for detection*

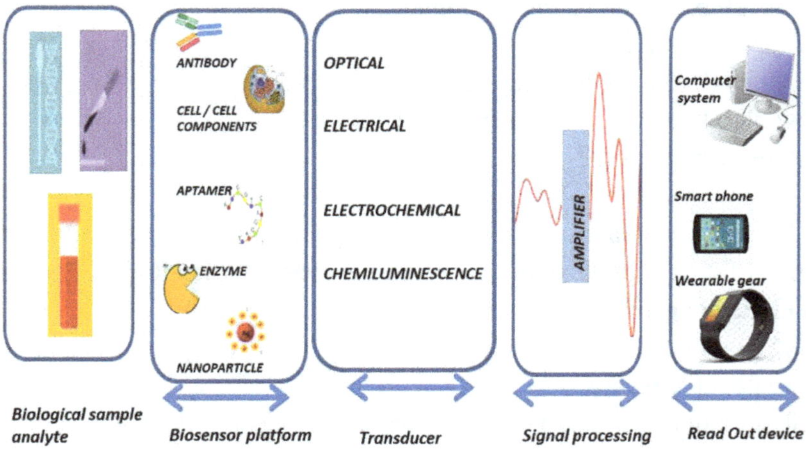


## CONCLUSION

The early diagnosis of infection still poses great challenge to clinicians. The functional imaging techniques offer the advantages to some extent, but for the specificity of the diagnosis, the healthcare community still relies on traditional laboratory investigations like blood culture tests.

The new developments in AI enabled molecular approach based diagnostics have shown great potential to compensate for intrinsic limitations of culture-based diagnostic tests. Further, the body fluid based biomarker detection can also aid in enhancing diagnostic capabilities.

Hence, these new-age diagnostic techniques are being designed with substantial focus on robustness of the tests, and comprehensive data interpretation methodologies so that there is substantial improvement in the patient–health.

## ACKNOWLEDGEMENT

A. K. Mishra served as the corresponding author of this chapter.

## REFERENCES

Aslam, B., Wang, W., Arshad, M. I., Khurshid, M., Muzammil, S., Rasool, M. H., & Baloch, Z. (2018). Antibiotic resistance: A rundown of a global crisis. *Infection and Drug Resistance*, *11*, 1645–1658. doi:10.2147/IDR.S173867 PMID:30349322

Baldoni, D. (2009). *Innovative methods for the diagnosis and treatment of implant-associated infections* (Doctoral dissertation). University of Basel.

Brant, W. E., & Helms, C. A. (Eds.). (2012). Fundamentals of diagnostic radiology. Academic Press.

CDC. (2019). Antibiotic Resistance Threats in the United States, 2019. CDC. DOI: :82532 doi:10.15620/cdc

Clancy, C. J., & Nguyen, M. H. (2018). T2 magnetic resonance for the diagnosis of bloodstream infections: Charting a path forward. *The Journal of Antimicrobial Chemotherapy*, *73*(suppl_4), iv2–iv5. doi:10.1093/jac/dky050 PMID:29608754

Conti, F., Dall'Agata, M., Gramenzi, A., & Biselli, M. (2015). Biomarkers for the early diagnosis of bacterial infection and the surveillance of hepatocellular carcinoma in cirrhosis. *Biomarkers in Medicine*, *9*(12), 1343–1351. doi:10.2217/bmm.15.100 PMID:26580585

Cootney, R. W. (2001). Ultrasound imaging: Principles and applications in rodent research. *ILAR Journal*, *42*(3), 233–247. doi:10.1093/ilar.42.3.233 PMID:11406722

Das, S. S., Hall, A. V., Wareham, D. W., & Britton, K. E. (2002). Infection imaging with radiopharmaceuticals in the 21st century. *Brazilian Archives of Biology and Technology, 45*(spe), 25-37.

Dellinger, R. P., Levy, M. M., Rhodes, A., Annane, D., Gerlach, H., Opal, S. M., & Osborn, T. M. (2013). Surviving sepsis campaign: International guidelines for management of severe sepsis and septic shock. *Critical Care Medicine*, *41*(2), 580–637. doi:10.1097/CCM.0b013e31827e83af PMID:23353941

Elgazzar, A. H. (2006). *The pathophysiologic basis of nuclear medicine*. Springer Science &Business Media. doi:10.1007/978-3-540-47953-6

Gemmel, F., Dumarey, N., & Welling, M. (2009). Future diagnostic agents. In *Seminars in nuclear medicine*. Elsevier. doi:10.1053/j.semnuclmed.2008.08.005

Gotthardt, M., Bleeker-Rovers, C. P., Boerman, O. C., & Oyen, W. J. (2010). Imaging of inflammation by PET, conventional scintigraphy, and other imaging techniques. *Journal of Nuclear Medicine*, *51*(12), 1937–1949. doi:10.2967/jnumed.110.076232 PMID:21078798

Haisch, C. (2012). Optical tomography. *Annual Review of Analytical Chemistry (Palo Alto, Calif.)*, *5*(1), 57–77. doi:10.1146/annurev-anchem-062011-143138 PMID:22524216

Kim, C., Lee, J., Kim, J. Y., Hwang, K. T., & Choi, Y. Y. (2016). Comparative Analysis of 99mTc-MDP Three-Phase Bone Scan with SPECT/CT and 99mTc-HMPAO-Labeled WBC SPECT/CT in the Differential Diagnosis of Clinically Suspicious Post-traumatic Osteomyelitis. *Nuclear Medicine and Molecular Imaging, 51*(1), 40–48. Advance online publication. doi:10.100713139-016-0441-x PMID:28250857

Kok, M., & Pechère, J. C. (2004). *Nature and pathogenicity of microorganisms*. Academic Press.

Langford, B. J., So, M., Raybardhan, S., Leung, V., Westwood, D., MacFadden, D. R., ... Daneman, N. (2020). Bacterial co-infection and secondary infection in patients with COVID-19: a living rapid review and meta-analysis. *Clinical Microbiology and Infection*.

Lopez, A. D., Mathers, C. D., Ezzati, M., Jamison, D. T., & Murray, C. J. (2006). Global and regional burden of disease and risk factors, 2001: Systematic analysis of population health data. *Lancet, 367*(9524), 1747–1757. doi:10.1016/S0140-6736(06)68770-9 PMID:16731270

Marková, M., Brodská, H., Malíčková, K., Válková, V., Cetkovský, P., Kolář, M., & Haluzík, M. (2013). Substantially elevated C-reactive protein (CRP), together with low levels of procalcitonin (PCT), contributes to diagnosis of fungal infection in immunocompromised patients. *Supportive Care in Cancer, 21*(10), 2733–2742. doi:10.100700520-013-1844-1 PMID:23712521

Massad-Ivanir, N., Shtenberg, G., Raz, N., Gazenbeek, C., Budding, D., Bos, M. P., & Segal, E. (2016). Porous silicon-based biosensors: Towards real-time optical detection of target bacteria in the food industry. *Scientific Reports, 6*(1), 1–12. doi:10.1038rep38099 PMID:27901131

Moorhead, R. (2002). William Budd and typhoid fever. *Journal of the Royal Society of Medicine, 95*(11), 561–564. doi:10.1177/014107680209501115 PMID:12411628

Mylonakis, E., Clancy, C. J., Ostrosky-Zeichner, L., Garey, K. W., Alangaden, G. J., Vazquez, J. A., & Zervou, F. N. (2015). T2 magnetic resonance assay for the rapid diagnosis of candidemia in whole blood: A clinical trial. *Clinical Infectious Diseases, 60*(6), 892–899. doi:10.1093/cid/ciu959 PMID:25586686

Neumann, H., & Keller, M. F. (2013). Advanced Endoscopic Imaging. *Diagnostic and Therapeutic Endoscopy*, 2.

Olsen, Ø. E. (2013). MRI: How to perform a pediatric scan. *Acta Radiologica, 54*(9), 991–997. doi:10.1177/0284185112474917 PMID:23390159

Palestro, C. J., Love, C., & Miller, T. T. (2007). Diagnostic imaging tests and microbial infections. *Cellular Microbiology*, *9*(10), 2323–2333. doi:10.1111/j.1462-5822.2007.01013.x PMID:17645552

Pelczar & Krieg. (1993). Microbiology. Tata McGraw Hill.

Pfaller, M., & Diekema, D. (2007). Epidemiology of invasive candidiasis: A persistent public health problem. *Clinical Microbiology Reviews*, *20*(1), 133–163. doi:10.1128/CMR.00029-06 PMID:17223626

Roi, A., Rusu, L. C., Roi, C. I., Luca, R. E., Boia, S., & Munteanu, R. I. (2019). A New Approach for the Diagnosis of Systemic and Oral Diseases Based on Salivary Biomolecules. *Disease Markers*, *2019*, 2019. doi:10.1155/2019/8761860 PMID:30906485

Rutanga, J. P., Van Puyvelde, S., Heroes, A. S., Muvunyi, C. M., Jacobs, J., & Deborggraeve, S. (2018). 16S metagenomics for diagnosis of bloodstream infections: Opportunities and pitfalls. *Expert Review of Molecular Diagnostics*, *18*(8), 749–759. doi:10.1080/14737159.2018.1498786 PMID:29985081

Scatliff, J. H., & Morris, P. J. (2014). From Roentgen to magnetic resonance imaging: The history of medical imaging. *North Carolina Medical Journal*, *75*(2), 111–113. doi:10.18043/ncm.75.2.111 PMID:24663131

Sharpe, J. (2004). Optical projection tomography. *Annual Review of Biomedical Engineering*, *6*(1), 209–228. doi:10.1146/annurev.bioeng.6.040803.140210 PMID:15255768

Simoska, O., & Stevenson, K. J. (2019). Electrochemical sensors for rapid diagnosis of pathogens in real time. *Analyst (London)*, *144*(22), 6461–6478. doi:10.1039/C9AN01747J PMID:31603150

Smith, K. P., & Kirby, J. E. (2020). Image analysis and artificial intelligence in infectious disease diagnostics. *Clinical Microbiology and Infection*, *24*(12), 1318–1323. doi:10.1016/j.cmi.2020.03.012 PMID:32213317

Stoma, I., Karpov, I., Uss, A., Krivenko, S., Iskrov, I., Milanovich, N. &Cherniak, V. (2019). Combination of sepsis biomarkers may indicate an invasive fungal infection in hematological patients. *Biomarkers*, 1-18.

Tallury, P., Malhotra, A., Byrne, L. M., & Santra, S. (2010). Nanobioimaging and sensing of infectious diseases. *Advanced Drug Delivery Reviews*, *62*(4-5), 424–437. doi:10.1016/j.addr.2009.11.014 PMID:19931579

Tkadlec, J., Peckova, M., Sramkova, L., Rohn, V., Jahoda, D., Raszka, D., & Drevinek, P. (2019). The use of broad-range bacterial PCR in the diagnosis of infectious diseases: A prospective cohort study. *Clinical Microbiology and Infection*, *25*(6), 747–752. doi:10.1016/j.cmi.2018.10.001 PMID:30321604

Tlili, C., Sokullu, E., Safavieh, M., Tolba, M., Ahmed, M. U., & Zourob, M. (2013). Bacteria screening, viability, and confirmation assays using bacteriophage-impedimetric/loop-mediated isothermal amplification dual-response biosensors. *Analytical Chemistry*, *85*(10), 4893–4901. doi:10.1021/ac302699x PMID:23510137

Tsalik, E. L., Bonomo, R. A., & Fowler, V. G. Jr. (2018). New molecular diagnostic approaches to bacterial infections and antibacterial resistance. *Annual Review of Medicine*, *69*(1), 379–394. doi:10.1146/annurev-med-052716-030320 PMID:29414265

Tsao, Y. T., Tsai, Y. H., Liao, W. T., Shen, C. J., Shen, C. F., & Cheng, C. M. (2020). Differential Markers of Bacterial and Viral Infections in Children for Point-of-Care Testing. *Trends in Molecular Medicine*, *26*(12), 1118–1132. PMID:33008730

Van der Does, Y., Rood, P., Ramakers, C., Schuit, S., Patka, P., van Gorp, E., & Limper, M. (2018). Identifying patients with bacterial infections using a combination of C-reactive protein, procalcitonin, TRAIL, and IP-10 in the emergency department: a prospective observational cohort study. *Clinical Microbiology and Infection, 24*(12), 1297–1304.

Wareham, D., Michael, J., & Das, S. (2005). Advances in bacterial specific imaging. *Brazilian Archives of Biology and Technology*, *48*(spe2), 145–152. doi:10.1590/S1516-89132005000700021

WHO Factsheet. (2019). *The top 10 causes of death*. https://www.who.int/en/news-room/fact-sheets/detail/the-top810-causes-of-death

Wickline, S. A., & Lanza, G. M. (2002). Molecular imaging, targeted therapeutics, and nanoscience. *Journal of Cellular Biochemistry*, *87*(S39), 90–97. doi:10.1002/jcb.10422 PMID:12552608

Wu, R., Zhou, S., Chen, T., Li, J., Shen, H., Chai, Y., & Li, L. S. (2018). Quantitative and rapid detection of C-reactive protein using quantum dot-based lateral flow test strip. *Analytica Chimica Acta*, *1008*, 1–7. doi:10.1016/j.aca.2017.12.031 PMID:29420938

Yang, S., & Rothman, R. E. (2004). PCR-based diagnostics for infectious diseases: Uses, limitations, and future applications in acute-care settings. *The Lancet. Infectious Diseases*, *4*(6), 337–348. doi:10.1016/S1473-3099(04)01044-8 PMID:15172342

Zarei, M. (2018). Infectious pathogens meet point-of-care diagnostics. *Biosensors & Bioelectronics*, *106*, 193–203. doi:10.1016/j.bios.2018.02.007 PMID:29428589

# Chapter 3
# Exploring the Potential of Peptides and Peptidomimetics in Biosensing

**Radhika R. Jaswal**
https://orcid.org/0000-0002-6364-495X
*Panjab University, India*

**Kanica Kaushal**
*Panjab University, India*

**Shubhi Joshi**
*Panjab University, India*

**Pratibha Sharma**
*Panjab University, India*

**Shweta Sharma**
https://orcid.org/0000-0002-1936-2947
*Panjab University, India*

**Simran Preet**
*Panjab University, India*

**Avneet Saini**
*Panjab University, India*

## ABSTRACT

*Biosensors are devices that capture the biological signal and convert it into a detectable electrical signal through transduction. Biological entities like DNA, RNA, and proteins/enzymes can be conjugated onto the biosensor surface to detect and observe certain biological analytes in environment, biomedical, and food industries. Peptides have been efficiently used in the fabrication of peptide-based biosensors due to their attractive properties like established synthesis protocols, diverse structures, and as highly enzyme-selective substrates. However, owing to their labile nature, peptidomimetics are the best alternatives at the bioreceptor interface due to their specificity and stability, relatively low cost and easy modifications, and capability*

DOI: 10.4018/978-1-7998-0307-2.ch003

*to form supramolecular assemblies like nanosheets. Such bioconjugation strategies efficiently convert interaction information into a measurable signal, thus highlighting the importance in the fabrication of next-generation novel robust biosensors desirable for detection and dissemination of pathogens causing infections in the living and non-living worlds.*

## INTRODUCTION

In the biochemical field, sensors are usually defined as a device which includes both a receptor (bio-recognition element) and a transducer, providing specific quantitative or semi quantitative analytical information. Biosensors are powerful tunable systems capable of switching between an ON/OFF status in response to an external stimulus. In general, *biosensing techniques* can be defined as any of a variety of procedures which use biomolecular probes to measure the presence or concentration of biological molecules, biological structures, microorganisms, etc., by translating a biochemical interaction at the probe surface into a measurable physical signal. The general function of biosensors involves a receptor in the most general sense recognizing an analyte, and then a transducer either triggers a quantifiable signal or catalyzes a reaction related to the analyte concentration to generate a signal (Griffin et al., 2009). In clinical diagnosis, a sensitive, quick, convenient and versatile molecular biosensor has been desired to simplify the testing process, reduce the cost and shorten testing time (Salazar-Salinas et al., 2009). Recent advances in both disciplines allow redesigning the configuration of the sensing elements – either by modifying toggle switches and gene networks, or by producing synthetic entities mimicking the key properties of natural molecules. The primary requirement in the selection of various substances/factors as the components of biosensors includes rapid responding ability, high specificity & sensitivity, reliability, portability, productivity and long-lasting stability. In addition to this, immobilization/fabrication of bioanalyte in its native conformation, high accessibility of the receptor's sites to the species of interest and effective adsorption of the analyte to the employed support (Marx, 2007) are the main crucial factors that should be considered during the engineering of high performance biosensors. These demands need to be ardently addressed when developing the design of biosensors. Peptides and peptide based analogs possess the potential candidature for fulfilling many of these requirements.

## BIOSENSORS: A GENERAL OVERVIEW

The word "sensor" is derived from the Latin word "sentire" which basically means 'to identify' anything. In terms of classification, physical sensors and chemical sensors are the two most primary and widely opted classes of sensors. Biosensors, an amalgam of physical and chemical sensing method, are the most recent type of sensors in a way that these have been recognized by only some eighteen years prior to today (Buenger et al., 2012). Basically, biosensors are receptor-transducer based devices which interpret some specific biophysical or biochemical property of the medium. Furthermore, the most interesting quality that sets them apart is the presence of biological/organic recognition element which enables the detection of particular biological molecules in the medium (Wang, 2006).

### i. Major Breakthrough in the Development of the Biosensors

Ever since the development of the first biosensor as a potentiometric enzyme based electrode that was invented for the detection and measurement of glucose in any medium (Clark and Lyons, 1962) the major focus has remained on reducing the size and making the laboratories economically stable towards the development of portable, small-sized/nanosized and multi-functional biosensors (Chao et al., 2016). The bioelement is principally any organic body which is able to detect any particular analyte from the medium of interest while remaining irresponsive towards any other potentially interfering species whereas sensing element consists of the signal transducing portion of the biosensor which could be in the form of any magnetic, optical, electrical or electrochemical etc. transducing mechanism (Vidal, 2013). Elaboration of these two components is given in Figure 1.

### ii. Types of Biosensors

Depending upon the type of transducer, biosensors can be categorized in the groups of electrochemical, mass dependent, optical, radiation sensitive and so on. On the basis of the choice of the bioelement such as enzyme, nucleic acid, proteins, saccharides, oligonucleotides, ligands etc. there are various sets of biosensors (Mohanty and Kougianos, 2006), that can be fabricated keeping in view the type of analyte under consideration (Turner, 2000). Peptide sequences play a substantial role in enzymatic and many other assays due to their characteristic properties such as good stability, regular synthetic protocol, and hence, these molecules are of utmost importance in the fabrication of sensitive, fast, and convenient biosensors.

*Figure 1. Schematic representation of different parts of a biosensor*

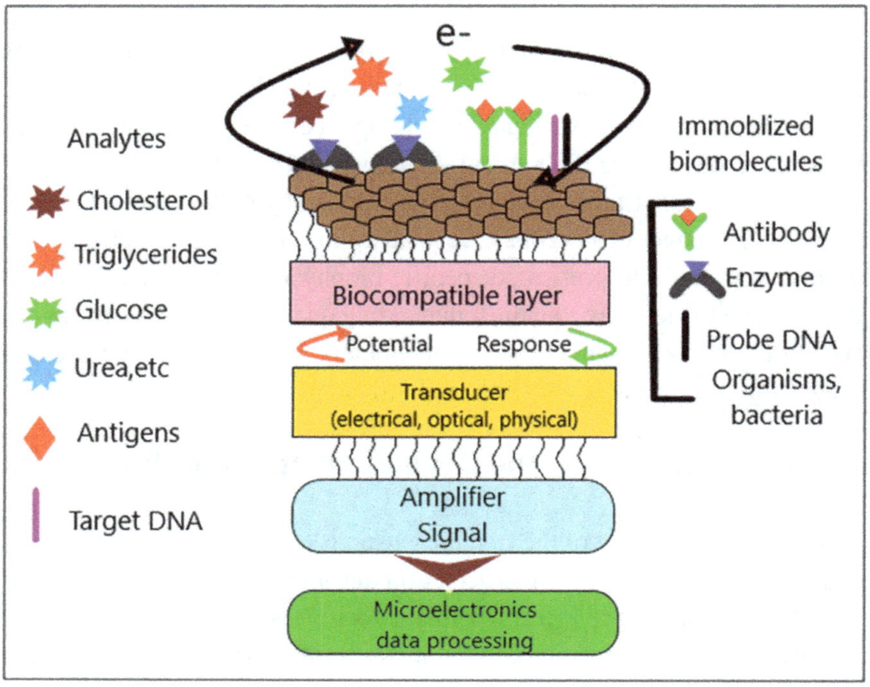


## PEPTIDE-BASED BIOSENSORS

Peptides are formed by natural or synthetic short polymers of amino acids which are linked by peptide bonds. Peptide sequences that are specific enzyme substrates play a crucial role in assays of enzymatic activity and screening of enzymatic inhibitors. Due to these unique properties, peptides are excellent candidates for developing sensitive, fast, and convenient biosensors. Generally, peptides do not generate a measurable signal directly in response to a binding event, hence conjugating them with a signal marker is an efficient strategy to convert the information of analyte/binding into a measurable signal and till date, several methods have been utilized to construct such peptide-based biosensors. These biosensors can be used for detecting various analytes including metallic ions, proteins, proteases, kinases, bacillus species, nucleic acids and antibodies.

### i. Fluorosence Based Biosensors

Environmentally sensitive fluorophores conjugated with peptides are most widely used common signal markers. Their fluorescence emission can be affected by changes in the local environment caused by the affinity or interaction between conjugated peptide and analytes. The method of conjugating peptides to environmentally-sensitive fluorophores has been utilized to develop various peptide-conjugated molecular probes/sensors (Pavan and Berti, 2012), including ion sensors, DNA sensors, redox sensors and protein sensors. Similar properties of such signal markers can also be attributed to near-infrared dyes (Peng et al., 2006), nanoparticles (Wang et al., 2010), quantum dots (Clarke et al., 2010), graphene (Feng et al., 2012; Feng et al., 2011), polydiacetylene (PDA)-liposomes (Jaworski et al., 2011), lanthanide chelators (Yan et al., 2011) and electrochemical markers (Zhao et al., 2010). Several biosensors have been widely reported in literature signifying the budding use of peptide rational design to overcome the challenges related to biosensor commercialization (Scognamiglio et al., 2013). For example, Enander et al. (2008) proved the ability of peptides combined with ratiometric fluorophores to ensure robustness in recognition and signaling in immobilized format quantification assays.

## Excimer-Type Biosensors

Similarly, excimer-type protein molecular biosensors comprise of a peptide and two identical fluorophores, which are attached to the opposite ends of the peptide (Figure 2). In the presence of the target analyte, the interaction between the analyte and peptide forces the fluorophores to separate, resulting in a shift of the emission peak of the fluorophores to that of the monomer. An excimer-type peptide beacon for detection of anti-HIV antibody has been developed (Plaxco et al., 2006). The peptide beacon concept was further extended to develop an HIV sensor based on the probe–quencher pair model. In probe-quencher pair protein-sensors, fluorophores and quencher units are attached at various residues of the peptide sequence. Affinity between a peptide sequence and target protein influences the distance between the fluorophore and quencher, resulting in a change in the fluorescence emission, thus generating a signal.

In another case, the use of CPPs (Cell Penetrating Peptides) has enabled the visualization of real-time viral infection of living cells (Yeh et al., 2008). The fact that molecular beacons are able to detect viral RNA in infected cells has already been revealed (Yeh et al., 2008). These single-stranded molecular beacon oligonucleotides were specifically chosen to target a noncoding region of a viral genome and were labeled with a fluorophore at one end and a quencher at the other. Prior to encountering the virus, these constructs existed in a stem-loop structure with the quencher and fluorophore in close range, thus resulting in an absence of fluorescence as shown in Figure 2. However, as the viral particles entered cells, the molecular beacons hybridized with the viral genome and changed conformation, thereby separating the quencher and fluorophore. As a result, fluorescence was detected allowing the viral infection to be visualized. While this approach is highly

*Figure 2. Working of a molecular beacon system*

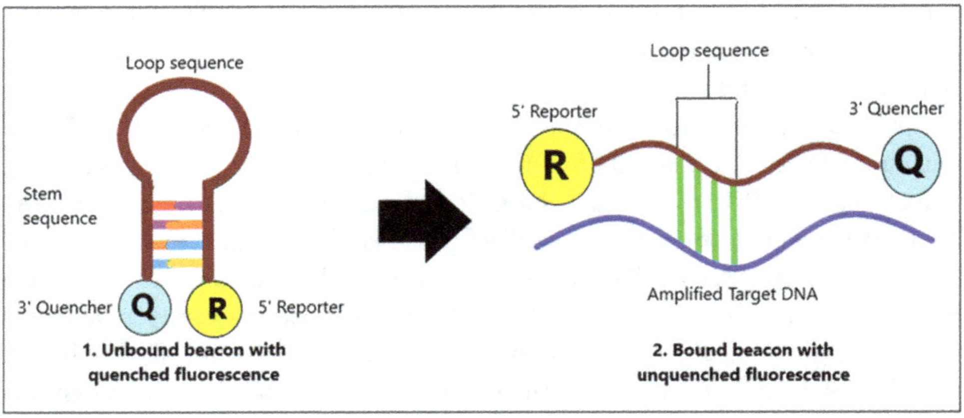


sensitive (detection of a single viral particle was possible), it was only successful in fixed cells permeabilized with Triton to allow uptake of the molecular beacons. In order to deliver these molecular beacons in a less invasive manner, Tat peptide was harnessed to deliver the constructs into uncompromised cells in a consecutive study which enabled real time detection of viral replication and infection (Yeh et al., 2008).

## LIMITATIONS OF PEPTIDE-BASED BIOSENSORS AND THE STRATEGIES TO OVERCOME

Despite innovations and improvements in this technology, peptide-based biosensors are still limited by a number of factors, such as long term stability, sensitivity, read-out time, miniaturization and cost & productivity related challenges (Scognamiglio et al., 2010). Lowering of the detection limits and the increase of robustness are the main confrontations of the future generation of biosensors. Synthetic biology can deal with these key issues, improving biosensor potential in terms of sensitivity and stability, as well as selectivity and detection in complex mixtures, by extending the perfect response of biological systems and/or engineering them to provide new desired features (Andrianantoandro et al., 2006). Recent advances in technologies like nanotechnology, bioinformatics and molecular engineering are capable of creating new biological components alongwith decoding and rewriting genes, enzymes, and networks, modifying them to alter their core function and structure (Effendi et al., 2008). Emerging strategies are being adopted to take the advantage of engineered organisms to generate a specific response to pollutants, or to design combinatorial library of bioinspired molecules, like peptidomimetics, molecular imprinting polymers, and aptamers, with customized features regarding specificity and stability (Checa et al., 2012).

## WHAT ARE PEPTIDOMIMETICS?

Peptidomimetics are the structural and functional mimics of bioactive peptides using non-canonical amino acids or non-peptide scaffolds which comprise an important class of drug molecules (Mizuno et al., 2017; Floris and Moro, 2012; Avan et al., 2014). These compounds have the ability to interact with the biological target and also produce the same biological effect as of protein (Vagner et al., 2008). Peptidomimetics are used to counter the undesirable effects of peptides while retaining their beneficial characteristics. Peptidomimetics have been subdivided into three types (Azzarito et al., 2013; Pelay-Gimeno et al., 2015) i.e. Type I mimetics or structural mimetics are short peptides that replicate the topography of a secondary structure and distinguish themselves from their parent peptide only by substitutions introduced to stabilize the desired conformation. Type II mimetics or functional mimetics are non-peptidic functional molecules that have a small molecular scaffold and do not recapitulate all the side chain interactions of the parent protein. Type III mimetics or functional-structural mimetics are the molecules which include nonpeptide templates that are topologically similar to the parent peptide but do not show atom-by-atom analogy. Further, these three categories have been sub-divided by Pelay-Gimeno et al. into four different classes based on the degree of their similarity to the natural peptide precursor i.e. Classes A–D where Class A mimetics are most similar to the parent peptide whereas Class D mimetics show the least similarities (Pelay-Gimeno et al., 2015). Classes A and B include peptide-like structures whereas classes C and D encompass small molecular scaffolds:

- Class A mimetics or Type I mimetics are defined as peptides that mainly consist of the parent peptide amino acid sequence. They have minimal alterations as the backbone a side chains of this class align closely with the bioactive conformation of the precursor peptide.
- Class B mimetics or Type II mimetics are the modified class A mimetics with various isolated non-natural amino acids acting as the building blocks and with major backbone alterations. These comprise of foldamers such as β- and α/β-peptides as well as peptoids in which their side chains align topologically similar to the precursor peptide.
- Class C mimetics or Type III mimetics include highly modified structures with small molecule replacing the peptide backbone completely.

- Class D mimetics or Type IV mimetics are molecules that mimic the mode of action of a bioactive peptide without a direct link to its side chain functionalities and they can be generated by affinity optimization of a class C molecule or they can be identified in screenings of compound libraries or by *in silico* screening of virtual libraries.

*Figure 3. Various classes of molecular modifications*

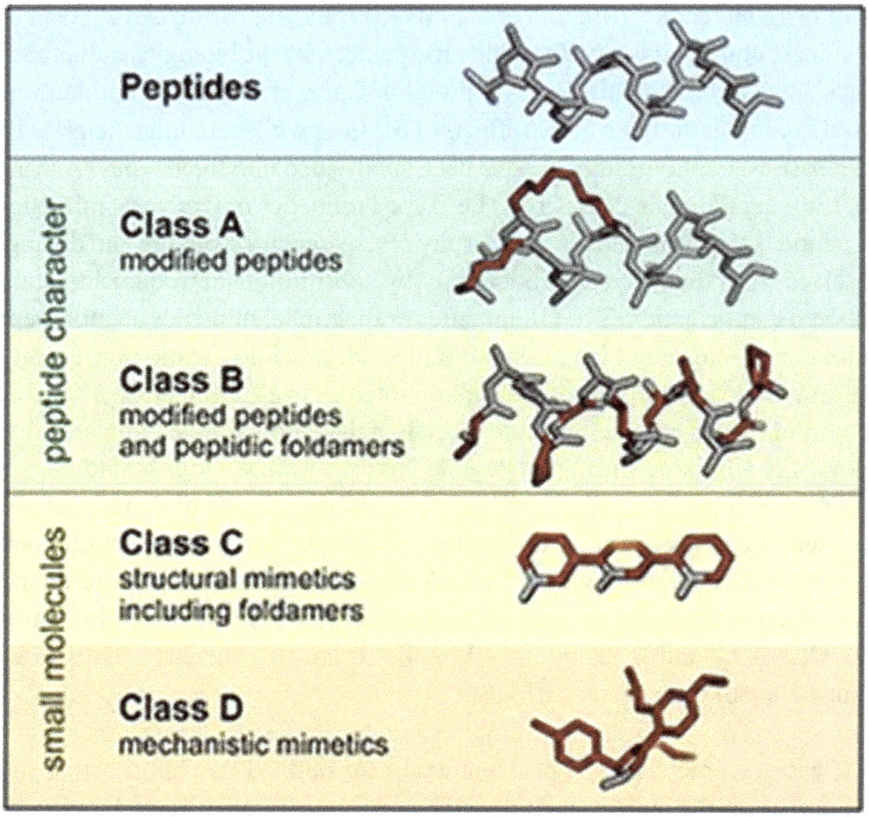


Peptidomimetics can be a promising class of molecules to regulate protein–protein interactions (PPI) in cells and hence, they play an important role in current drug-development research (Whitby and Boger, 2012). They have various properties like antimicrobial, antiviral, anticancer, antihypersensitive, antimalarial and antifouling that make them attractive in the field of therapeutics as well as diagnostics (Tenenbaum & Segal, 2015; Zhang, 2015).

## TYPES OF PEPTIDOMIMETICS

### a. Global Restrictions

Restricting the peptide globally means altering the final conformation of the peptide. This can be achieved by numerous ways. For instance, cyclization is the simplest way to introduce a conformational constraint into a peptide sequence. This results in the modification of the overall conformational profile of the target peptide compound and also increases the *in vivo* stability of the cyclic peptides compared to their linear analogs. Cyclization improves the quality of the bioactive compound i.e. bioavailability and potency as the high proportion of *cis* amide bonds and the absence of free *C*- and *N*-termini confer higher metabolic resistance. The limited conformational freedom results in higher receptor selectivity and binding affinity by reducing entropic effects.

Backbone cyclization combines with *N*-alkylation to enhance the stability of peptides (Gilon et al., 1991). The most popular approach which has been also extensively studied is the head-to-tail generation of cyclic peptide via amide bond formation according to standard peptide chemistry (Li et al., 2002). Another popular approach is the generation of cyclic peptidomimetics through a chemical bond involving two side-chains and that too by utilizing basic amino acid residues to form an amide bond through disulfide bridges between the two side-chains.

### b. Local Modifications

The local modifications are restricted to single amino acids and are local alterations of the peptide structure. They can be grouped into side-chain and backbone modifications with the aim of introducing conformational restrictions and to stabilize the molecule towards protease-mediated degradation as shown in Figure 4.

The backbone modification is introduced by amide bond surrogates which improves the stability of the peptide *in vivo* (Figure 5). The amide bond isosteres mimic the structural features of the peptide bond, modify the conformational profile and the hydrogen-bonding capability (Wolff, 1995; Choudhary and Raines, 2011).

The local modifications around side-chains are achieved by modulating the conformational profile of the peptide and thus altering all the rotatable bonds present in the amino acid unit. The side-chain modifications have explored pharmacophoric steric and electronic interactions such as modulation of the hydrophobic content by adding aromatic moieties or the introduction of polar appendages to address any polar or hydrogen-bonding interactions with the target receptor. Such localized modifications give rise to unique secondary structures which can be exploited to deliver compounds of utmost importance in therapeutics (Nandel & Jaswal, 2007).

*Figure 4. Isosteric local modification*

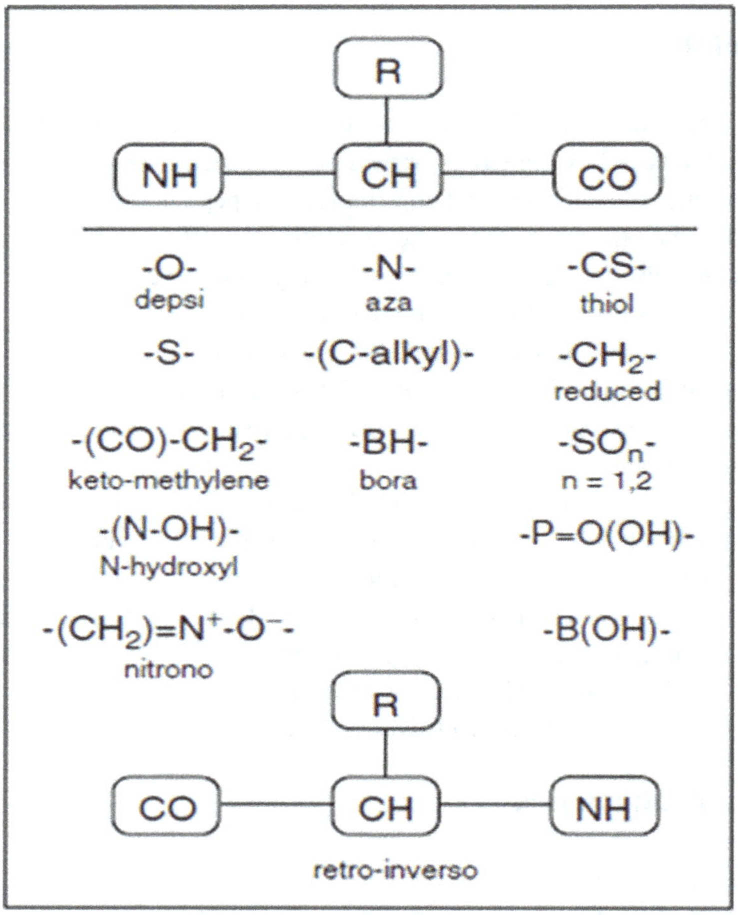


## Peptoids

Peptoids can be defined as mimetics of α-peptides in which the side chain is attached to the backbone amide nitrogen instead of the α-carbon and results in the formal shift of the position of the side chain with respect to the parent peptide backbone. They were first reported by Bartlett and coworkers in 1992 as oligomers of N-substituted glycine (Simon et al., 1992). Peptoids were initially recommended as an accessible class of molecules from which lead compounds could be identified for drug discovery. The sequence-specific peptoid oligomers are easily constructed from primary amines by the solid-phase submonomer method and their applications are of particular interest to study the peptoid secondary structures and drug design. Peptoids are advantageous as research and pharmaceutical tools which include the ease and economy of synthesis, highly variable backbone and side-chain chemistry possibilities (Nandel & Saini, 2007; Nandel & Saini, 2007; Nandel & Saini, 2011; Saini et al., 2013; Nandel et al., 2014; Saini et al., 2015; Saini, 2016; Saini & Verma, 2017; Tripathi et al., 2017).

*Figure 5. Isosteres of the peptide bond*

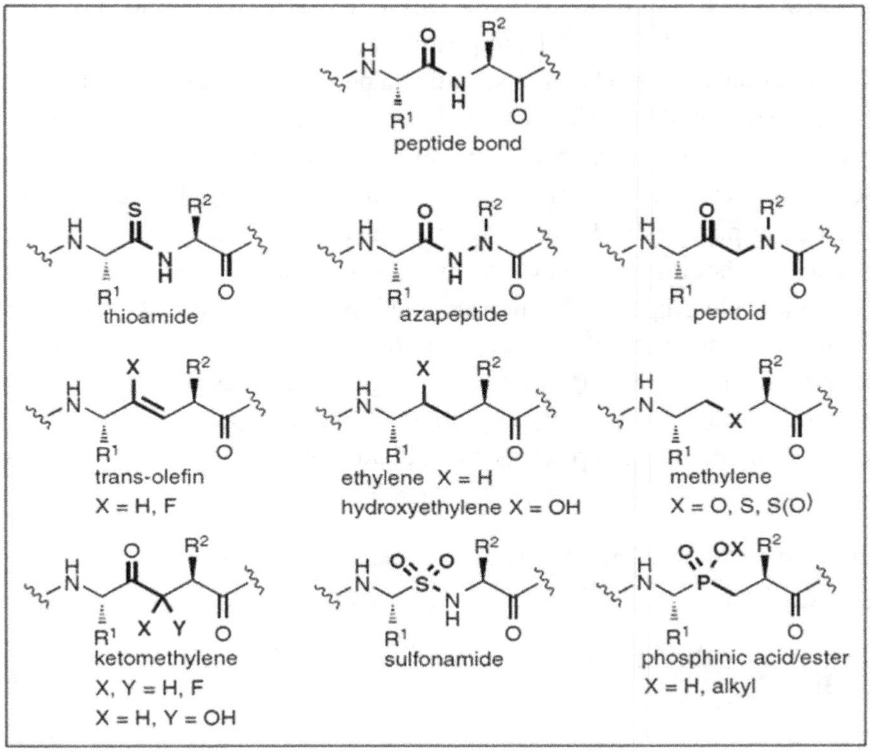


*Figure 6. Schematic representation of a peptide and a peptoid monomer*

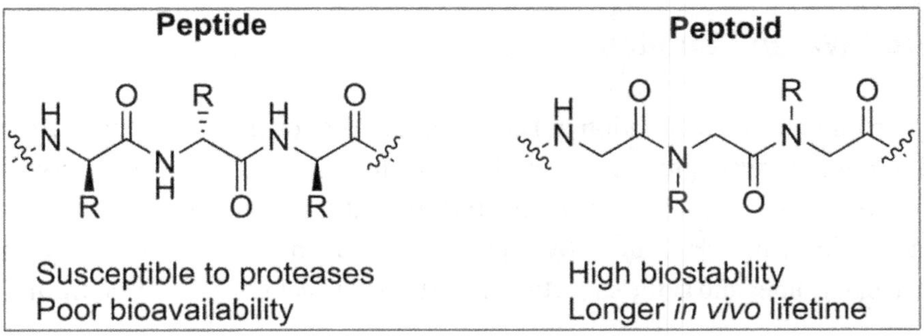

## N-Methylated Peptides (NMe Peptides)

It is a well-established approach for understanding the role of each amino acid constituting the bioactive peptide and strategic in identifying which amino acids act as hydrogen-bonding donors in the interaction with the target enzyme/receptor. This approach provides insight into the relationship between the conformational preferences of the peptide and bioactivity. The group of Kessler envisaged a simple approach of N-methylation to overcome various obstacles of peptides as a 'rational' way toward drug development (Chatterjee et al., 2013) and also to explore the reduction of conformational space while transforming a peptide into a peptidomimetic drug. Mono- and Multiple-N-methylations (Figure 7) of cyclic peptides were investigated to elucidate their remarkable conformational modulation ability by imparting steric constraints in the peptidic backbone and in improving the pharmacokinetic profile of the peptides to be used as drug leads. In this respect, N-methylation introduces another dimension to this 'spatial screening' (Kessler et al., 1996) due to the remarkable property of conformational modulation. In fact, N-methylation facilitates the occurrence of a cis-peptide bond and allows us to study the role of amide protons in establishing potential hydrogen bonds, resulting in optimization of the conformational and structural profile of a peptidomimetic which can be used to design potential drugs in pharmaceutical industry (Nandel & Jaswal, 2014)

*Figure 7. Structural representation of N-methylated peptide*

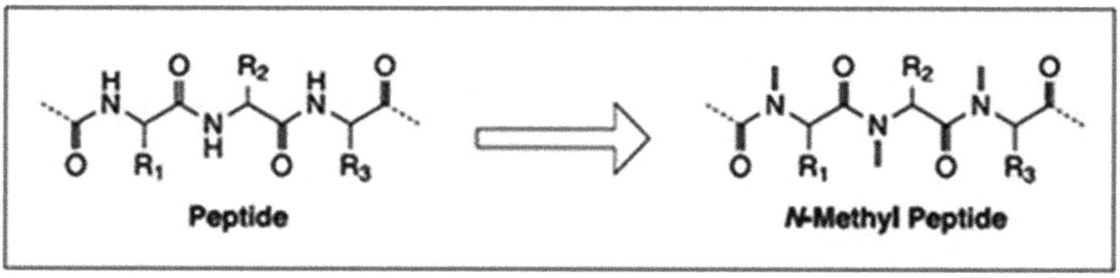


## Retro-Inverso Peptides

The retro-inverso isomerization is a common method for chemically modifying the peptidomimetic structure around amide bonds to prevent the protease recognizing the peptide-based inhibitor as a substrate inhibiting protease degradation. This isomerization can be achieved by replacing one or more l-amino acids with the parent enantiomer and at the same time inverting the backbone direction from N→C to C→N. The retro-inverso modification increases the *in vivo* stability due to the modification of amide bonds which are recognized by proteases for their hydrolysis.

## SYNTHESIS OF PEPTIDOMIMETICS AND THEIR ROLE IN BIOSENSING

The conversion of peptides to peptidomimetics is an attractive strategy where peptides are transformed to stable mimics that represent similar effects to their peptide analog in addition to increased uniformity in structure, target specificity, improved stability to proteolytic digestion and better cell membrane permeability. Consequently, incorporation of unusual amino acid substitutions, backbone amide bond modifications, stiff scaffolds or the addition of hydrophobic residues in such peptide analogs has been studied (Vagner et al., 2008; Vlieghe et al., 2010).

### Synthesis of Peptoids

Peptoids are normally synthesized using the highly flexible submonomer method (Zuckermann et al., 1992). This is a solid phase synthesis methodology (Figure 8) which includes two steps: Acylation using a halo-acetic acid also called as haloacetylation of an amine functionality attached to a solid support and then displacement of a halogen using a primary amine. This method allows high coupling efficiency and good final product yields and synthesis on solid-phase allows excess reactants to be removed easily at the end of each step.

*Figure 8. Schematic representation of synthesis process of peptoids*

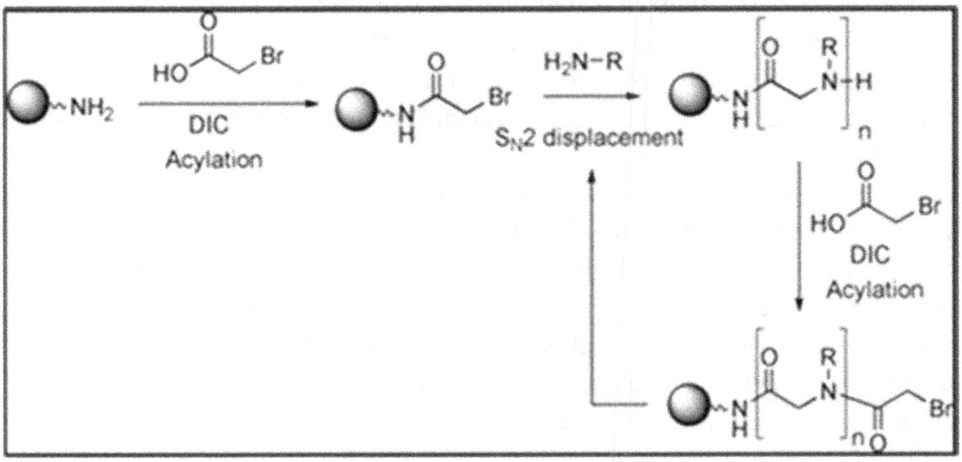

## Synthesis of N-Methylated Amino Acids

Although several efficient methods are available for the synthesis of N-methylated amino acids, the most commonly used conventional method is the one developed by Freidinger et al., involving the reductive cleavage of 5-oxazolidinones to obtain the enantiomerically pure N-methyl amino acids (Freidinger et al., 1983). However, this method is not applicable for some side-chain functionalized amino acids, such as Trp, His, or Cys and showed that the base-catalyzed N-alkylation of o-nitrobenzenesulfonamide (o-NBS) protected ⟨-amino groups using dimethylsulphate as the methylating agent is a better alternative (Miller and Scanlan, 1997). A further modification of this method employs a Mitsunobu reaction, which was optimized to obtain all possible N-methylated amino acids in solution (Biron and Kessler, 2005) or on solid support (Chatterjee et al., 2012).

## Synthesis of Specific Peptidomimetics

Similarly, N-protected amino alkyl thiols were treated with carbon disulphide in the presence of triethylamine (TEA) to generate trithiocarbonate salt which upon reaction with appropriate halides afforded dipeptidomimetics in good yields. This procedure was extended for the synthesis of N, N′-orthogonally protected trithiocarbonate-linked dipeptidomimetics (Marchiani et al., 2009). Pyrrolidinones were synthesized by the oxidation of the hydroxy function with tetrapropylammonium perruthenate/N-methylmorpholine-N-oxide and concomitant cyclization. Pyrrolidines were synthesized by tosylation of the hydroxy group and subsequent intramolecular nucleophilic substitution. Further, accessible substrates were transferred into peptidomimetics by attachment of amino acid moieties at both termini (Mayer et al., 1988).

The mono- and bis-glyoxylamide peptidomimetics were prepared via the facile ring-opening of N-acylisatins with amino acids and peptide derivatives. The ring-opening of N-acylisatins with dipeptides and tripeptides was one of the most effective strategies for the synthesis of second and third generation glyoxylamides (Merrifield, 2006).

Artificial peptides and their peptidomimetics is a great prospective that can be harnessed to develop the desired molecular biosensor because of their desirable properties such as diversified structure, high protein affinity, established synthesis protocol and modified approach (Choulier and Enander, 2010; Pazos et al., 2009). A range of peptidomimetics with specific sequences can be used to deliver high affinity to particular analytes, and can be acquired by screening and optimization of such artificial libraries. In addition, peptidomimetics like peptoids are blessed with high stability, easy modification and large chemical flexibility. For example, peptoids can be prepared with arbitrary sequences according to standard solid-phase submonomer synthesis method (Zuckermann et al., 1992). Also, peptidomimetics alongwith mass spectrometry can be used to design biosensors which can better analyze reaction mechanisms, assign enzyme substrates, detect ligand interactions etc. (Marholz et al., 2019).

The clinical potential of peptidomimetics was also confirmed by authors who projected a protein/peptide array to detect anti-Delta genotype 1,6,8 specific antibodies among Hepatitis D Virus infected patients by surface plasmon resonance imaging. An array of recombinant proteins and mimetic peptides has been defined which provided a high throughput viral hepatitis diagnostic, combining the potential of artificial molecules with the Surface Plasmon Resonance Imaging (SPRi), a technique to harness a wide sample screening. The peptidomimetics of small analytes isolated from phage display libraries were also synthesized as chimeric constructs with the NanoLuc to produce recombinant reagents for producing phage-free immunoassays, a significant contribution to the field of small-molecule immunoassays with no need of chemical conjugation of the actual analyte, the stoichiometry (peptide/enzyme) is fixed with no batch-to-batch variations, and the final tracer helps in the development of single-step immunoassays with fast readouts (Scognamiglio et al., 2015).

## PEPTOIDS AT THE BIORECEPTOR INTERFACE

The detection of bacteria in complex matrices containing many undesired interferents like human cells, commensal bacteria alongwith many proteins and metabolites requires the development of novel bioreceptors, including bacteriophages, non-antibody binding proteins, half-antibodies, and single-chain (camelid) antibodies which offer higher specificity. While antibodies are the most widely used bioreceptors in affinity biosensor research, but problems in their production, purification costs, stability during & after immobilization on sensor surface and decreased binding efficiency with time are the main limiting factors although recently antimicrobial peptides has also been exploited for the detection of whole bacteria (Pardoux et al., 2020). For the simultaneous detection of many bacteria, regeneration of the sensor surface and multiplexing are the two other crucial aspects.

To overcome some of these deficiencies, recent advances in engineered antibody mimetics include peptoid nanosheets (Olivier et al., 2013), where antibody mimetic peptoids self-assemble to form 3 to 5 nm thick sheets with surface loops expressing antigen binding sites. Such nanosheets exhibit remarkable chemical and biological stablility and can be produced with ease and precise control. Other notable designed antibody alternatives include single-chain variable fragments (ScFv) (Ahmad et al., 2012), camelid-derived heavy variable chain (VHH) antibodies (nanobodies)

(Muyldermans, 2013; Hassanzadeh-Ghassabeh et al, 2013), and single-chain antibodies expressed via yeast surface display (Richman et al., 2009), DARPins (Stumpp and Amstutz, 2007), and other artificial proteins such as adhirons (Tiede et al., 2014). These alternatives are comparatively small, easy to customize, and convenient to produce in bulk in bacterial systems, avoiding the traditional antibody production route in mammals or birds. Successful regeneration is possible with stable bioreceptors, since they can often withstand harsh regeneration buffers without compromising binding capacity. Parallel multiplexing on a single chip also reduces detection costs, providing multiple items of information from a single-shot analysis (Ahmed et al., 2014).

Two-dimensional (2D) atomically defined organic nanomaterials are an essential class of materials with broad applications. The nanostructured carbons such as carbon nanotubes (CNTs) or graphene (GF) are widely employed as electronic or electrochemical transducer in biosensor devices (Valentini et al., 2013). Graphene (GF) is the emerging choice of the nanomaterial in this perspective in biosensors because of its remarkable two-dimensional (2D) layered character. Similarly, emerging graphene-like 2D materials comprise of multiple elements that possess more versatility, greater flexibility and better functionality with a wide range of potential applications. Recently, a graphene composite biosensor has been designed to detect microcystin-LR (MC-LR), a seriously hazardous and bioaccumulative cyanotoxin for water quality monitoring assessment. A three-step linking protocol was used to immobilize MC-LR onto the GF electrodes and then incubating with conjugated monoclonal antibodies to detect MC-LR toxin. The increase of electron transfer resistance upon bioconjugation of MC-LR and antibodies on GF electrodes was used to detect the change of MC-LR concentration. A great linear sensing response ($R2 = 0.99$) of EIS change was established over a wide MC-LR concentration range of 0.005 to 10 μg/L with good reproducibility. Environmental water samples were collected from different locations like Tokyo metropolitan tap, Sanshiro pond (Tokyo, Japan), Shinobazu pond (Tokyo, Japan) and Inba lake water (Chiba prefecture, Japan). This discovery is one step close to the *in-situ* detection of even very less amount of MC-LR (i.e. 2.3 ng/L) in assessing the severity of contamination of the water source with very low detection limits (Zhang et al., 2018). Such graphene film composite biosensors deliver a promising prospect of large-scale manufacture of sensor tips due to their macroscopic free-standing nature, scalable fabrication route and easily tunable size. Similarly, this attribute of orderdness can also be extended other molecules.

One of the ways to design ordered 2D organic nanomaterials is through the supramolecular assembly of sequence-defined synthetic polymers. Individual peptoid polymers with a simple sequence of alternating hydrophobic and ionic monomers can self-assemble into highly ordered, free-floating nanosheets. A wide range of experimental characterization techniques (e.g., scanning probe, electron, and optical microscopy, X-ray diffraction, surface-selective vibrational spectroscopy, and surface tensiometry) and computational techniques (coarse-grained and atomistic modeling) have established the fundamental properties of peptoid nanosheets, their mechanism of formation, and their application as robust scaffolds for molecular recognition and as templates for the growth of inorganic minerals. Peptoid nanosheets are supramolecular assemblies of 16-42-mer chains freely floating in water that form molecular bilayers spanning many microns in lateral dimensions. Hydrophobic and electrostatic interactions result in the tight packing of the highly ordered, fully extended and packed parallel to one another component chains. Nanosheets form via a novel interface-catalyzed monolayer collapse mechanism. These nanosheets are readily engineerable, as functional monomers can be easily integrated onto the nanosheet surface or into the interior. For instance, functional hydrophilic "loops" have been displayed on the surfaces of nanosheets which can interact with specific protein targets, providing a potential platform for the molecular recognition events. Such nanosheets can also bind metal ions and serve as 2D templates for mineral growth (Robertson et al., 2016).

The performance of biosensors lies in the efficient signal capture of the biological recognition event (transduction). Transducers translate the interaction of the analyte with the biological element and generate electrochemical, electrochemiluminescent, magnetic, gravimetric, or optical signals. Nanomaterials can immobilize an enhanced quantity of bioreceptor units at reduced volumes and even act as transduction element thus increasing sensitivity and lowering the detection limits (Holzinger et al., 2014). Because of their trending potential, Olivier et al. has developed peptoid nanosheets in a way identical to the action of Human immune system. These nanosheets resembling tiny sheets of Velcro can be designed to identify varied kind of molecules, each just one-hundred nanometers across and could lead to a new class of biosensors. Antibodies have a typical architectural design: a structural scaffold which remains the same irrespective of the severity of the disease, and incessantly variable functional loops which is mimicked in a two-dimensional nanosheet scaffold covered with little functional loops like Velcro.

Functional loops on nanosheets were achieved by inserting short molecular segments into nanosheet-forming peptoid polymers. As the peptoids knit themselves together into sheets, the inserted segments expelled from the fold. These functional loops can be engineered to selectively bind certain enzymes or inorganic materials, which impart these novel materials with promising chemical sensing and catalysis with an additional advantage of a very high yield. Nanosheets with loops of varying composition, length, and density were designed to pick and sense specific enzymes out of a solution, tolerating much harsher conditions than peptides, and hence becoming the choice of material for building of a diagnostic device. This finding also represents an important step towards extending the rules of protein folding to the world of synthetic materials (Olivier et al., 2013).

In the biosensing area, small sized coumarins are one of the broadly used fluorophores because their high quantum yield, and efficient membrane permeability. Metal-complexed coumarin derivatives are multipurpose sensor molecules (Hou et al., 2013; Yeh et al., 2014; Garcia-Beltran et al., 2014). Recently, increasing numbers of triazolyl coumarin motifs have been utilized for various biological applications, such as the development of anti-inflammatory (Stefani et al., 2012), anticancer (Zhang et al., 2014), and antibacterial agents (Shi and Zhou, 2011). These compounds have also been explored as selective chemosensors (Stefani et al., 2012; Ho et al., 2012; Shi et al., 2014). Among the anions, efficient detection of the cyanide anion is particularly useful due to its severe toxic issues related to human health (Li et al., 2014; Peng et al., 2014) and upon release to environment. Lim and Lee explored the development and utility of peptoid-derived fluorescence sensors (Fuller et al., 2013) and reported the synthesis and characterization of a novel peptoid-based fluorescence probe. Authors incorporated the coumarin moiety into the peptoid backbone via copper (I) catalyzed azide-alkyne [3+2] cycloaddition reaction and the quenching of the complex was achieved by coordination with $Cu^{2+}$ ion. The peptoid-$Cu^{2+}$ complex exhibited significant Turn-ON fluorescence upon the addition of $CN^-$. In view of its utility in rapid, selective, and reversible detection of $CN^-$, novelty of the newly-generated peptoid molecule as a potential $CN^-$ sensor was established (Lim and Lee, 2016).

Peptoids can also be exploited for imaging applications to add specificity to contrast and imaging agents. For example, a potent peptoid antagonist for vascular endothelial growth factor (VEGF) receptor 2 was labeled with a positron emitter for successful positron emission tomography (PET) imaging of the receptor *in vivo*. Although, this technique was applied for tumor detection, similar approaches can be developed using peptoid ligands for neurodegeneration biomarkers, which may range from altered receptor expression to amyloid deposition (Hao et al., 2011).

## OTHER BIOMOLECULAR ANALOGS IN RANGE

Immobilized molecular receptors on self-assembled monolayers are generally used to detect biomolecules with surface plasmon resonance (SPR) based sensors (Wink et al., 1997). SPR has evolved into a useful bioanalytical tool to detect proteins, DNA, enzymes, and other biomolecules because of its excellent sensitivity to detect proteins. However, the potential of these biosensors is limited by the nonspecific interactions of biomolecules in complex matrixes (such as cell lysate, serum, and blood). Nonspecific proteins interact with the surface of biosensors creating a false positive response, obstructing the detection of analytes in crude biological fluids. Therefore, it necessitates the need to reduce nonspecific interactions so as to allow the use of biosensors for direct monitoring of biomolecules in biological fluids. This will eliminate the need of cleaning steps, signal amplification, labeling, or indirect detection of the analyte of interest. Preferably, it is the role of the monolayer to protect the SPR surface from nonspecific adsorption and provide an anchoring point for the molecular receptor. Hence, extensive research in surface chemistry has been undertaken in the past decade to overcome nonspecific adsorption. In fact, the high sensitivity of SPR to protein adsorption also makes it an excellent tool to monitor nonspecific adsorption on self-assembled monolayers (Mrksich et al., 1995). Among layers studied polyethyleneglycol (PEG), also known as polyethylene oxide (PEO), exhibits optimal performances in limiting nonspecific adsorption (Ostuni et al., 2001). Numerous variants of PEG monolayers have been investigated for nonspecific adsorption (Trmcic-Cvitas et al., 2009). Although PEG monolayers have been used for detection of proteins in serum (Teramura and Iwata, 2007), oxidative damage and the need for carboxylic acid functionalization of PEG to immobilize the molecular receptor are their two main disadvantages. To overcome these limitations, hybrid materials have been engineered to improve on the properties of PEG. For example, PEG copolymers of maleimide-PEG monolayers (Lee et al., 2007), polypropylene sulfide-PEG (Feller et al., 2005) and poly (lysine)-PEG18 were synthesized and successfully resisted the nonspecific adsorption. This approach was further modified by Messersmith and coworkers, who successfully developed a peptidomimetic polymer with ethylene glycol brushes (Statz et al., 2008) resulting in the reduced nonspecific adsorption of cells on a metallic surface. Similarly, a poly (aspartic acid) peptide with ethylene glycol (EG)-biotin brushes showed decreased nonspecific adsorption and can immobilize molecules via biotin-streptavidin-modified EG and biotinylated antibodies (Jeong et al., 2004; Jeong et al., 2006).

Similarly, phage display peptide libraries are useful tools for rapid selection of specific ligands for the development of immunoassay, pathogenic bacteria detection, cells targeted imaging, and so on (Arola et al., 2016; Ono et al., 2014; Ma et al., 2013; Gunay et al., 2017; Arevalo et al., 2012). For small molecular immunoassays, phage borne peptides are lucrative reagents for the development of competitive tests, in which an analyte peptidomimetic, a mimotope, substitutes for the chemical hapten and/or the analyte (Arevalo et al., 2012; Kim et al., 2008) or the noncompetitive tests, where these peptides react specifically with the analyte-antibody immunocomplex but not with the uncomplexed antibody (Arola et al., 2016). In these applications, the strong signal associated with the detection of the large phage surface and the possibility of noncompetitive detection in general result in assays with improved sensitivity, in the latter, the two-site recognition also provides better specificity (Hua et al., 2015).

However, because of their large size, low diffusion rate, and biological nature, there are some limitations to the use of phage particles and thus, require secondary reagents for its detection. Recently, the use of the Nano luciferase (NanoLuc) as a fusion partner to generate recombinant tracers has been explored for the development of such immunoassay. The imidaclothiz peptidomimetic C2-15 that specifically binds to the anti-imidaclothiz monoclonal antibody (mAb) 1E7 was fused to NanoLuc, both at the N terminus (C2-15-NanoLuc) and C terminus (NanoLuc-C2-15). The conjugation at C-terminus showed better performance than that at the N-terminus and was exploited to develop a bioluminescent enzyme immunoassay (BLEIA) and a bioluminescence lateral flow immunoassay (BLLFIA) for imidaclothiz. The luminescence signal of NanoLuc-C2-15 rapidly reaches high intensity with slow attenuation, which enabled one to capture the BLLFIA readout by using a smartphone without an external light source. The $IC_{50}$ of the BLEIA and BLLFIA were $3.3 \pm 0.2$ and $6.4 \pm 0.4$ ng $mL^{-1}$, respectively. Both immunoassays exhibited good accuracy for the detection of imidaclothiz in environmental and agricultural samples (Ding et al., 2018).

## CONCLUSION

The field of biosensing has emerged at a very rapid pace particularly in the last two decades and proved its potential through single analyte or multiple array based detection technology. Synthetic biology has profoundly affected the strategies to synthesize novel bioinspired materials. Furthermore, the application of nanotechnology in biosensing technology in combination with synthetic biology, is creating a deep impact on the design of biological circuits, bioproduction systems, biopolymers, biological machines, and consequently on the development of diagnostic tools for agri food, environmental, pharmaceutical and biomedical application. However, commercially robust biosensors are constantly being researched on. Therefore, various peptidomimetics together with nanotechnology offer more reliable biological recognition elements with enhanced sensitivity and stability, which actually represents the main challenges to fabricate a commercially successful biosensor.

## FUTURE PERSPECTIVES

Peptidomimetics has arisen as an exciting and promising research field, receiving a noteworthy attention from the scientific community. This consideration of various peptidomimetics, particularly peptoids arises from a number of remarkable characteristics of easy synthesis, tunable properties with scaffolds that can tolerate tough laboratory conditions and thus can serve as a solution for challenges in different application fields. More studies need to focus on the potentiality of peptidomimetics like peptoids that can be further harnessed for the development of various kinds of engineered biomaterials. The design and production of such tunable systems, based on the use of interdisciplinary technologies, has created important technological breakthroughs, allowing the expansion in the development of a wide range of research and application fields, from pharmaceutical to biomedical diagnosis and therapeutics, as well as in the biotechnological and industrial sector.

## REFERENCES

Ahmad, Z. A., Yeap, S. K., Ali, A. M., Ho, W. Y., Alitheen, N. B. M., & Hamid, M. (2012). scFv antibody: Principles and clinical application. *Clinical & Developmental Immunology*. PMID:22474489

Ahmed, A., Rushworth, J. V., Hirst, N. A., & Millner, P. A. (2014). Biosensors for whole-cell bacterial detection. *Clinical Microbiology Reviews*, *27*(3), 631–646. doi:10.1128/CMR.00120-13 PMID:24982325

Allemann, E., Leroux, J. C., & Gurny, R. (1998). Polymeric nano- and microparticles for the oral delivery of peptides and peptidomimetics. *Advanced Drug Delivery Reviews*, *34*(2-3), 171–189. doi:10.1016/S0169-409X(98)00039-8 PMID:10837677

Andrianantoandro, E., Basu, S., Karig, D. K., & Weiss, R. (2006). Synthetic biology: New engineering rules for an emerging discipline. *Molecular Systems Biology*, *2*(1). Advance online publication. doi:10.1038/msb4100073 PMID:16738572

Arevalo, F. J., Gonzalez-Techera, A., Zon, M. A., González-Sapienza, G., & Fernández, H. (2012). Ultra-sensitive electrochemical immunosensor using analyte peptidomimetics selected from phage display peptide libraries. *Biosensors & Bioelectronics*, *32*(1), 231–237. doi:10.1016/j.bios.2011.12.019 PMID:22226409

Arola, H. O., Tullila, A., Kiljunen, H., Campbell, K., Siitari, H., & Nevanen, T. K. (2016). Specific noncompetitive immunoassay for HT-2 mycotoxin detection. *Analytical Chemistry*, *88*(4), 2446–2452. doi:10.1021/acs.analchem.5b04591 PMID:26785138

Avan, I., Hall, C. D., & Katritzky, A. R. (2014). Peptidomimetics via modifications of amino acids and peptide bonds. *Chemical Society Reviews*, *43*(10), 3575–3594. doi:10.1039/c3cs60384a PMID:24626261

Azzarito, V., Long, K., Murphy, N. S., & Wilson, A. J. (2013). Inhibition of α-helix-mediated protein-protein interactions using designed molecules. *Nature Chemistry*, *5*(3), 161–173. doi:10.1038/nchem.1568 PMID:23422557

Biron, E., & Kessler, H. (2005). Convenient synthesis of N-methylamino acids compatible with Fmoc solid-phase peptide synthesis. *The Journal of Organic Chemistry*, *70*(13), 5183–5189. doi:10.1021/jo050477z PMID:15960522

Bodanszky, M. (1985). In search of new methods in peptide synthesis. A review of the last three decades. *International Journal of Peptide and Protein Research*, *25*(5), 449–474. doi:10.1111/j.1399-3011.1985.tb02200.x PMID:3894264

Buenger, D., Topuz, F., & Groll, J. (2012). Hydrogels in sensing applications. *Progress in Polymer Science*, *37*(12), 1678–1719. doi:10.1016/j.progpolymsci.2012.09.001

Chao, J., Zhu, D., Zhang, Y., Wang, L., & Fan, C. (2016). DNA nanotechnology-enabled biosensors. *Biosensors & Bioelectronics*, *76*, 68–79. doi:10.1016/j.bios.2015.07.007 PMID:26212206

Chatterjee, J., Laufer, B., & Kessler, H. (2012). Synthesis of N-methylated cyclic peptides. *Nature Protocols*, *7*(3), 432–444. doi:10.1038/nprot.2011.450 PMID:22322216

Chatterjee, J., Rechenmacher, F., & Kessler, H. (2013). N-methylation of peptides and proteins: An important element for modulating biological functions. *Angewandte Chemie International Edition*, *52*(1), 254–269. doi:10.1002/anie.201205674 PMID:23161799

Checa, S. K., Zurbriggen, M. D., & Soncini, F. C. (2012). Bacterial signaling systems as platforms for rational design of new generations of biosensors. *Current Opinion in Biotechnology*, *23*(5), 766–772. doi:10.1016/j.copbio.2012.05.003 PMID:22658939

Choudhary, A., & Raines, R. T. (2011). An evaluation of peptide-bond isosteres. *ChemBioChem*, *12*(2), 1801–1807. doi:10.1002/cbic.201100272 PMID:21751326

Choulier, L., & Enander, K. (2010). Environmentally Sensitive Fluorescent Sensors Based on Synthetic Peptides. *Sensors (Basel)*, *10*(4), 3126–3144. doi:10.3390100403126 PMID:22319290

Clark, L. C. Jr, & Lyons, C. (1962). Electrode systems for continuous monitoring in cardiovascular surgery. *Annals of the New York Academy of Sciences*, *102*(1), 29–45. doi:10.1111/j.1749-6632.1962.tb13623.x PMID:14021529

Clarke, S., Pinaud, F., Beutel, O., You, C. J., Piehler, J., & Dahan, M. (2010). Covalent monofunctionalization of peptide-coated quantum dots for single-molecule assays. *Nano Letters*, *10*(6), 2147–2154. doi:10.1021/nl100825n PMID:20433164

Ding, Y., Hua, X., Chen, H., Liu, F., González-Sapien, G., & Wang, M. (2018). Recombinant Peptidomimetic-Nano Luciferase Tracers for Sensitive Single-Step Immunodetection of Small Molecules. *Analytical Chemistry*, *90*(3), 2230–2237. doi:10.1021/acs.analchem.7b04601 PMID:29280616

Effendi, L., Nielsen, D., Solomon, K., & Prather, K. J. (2008). Protein promiscuity and its implications for biotechnology. *Trends in Biotechnology*, *26*(12), 674–681. PMID:18977048

Feller, L. M., Cerritelli, S., Textor, M., Hubbell, J. A., & Tosatti, S. G. P. (2005). Influence of poly (propylene sulfide-block-ethylene glycol) di- And triblock copolymer architecture on the formation of molecular adlayers on gold surfaces and their effect on protein resistance: A candidate for surface modification in biosensor research. *Macromolecules*, *38*(25), 10503–10510. doi:10.1021/ma051424m

Feng, D., Zhang, Y. Y., Feng, T. T., Shi, W., Li, X. M., & Ma, H. M. (2011). A graphene oxide–peptide fluorescence sensor tailor-made for simple and sensitive detection of matrix metalloproteinase 2. *Chemical Communications*, *47*(38), 10680–10682. doi:10.1039/c1cc13975d PMID:21892449

Feng, T. T., Feng, D., Shi, W., Li, X. H., & Ma, H. M. (2012). A graphene oxide-peptide fluorescence sensor for proteolytically active prostate-specific antigen. *Molecular BioSystems*, *8*(5), 1441–1445. doi:10.1039/c2mb05379a PMID:22334395

Fischer, P. M. (2003). The Design, Synthesis and Application of Stereochemical and Directional Peptide Isomers: A Critical Review. *Current Protein & Peptide Science*, *4*(5), 339–356. doi:10.2174/1389203033487054 PMID:14529528

Floris, M., & Moro, S. (2012). Mimicking Peptides… In Silico. *Molecular Informatics*, *31*(1), 12–20. doi:10.1002/minf.201100093 PMID:27478173

Freidinger, R. M., Hinkle, J. S., & Perlow, D. S. (1983). Synthesis of 9-fluorenylmethyloxycarbonyl-protected N-alkyl amino acids by reduction of oxazolidinones. *The Journal of Organic Chemistry*, *48*(1), 77–81. doi:10.1021/jo00149a016

Fukuyama, T., Jow, C. K., & Cheung, M. (1995). 2- and 4-Nitrobenzenesulfonamides: Exceptionally versatile means for preparation of secondary amines and protection of amines. *Tetrahedron Letters*, *36*(36), 6373–6374. doi:10.1016/0040-4039(95)01316-A

Fuller, A. A., Holmes, C. A., & Seidl, F. J. (2013). A fluorescent peptoid ph-sensor. *Biopolymers*, *100*(4), 380–386. doi:10.1002/bip.22242 PMID:23526424

Gante, J. (1994). Peptidomimetics—Tailored Enzyme Inhibitors. *Angewandte Chemie International Edition in English*, *33*(17), 1699–1720. doi:10.1002/anie.199416991

Garcia-Beltran, O., Cassels, B. K., Perez, C., Mena, N., Nunez, M. T., Martinez, N. P., Pavez, P., & Aliaga, M. E. (2014). Coumarin-based fluorescent probes for dual recognition of copper (II) and iron (III) ions and their application in bio-imaging. *Sensors (Basel)*, *14*(1), 1358–1371. doi:10.3390140101358 PMID:24419164

Gilon, C., Halle, D., Chorev, M., Selincer, Z., & Byk, G. (1991). Backbone cyclization: A new method for conferring conformational constraint on peptides. *Biopolymers*, *31*(6), 745–750. doi:10.1002/bip.360310619 PMID:1718473

González-Techera, A., Kim, H. J., Gee, S. J., Last, J. A., Hammock, B. D., & González-Sapienza, G. (2007). Polyclonal antibody-based noncompetitive immunoassay (PHAIA) for small analytes developed with short peptide loops isolated from phage libraries. *Analytical Chemistry*, *79*(23), 9191–9196. doi:10.1021/ac7016713 PMID:17973501

González-Techera, A., Umpiérrez-Failache, M., Cardozo, S., Obal, G., Pritsch, O., Last, J. A., Gee, S. J., Hammock, B. D., & González-Sapienza, G. (2008). High-Throughput Method for Ranking the Affinity of Peptide Ligands Selected from Phage Display Libraries. *Bioconjugate Chemistry*, *19*(5), 993–1000. doi:10.1021/bc700279y PMID:18393454

Goodman, M., & Ro, S. (1995). *Burger's Medicinal Chemistry and Drug Discovery*. John Wiley & Sons Inc.

Griffin, G. D., & Stratis-Cullum, D. N. (2009). *Biosensors, Encyclopedia of Microbiology*. Academic Press.

Gunay, K. A., Benczedi, D., Herrmann, A., & Klok, H. A. (2017). Peptide-Enhanced Selective Surface Deposition of Polymer-Based Fragrance Delivery Systems. *Advanced Functional Materials*, *27*(2), 1603843. doi:10.1002/adfm.201603843

Gunning, P. T., Katt, W. P., Glenn, M., Siddique, K., Kim, J. S., Jove, R., Sebti, S. M., Turkson, J., & Hamilton, A. D. (2007). Isoform selective inhibition of STAT1 or STAT3 homo-dimerization via peptidomimetic probes: Structural recognition of STAT SH2 domains. *Bioorganic & Medicinal Chemistry Letters*, *17*(7), 1875–1878. doi:10.1016/j.bmcl.2007.01.077 PMID:17336521

Hao, G., Hajibeigi, A., León-Rodríguez, L. M., Oz, O. K., & Sun, X. (2011). Peptoid-based PET imaging of vascular endothelial growth factor receptor (VEGFR) expression. *American Journal of Nuclear Medicine and Molecular Imaging*, *1*(1), 65–75. PMID:23133797

Hassanzadeh-Ghassabeh, G., Devoogdt, N., De Pauw, P., Vincke, C., & Muyldermans, S. (2013). Nanobodies and their potential applications. *Nanomedicine (London)*, *8*(6), 1013–1026. doi:10.2217/nnm.13.86 PMID:23730699

Ho, I. T., Lai, T. L., Wu, R. T., Tsai, M. T., Wu, C. M., Lee, G. H., & Chung, W. S. (2012). Design and synthesis of triazolyl coumarins as Hg2+ selective fluorescent chemosensors. *Analyst (London)*, *137*(24), 5770–5776. doi:10.1039/c2an36076d PMID:23092986

Holzinger, M., Le Goff, A., & Cosnier, S. (2014). Nanomaterials for biosensing applications: A review. *Frontiers in Chemistry*, *2*, 63. doi:10.3389/fchem.2014.00063 PMID:25221775

Homola, J., Yee, S. S., & Gauglitz, G. (1999). Surface Plasmon Resonance Sensors. *Sensors and Actuators. B, Chemical*, *54*(1-2), 3–15. doi:10.1016/S0925-4005(98)00321-9

Hou, J. T., Li, K., Yu, K. K., Wu, M. Y., & Yu, X. Q. (2013). Coumarin-dpa-cu (II) as a chemosensing ensemble towards histidine determination in urine and serum. *Organic & Biomolecular Chemistry*, *11*(5), 717–720. doi:10.1039/C2OB26955D PMID:23238391

Hua, X. D., Zhou, L. L., Feng, L., Ding, Y., Shi, H. Y., Wang, L. M., Gee, S. J., Hammock, B. D., & Wang, M. H. (2015). Competitive and noncompetitive phage immunoassays for the determination of benzothiostrobin. *Analytica Chimica Acta*, *890*, 150–156. doi:10.1016/j.aca.2015.07.056 PMID:26347177

Jaworski, J., Yokoyama, K., Zueger, C., Chung, W. J., Lee, S. W., & Majumdar, A. (2011). Polydiacetylene incorporated with peptide receptors for the detection of trinitrotoluene explosives. *Langmuir*, *27*(6), 3180–3187. doi:10.1021/la104476p PMID:21275406

Jeong, J. H., Kim, B. Y., Cho, S. H., & Kim, J. D. J. (2004). Surface Fabrication Of Biotinylated Polypeptide Multi-Layer Designed For Surface Plasmon Resonance Sensing. *Nonlinear Opt. Phys. Mater.*, *13*(03n04), 525–534. doi:10.1142/S0218863504002213

Jeong, J. H., Kim, B. Y., Lee, S. J., & Kim, J. D. (2006). A novel immobilization technique using a poly(amino acid) multilayer designed for surface plasmon resonance sensing. *Chemical Physics Letters*, *421*(4-6), 373–377. doi:10.1016/j.cplett.2006.01.098

Kessler, H., Gratias, R., Hessler, G., Gurrath, M., & Müller, G. (1996). Conformation of cyclic peptides. Principle concepts and the design of selectivity and superactivity in bioactive sequences by 'spatial screening'. *Pure and Applied Chemistry*, *68*(6), 1201–1205. doi:10.1351/pac199668061201

Kim, H. J., González-Techera, A., González-Sapienza, G., Ahn, K. C., Gee, S. J., & Hammock, B. D. (2008). Phage-borne Peptidomimetics Accelerate the Development of Polyclonal Antibody-based Heterologous Immunoassays for the Detection of Pesticide Metabolites. *Environmental Science & Technology*, *42*(6), 2047–2053. doi:10.1021/es702219a PMID:18409635

Knight, A. S., Zhou, E. Y., Francis, M. B., & Zuckermann, R. N. (2015). Sequence Programmable Peptoid Polymers for Diverse Materials Applications. *Advanced Materials*, *27*(38), 5665–5691. doi:10.1002/adma.201500275 PMID:25855478

Lee, C. Y., Gamble, L. J., Grainger, D. W., & Castner, D. G. (2006). Mixed DNA/oligo (ethylene glycol) functionalized gold surfaces improve DNA hybridization in complex media. *Biointerphases*, *1*(2), 82–92. doi:10.1116/1.2219110 PMID:20408620

Li, J., Gao, J., Xiong, W. W., Li, P. Z., Zhang, H., Zhao, Y., & Zhang, Q. (2014). Pyridinium-fused pyridinone: A novel "turn-on" fluorescent chemodosimeter for cyanide. *Chemistry, an Asian Journal*, *9*(1), 121–125. doi:10.1002/asia.201301144 PMID:24347071

Li, P., Roller, P. P., & Xu, J. (2002). Current synthetic approaches to peptide and peptidomimetic cyclization. *Current Organic Chemistry*, *6*(5), 411–440. doi:10.2174/1385272024604970

Lim, B., & Lee, J. (2016). A Peptoid-Based Fluorescent Sensor for Cyanide Detection. *Molecules (Basel, Switzerland)*, *21*(3), 339. doi:10.3390/molecules21030339 PMID:26978334

Ma, K., Wang, D. D., Lin, Y., Wang, J., Petrenko, V., & Mao, C. (2013). Synergetic targeted delivery of sleeping-beauty transposon system to mesenchymal stem cells using LPD nanoparticles modified with a phage-displayed targeting peptide. *Advanced Functional Materials*, *23*(9), 1172–1181. doi:10.1002/adfm.201102963 PMID:23885226

Marchiani, A., Antolini, N., Calderan, A., & Ruzza, P. (2009). A typical SH3 domain binding motifs: Features and properties. *Current Top Pept, 10*, 45–56.

Marholz, L. J., Yang, T-Y., & Parker, L. L. (2019) Probing Biological Activities with Peptide and Peptidomimetic Biosensors. In Mass Spectrometry-Based Chemical Proteomics. John Wiley & Sons.

Marner, W. D. II. (2009). Practical application of synthetic biology principles. *Biotechnology Journal, 4*(10), 1406–1419. doi:10.1002/biot.200900167 PMID:19830715

Marx, K. A. (2006). Toward understanding the intelligent properties of biological macromolecules-implications for their design into biosensors. In *Smart Biosensor Technology* (pp. 25–104). CRC Press. doi:10.1201/9781420019506.pt1

Mayer, B. J., Hamaguchi, M., & Hanafusa, H. (1988). A novel viral oncogene with structural similarity to phospholipase C. *Nature, 332*(6161), 272–275. doi:10.1038/332272a0 PMID:2450282

Merrifield, R. B. (1969). Solid-Phase Peptide Synthesis. *Advances in Enzymology and Related Areas of Molecular Biology, 32*, 221–296. PMID:4307033

Miller, S. C., & Scanlan, T. S. (1997). Site-Selective N-Methylation of Peptides on Solid Support. *Journal of the American Chemical Society, 119*(9), 2301–2302. doi:10.1021/ja9635443

Miller, S. M., Simon, R. J., Ng, S., Zuckermann, R. N., Kerr, J. M., & Moos, W. H. (1995). Comparison of the Proteolytic Susceptibilities of Homologous L-Amino Acid, D-Amino Acid, and N-Substituted Glycine Peptide and Peptoid Oligomers. *Drug Development Research, 35*(1), 20–32. doi:10.1002/ddr.430350105

Mitchell, A. R. (2008). Bruce Merrifield and solid-phase peptide synthesis: A historical assessment. *Peptide Science (Hoboken, N.J.), 90*(3), 175–184. PMID:18213693

Mizuno, A., Matsui, K., & Shuto, S. (2017). From Peptides to Peptidomimetics: A Strategy Based on the Structural Features of Cyclopropane. *Chemistry (Weinheim an der Bergstrasse, Germany), 23*(58), 14394–14409. doi:10.1002/chem.201702119 PMID:28632330

Mohanty, S. P., & Kougianos, E. (2006). Biosensors: A tutorial review. *IEEE Potentials, 25*(2), 35–40. doi:10.1109/MP.2006.1649009

Mrksich, M., Sigal, G. B., & Whitesides, G. M. (1995). Surface Plasmon Resonance Permits in Situ Measurement of Protein Adsorption on Self-Assembled Monolayers of Alkanethiolates on Gold. *Langmuir, 11*(11), 4383–4385. doi:10.1021/la00011a034

Muyldermans, S. (2013). Nanobodies: Natural single-domain antibodies. *Annual Review of Biochemistry*, *82*(1), 775–797. doi:10.1146/annurev-biochem-063011-092449 PMID:23495938

Nam, K. T., Shelby, S. A., Choi, P. H., Marciel, A. B., Chen, R., Tan, L., Chu, T. K., Mesch, R. A., Lee, B.-C., Connolly, M. D., Kisielowski, C., & Zuckermann, R. N. (2010). Free-Floating Ultrathin Two-Dimensional Crystals from Sequence-Specific Peptoid Polymers. *Nature Materials*, *9*(5), 454–460. doi:10.1038/nmat2742 PMID:20383129

Nandel, F. S., & Jaswal, R. (2007). New Type of Helix and 27 Ribbon Structure Formation in Poly ΔLeu Peptides: Construction of a Single-Handed Template. *Biomacromolecules*, *8*(10), 3093–3101. doi:10.1021/bm700504h PMID:17883275

Nandel, F. S., & Jaswal, R. R. (2014). Conformational study of N-methylated alanine inhibitor. *Indian Journal of Biochemistry peptides and design of A & Biophysics*, *51*, 7-18.

Nandel, F. S., Jaswal, R. R., Saini, A., Nandel, V., & Shafique, M. (2014). Construction and conformational behavior of peptoids with cis-amide bond geometry: Design of a peptoid with alternate φ, ψ values of inverse PP-II/PP-II and PP-I structures. *Journal of Molecular Modeling*, *20*(9), 2429. doi:10.100700894-014-2429-8 PMID:25173871

Nandel, F. S., & Saini, A. (2007). Conformational Study of Short peptoid models for future applications as potent antimicrobial compounds. *Macromolecular Theory and Simulations*, *16*(3), 295–303. doi:10.1002/mats.200600080

Nandel, F. S., & Saini, A. (2007). Construction and Design of Single Strand Collagen like Structure. *Indian Journal of Biochemistry & Biophysics*, *44*, 106–113. PMID:17536339

Nandel, F. S., & Saini, A. (2011). Peptoids with aliphatic sidechains as helical structures without hydrogen bonds and collagen/ inverse-collagen type structures. *Journal of Biophysical Chemistry*, *2*(01), 37–48. doi:10.4236/jbpc.2011.21006

Oh, K. J., Cash, K. J., & Plaxco, K. W. (2006). Excimer-Based Peptide Beacons: A Convenient Experimental Approach for Monitoring Polypeptide-Protein and Polypeptide-Oligonucleotide Interactions. *Journal of the American Chemical Society*, *128*(43), 14018–14019. doi:10.1021/ja0651310 PMID:17061871

Olivier, G. K., Cho, A., Sanii, B., Connolly, M. D., Tran, H., & Zuckermann, R. N. (2013). Antibody-Mimetic Peptoid Nanosheets for Molecular Recognition. *ACS Nano*, *7*(10), 9276–9286. doi:10.1021/nn403899y PMID:24016337

Ono, K., Takeuchi, K., Ueda, H., Morita, Y., Tanimura, R., Shimada, I., & Takahashi, H. (2014). Structure-based approach to improve a small-molecule inhibitor by the use of a competitive peptide ligand. *Angewandte Chemie International Edition*, *53*(10), 2597–2601. doi:10.1002/anie.201310749 PMID:24481871

Pardoux, É., Boturyn, D., & Roupioz, Y. (2020). Antimicrobial Peptides as Probes in Biosensors Detecting Whole Bacteria: A Review. *Molecules (Basel, Switzerland)*, *25*(8), 1998–2018. doi:10.3390/molecules25081998 PMID:32344585

Patch, J. A., Kirshenbaum, K., Seurynck, S. L., Zuckermann, R. N., & Barron, A. E. (2004). *Versatile oligo (N-substituted) glycines: the many roles of peptoids in drug discovery*. John Wiley & Sons.

Pavan, S., & Berti, F. (2012). Short peptides as biosensor transducers. *Analytical and Bioanalytical Chemistry*, *402*(10), 3055–3070. doi:10.100700216-011-5589-8 PMID:22169951

Pazos, E., Vazquez, O., Mascarenas, J. L., & Vazquez, M. E. (2009). Peptide-based fluorescent biosensors. *Chemical Society Reviews*, *38*(12), 3348–3359. doi:10.1039/b908546g PMID:20449054

Pelay-Gimeno, M., Glas, A., Koch, O., & Grossmann, T. N. (2015). Structure-Based Design of Inhibitors of Protein-Protein Interactions: Mimicking Peptide Binding Epitopes. Angewandte Chemie International Edition, 54(31), 8896-8927.

Peng, M. J., Guo, Y., Yang, X. F., Suzenet, F., Li, J., Li, C. W., & Duan, Y. W. (2014). Coumarin–hemicyanine conjugates as novel reaction-based sensors for cyanide detection: Convenient synthesis and ict mechanism. *RSC Advances*, *4*(37), 19077–19085. doi:10.1039/c4ra01598c

Peng, X. Z., Draney, D. R., & Volcheck, W. M. (2006). Article. *Opt. Mol. Probes Biomed. Appl.*, *6097*, 150–F970.

Richman, S., Kranz, D. M., & Stone, J. D. (2009). Biosensor detection systems: engineering stable, high-affinity bioreceptors by yeast surface display. *Biosensors and Biodetection, 504*, 323-350.

Robertson, E. J., Battigelli, A., Proulx, C., Mannige, R. V., Haxton, T. K., Yun, L., Whitelam, S., & Zuckermann, R. N. (2016). Design, Synthesis, Assembly, and Engineering of Peptoid Nanosheets. *Accounts of Chemical Research*, *49*(3), 379–389. doi:10.1021/acs.accounts.5b00439 PMID:26741294

Saini, A., Jaswal, R. R., Negi, R., & Nandel, F. S. (2013). Insights on the structural characteristics of Vim-TBS (58-81) peptide for future applications as a cell penetrating peptide. *Bioscience Trends*, *7*(5), 209–220. doi:10.5582/bst.2013.v7.5.209 PMID:24241171

Saini, A., & S, F. (2016). Structural Modeling and Conformational Analysis of Aromatic Polypeptoid Models confined to different environmental conditions. *International Journal of Computers and Applications*, *143*(7), 46–56. doi:10.5120/ijca2016910265

Saini, A., Sharma, S., & Jaswal, R. R. (2015). Non-covalent interactions guide the structural plasticity of Desmin Tubulin binding peptides: A Molecular Mechanics and Molecular Dynamics Study. *International Journal of Peptide Research and Therapeutics*, *21*(4), 459–478. doi:10.100710989-015-9474-8

Saini, A., & Verma, G. (2017). Peptoids: Tomorrow's Therapeutics. In Nanostructures for Novel Therapy: Synthesis, Characterization and Applications (pp. 251-280). Elsevier Publications.

Salazar-Salinas, K., Jauregui, L. A., Kubli-Garfias, C., & Seminario, J. M. (2009). Molecular biosensor based on a coordinated iron complex. *The Journal of Chemical Physics, 130*(10), 03B608.

Sanii, B., Kudirka, R., Cho, A., Venkateswaran, N., Olivier, G. K., Olson, A. M., Tran, H., Harada, R. M., Tan, L., & Zuckermann, R. N. (2011). Shaken, not stirred: Collapsing a peptoid monolayer to produce free-floating, stable nanosheets. *Journal of the American Chemical Society*, *133*(51), 20808–20815. doi:10.1021/ja206199d PMID:21939206

Scognamiglio, V., Antonacci, A., Lambreva, M. D., Litescu, S. C., & Rea, G. (2015). Synthetic biology and biomimetic chemistry as converging technologies fostering a new generation of smart biosensors. *Biosensors & Bioelectronics*, *74*, 1076–1086. doi:10.1016/j.bios.2015.07.078 PMID:26277908

Scognamiglio, V., Pezzotti, G., Pezzotti, I., Cano, J., Buonasera, K., Giannini, D., & Giardi, M. T. (2010). Biosensors for effective environmental and agrifood protection and commercialization: From research to market. *Mikrochimica Acta, 170*(3-4), 215–225. doi:10.100700604-010-0313-5

Shi, D. T., Wei, X. L., Sheng, Y., Zang, Y., He, X. P., Xie, J., Liu, G., Tang, Y., Li, J., & Chen, G. R. (2014). Substitution pattern reverses the fluorescence response of coumarin glycoligands upon coordination with silver (I). *Scientific Reports*, *4*(1), 4252. doi:10.1038rep04252 PMID:24584644

Shi, Y. A., & Zhou, C. (2011). Synthesis and evaluation of a class of new coumarin triazole derivatives as potential antimicrobial agents. *Bioorganic & Medicinal Chemistry Letters*, *21*(3), 956–960. doi:10.1016/j.bmcl.2010.12.059 PMID:21215620

Simon, R. J., Kania, R. S., Zuckermann, R. N., Huebner, V. D., Jewell, D. A., Banville, S., & Marlowe, C. K. (1992). Peptoids: A modular approach to drug discovery. *Proceedings of the National Academy of Sciences of the United States of America*, *89*(20), 9367–9371. doi:10.1073/pnas.89.20.9367 PMID:1409642

Statz, A. R., Barron, A. E., & Messersmith, P. B. (2008). Protein, cell and bacterial fouling resistance of polypeptoid-modified surfaces: Effect of side-chain chemistry. *Soft Matter*, *4*(1), 131–139. doi:10.1039/B711944E PMID:21472038

Stefani, H. A., Gueogjan, K., Manarin, F., Farsky, S. H., Zukerman-Schpector, J., Caracelli, I., Pizano Rodrigues, S. R., Muscara, M. N., Teixeira, S. A., Santin, J. R., Machado, I. D., Bolonheis, S. M., Curi, R., & Vinolo, M. A. (2012). Synthesis, biological evaluation and molecular docking studies of 3-(triazolyl)-coumarin derivatives: Effect on inducible nitric oxide synthase. *European Journal of Medicinal Chemistry*, *58*, 117–127. doi:10.1016/j.ejmech.2012.10.010 PMID:23123728

Steinman, D. H., Curtin, M. L., Garland, R. B., Davidsen, S. K., Heyman, H. R., Holms, J. H., & Li, J. (1998). The design, synthesis, and structure-activity relationships of a series of macrocyclic MMP inhibitors. *Bioorganic & Medicinal Chemistry Letters*, *8*(16), 2087–2092. doi:10.1016/S0960-894X(98)00396-5 PMID:9873491

Stumpp, M. T., & Amstutz, P. (2007). DARPins: A true alternative to antibodies. *Current Opinion in Drug Discovery & Development*, *10*(2), 153–159. PMID:17436550

Tenenbaum, E., & Segal, E. (2015). Optical biosensors for bacteria detection by a peptidomimetic antimicrobial compound. *Analyst (London)*, *140*(22), 7726–7733. doi:10.1039/C5AN01717C PMID:26456237

Teramura, Y., & Iwata, H. (2007). Label-free immunosensing for alpha-fetoprotein in human plasma using surface plasmon resonance. *Analytical Biochemistry*, *365*(2), 201–207. doi:10.1016/j.ab.2007.03.022 PMID:17459320

Tiede, C., Tang, A. A. S., Deacon, S. E., Mandal, U., Nettleship, J. E., Owen, R. L., George, S. E., Harrison, D. J., Owens, R. J., Tomlinson, D. C., & McPherson, M. J. (2014). Adhiron: A stable and versatile peptide display scaffold for molecular recognition applications. *Protein Engineering, Design & Selection*, *27*(5), 145–155. doi:10.1093/protein/gzu007 PMID:24668773

Tripathi, R., Sharma, P., Saini, A., & Verma, G. (2017). Selfhealed Materials from Elastomeric Composites: Concepts, Strategies and Developments. In Smart Polymer Nanocomposites (pp. 219-242). Springer International Publishing.

Trmcic-Cvitas, J., Hasan, E., Ramstedt, M., Li, X., Cooper, M. A., Abell, C., Huck, W. T. S., & Gautrot, J. E. (2009). Biofunctionalized protein resistant oligo (ethylene glycol)-derived polymer brushes as selective immobilization and sensing platforms. *Biomacromolecules*, *10*(10), 2885–2894. doi:10.1021/bm900706r PMID:19761181

Valentini, F., Carbone, M., & Palleschi, G. (2013). Carbon nanostructured materials for applications in nanomedicine, cultural heritage, and electrochemical biosensors. *Analytical and Bioanalytical Chemistry*, *405*(2-3), 451–465. doi:10.100700216-012-6351-6 PMID:23064705

Vidal, J. C., Bonel, L., Ezquerra, A., Hernández, S., Bertolín, J. R., Cubel, C., & Castillo, J. R. (2013). Electrochemical affinity biosensors for detection of mycotoxins: A review. *Biosensors & Bioelectronics*, *49*, 146–158. doi:10.1016/j.bios.2013.05.008 PMID:23743326

Wang, C. K., Wang, J. E., Liu, D. J., & Wang, Z. (2010). Gold nanoparticle-based colorimetric sensor for studying the interactions of beta-amyloid peptide with metallic ions. *Talanta*, *80*(5), 1626–1631. doi:10.1016/j.talanta.2009.09.052 PMID:20152387

Wang, J. X., Sun, X. W., Wei, A., Lei, Y., Cai, X. P., Li, C. M., & Dong, Z. L. (2006). Zinc oxide nanocomb biosensor for glucose detection. *Applied Physics Letters*, *88*(23), 233106. doi:10.1063/1.2210078

Whitby, L. R., & Boger, D. L. (2012). Comprehensive Peptidomimetic Libraries Targeting Protein–Protein Interactions. *Accounts of Chemical Research*, *45*(10), 1698–1709. doi:10.1021/ar300025n PMID:22799570

Wink, T., Van Zuilen, S. J., Bult, A., & Van Bennekom, W. P. (1997). Self-assembled monolayers for biosensors. *Analyst (London)*, *122*(4), R43–R50. doi:10.1039/a606964i PMID:9177074

Xu, Z., Liu, X., Pan, J., & Spring, D. R. (2012). Coumarin-derived transformable fluorescent sensor for Zn2+. *Chemical Communications*, *48*(39), 4764–4766. doi:10.1039/c2cc30963g PMID:22472965

Yan, X. W., Yang, L. M., & Wang, Q. Q. (2011). Lanthanide-Coded Protease-Specific Peptide–Nanoparticle Probes for a Label-Free Multiplex Protease Assay Using Element Mass Spectrometry: A Proof-of-Concept Study. *Angewandte Chemie International Edition*, *50*(22), 5130–5133. doi:10.1002/anie.201101087 PMID:21504034

Yeh, H. Y., Hwang, Y. C., Yates, M. V., Mulchandani, A., & Chen, W. (2008). Detection of hepatitis a virus by using a combined cell culture-molecular beacon assay. *Applied and Environmental Microbiology*, *74*(7), 2239–2243. doi:10.1128/AEM.00259-08 PMID:18263747

Yeh, H. Y., Yates, M. V., Mulchandani, A., & Chen, W. (2008). Visualizing the dynamics of viral replication in living cells via Tat peptide delivery of nuclease-resistant molecular beacons. *Proceedings of the National Academy of Sciences of the United States of America*, *105*(45), 17522–17525. doi:10.1073/pnas.0807066105 PMID:18988730

Yeh, J. T., Chen, W. C., Liu, S. R., & Wu, S. P. (2014). A coumarin-based sensitive and selective fluorescent sensor for copper (II) ions. *New Journal of Chemistry*, *38*(9), 4434–4439. doi:10.1039/C4NJ00695J

Zhang, W., Jia, B., & Furumai, H. (2018). Fabrication of graphene film Composite electrochemical Biosensor as a pre-screening algal Toxin detection tool in the event of Water contamination. *Scientific Reports*, *8*(1), 10686–10696. doi:10.103841598-018-28959-w PMID:30013209

Zhang, W. J., Li, Z., Zhou, M., Wu, F., Hou, X. Y., Luo, H., Liu, H., Han, X., Yan, G. Y., Ding, Z. Y., & Li, R. (2014). Synthesis and biological evaluation of 4-(1,2,3-triazol-1-yl) coumarin derivatives as potential antitumor agents. *Bioorganic & Medicinal Chemistry Letters*, *24*(3), 799–807. doi:10.1016/j.bmcl.2013.12.095 PMID:24418772

Zhao, N., He, Y. Q., Mao, X., Sun, Y. H., Zhang, X. B., Li, C. Z., Lin, Y. H., & Liu, G. D. (2010). Electrochemical assay of active prostate-specific antigen (psa) using ferrocene-functionalized peptide probes. *Electrochemistry Communications*, *12*(3), 471–474. doi:10.1016/j.elecom.2010.01.022

Zhou, Y., Liu, K., Li, J. Y., Fang, Y., Zhao, T. C., & Yao, C. (2011). Visualization of nitroxyl in living cells by a chelated copper (II) coumarin complex. *Organic Letters*, *13*(6), 1290–1293. doi:10.1021/ol103077q PMID:21322578

Zuckermann, R. N., Kerr, J. M., Kent, S. B. H., & Moos, W. H. (1992). Efficient Method for the Preparation of Peptoids [oligo(N-substituted glycines)] by Submonomer Solid-Phase Synthesis. *Journal of the American Chemical Society*, *114*(26), 10646–10647. doi:10.1021/ja00052a076

# Chapter 4
# Advances in Clinical Diagnosis of Tuberculosis:
## Past, Present, and Future

**Ankush Parmar**
*Panjab University, India*

**Tanzeer Kaur**
https://orcid.org/0000-0003-2935-7575
*Panjab University, India*

**Shweta Sharma**
https://orcid.org/0000-0002-1936-2947
*Panjab University, India*

## ABSTRACT

*Tuberculosis (TB) holds a central and deadly platform around the globe, affecting mankind with around one-third of the world being affected by latent TB. TB progresses in the body through inhalation process and has a critical discrimination in terms of affecting individuals depending upon age, sex, socio-economic status, and even the stature of nation (developed or developing). The biggest challenge in TB management is accurate, direct, early diagnosis, and an ability to differentiate the type of mycobacterium. The most common and reliable direct methods include tuberculosis skin test (TST), smear microscopy, nucleic acid amplification tests (NAAT), and immuno-chromatographic-based methods. However, culturing the specimen on a mycobacterium specific media is considered the 'gold standard' for diagnosis of TB by the WHO. Mycobacterium cultures are used extensively for bacilli differentiation and also for predicting drug susceptibility testing in multi-drug-resistant TB. This chapter discusses the merits and demerits of many approaches to distinguish and identify the type of mycobacterium.*

DOI: 10.4018/978-1-7998-0307-2.ch004

## INTRODUCTION

Tuberculosis (TB) disease is a global public health malady, claiming almost 1.5 million lives annually . Putting India on a global scenario, it is found that one-fifth of the global incidence of TB occurs in India (Central TB division 2010 ). Tuberculosis is caused by members of the *Mycobacterium tuberculosis* (Fig. 1) complex [MTBC], which includes *M. tuberculosis, M. bovis, M. africanum, M. canetti, and M. microti.* TB is a major concern in industrialized countries due to their socio-economic factors such as immigration, a rapid increase in poverty, malnutrition, war, and limited medication access. To address this grave disease, the most important aspects are its effective treatment, and its proper diagnosis. The management of TB includes a 4-drug regime, and the treatment period is about six months long. To reduce transmission of this disease, and to achieve disease elimination, early, and accurate diagnosis followed by proper treatment is vital. However, there are still a lot of challenges associated with the diagnosis of TB with specificity, and sensitivity.

*Figure 1. Structure of Mycobacterium*

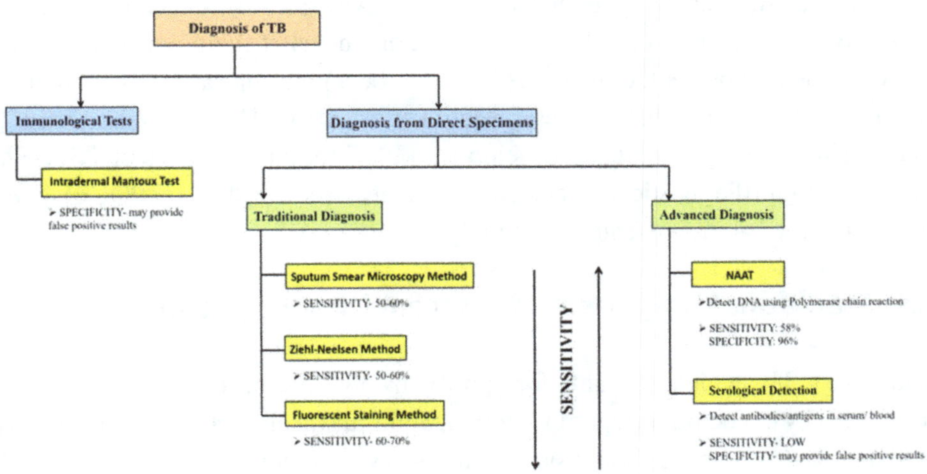


## 1. MOST COMMON IMMUNOLOGICAL TESTS OF TB

The most commonly used method for TB diagnosis is the tuberculosis Skin Test. This test is popularly known as the intradermal Mantoux test. It is the oldest diagnostic test, which is still included in the WHO latest recommendations for TB control (Lalvani 2007). The tuberculin fraction most widely used in this test is purified protein derivative (PPD), derived from cultures of *M. tuberculosis*. The reaction to intra-cutaneous injected tuberculin is the classic example of a delayed hypersensitivity reaction. A person who has been exposed to this bacterium is expected to mount an immune response in the skin. However, various factors both in the host, and inherent in the test lower its specificity, and sensitivity. PPD contains over 200 antigens present in the bacilli Chalmette-Guerin vaccine (BCG) and most non-tuberculosis bacteria also. Moreover, this test also has a poor ability to distinguish latent TB infection (Nayak and Acharjya 2012). This test gives both false-positive and false-negative results. It is found that about 20% of immuno-compromised children with culture-confirmed TB disease do not react initially to the TST (Dunn, Starke et al. 2016) thus giving false-negative results. Moreover, individuals vaccinated with BCG tend to show false positive results with this test. Therefore better alternatives are followed in high incidence TB countries.

## 2. DIAGNOSIS OF TB FROM DIRECT SPECIMENS

For the most effective treatment of TB, an early diagnosis becomes a pre-requisite. In patients with active TB, the infection of *M. tuberculosis* is high in the samples which can be directly analyzed for its presence. Any diagnostic method which can give the result on the same day and make a diagnosis at the point where patient consultation and presentation occurs is said to be a point-of-care (POC) diagnosis (Gondil and Chhibber 2018, Doctor, Mehra et al. 2019). The WHO, and Stop TB Partnership are emphasizing on developing a POC for the diagnosis of TB. So far, there are certain diagnostic methods available that can be translated as POC diagnosis of TB and they are represented in Figure 1.

### 2.1. Traditional Diagnosis of TB from Direct Specimens

Pulmonary TB, which accounts for approximately 85% of TB burden, is mainly diagnosed by expectorated sputum sample. Sputum smear microscopy has been the primary method for the diagnosis of pulmonary tuberculosis in low and middle-income countries . The smear of sputum is stained to identify acid-fast bacilli which further confirm the presence of Mycobacteria in a given clinical specimen. For direct microscopy, the Ziehl-Neelsen method is commonly used in low-income countries because of its low cost, and high specificity. However, this method has low sensitivity (50-60%) and cannot detect extra pulmonary TB cases. Another similar smear staining method is fluorescent staining with auramine O stain. This method is also rapid, specific, and has an enhanced sensitivity than the Ziehl-Neelsen method (by ~10% better). However, the usage of this assay is limited owing to its cost affectivity as it requires the employment of a sophisticated laboratory infrastructure (*viz.* fluorescent microscope, and a darkroom) which adds a toll on the existing cost (Steingart, Henry et al. 2006).

*Figure 2. Schematic representation showing varied POC diagnostic methods for TB detection*

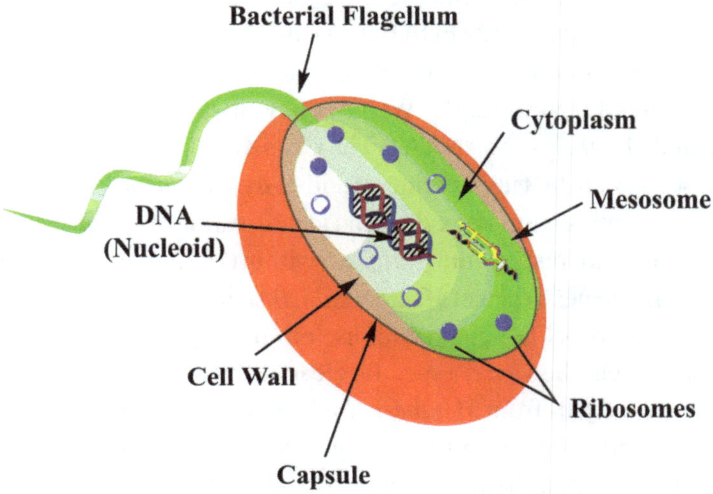


## 2.2. Advancements in the Diagnosis of TB from Direct Specimens: NAAT

NAAT is one of the most accurate known methods of detecting TB in direct samples and is an excellent choice for point-of-care diagnosis of TB. This method allows rapid detection of bacillus DNA (even in minuscule amounts) using polymerase chain reaction. Amidst all the NAAT available to date, Amplified Mycobacterium direct test (AMTD) (Hologic, San Diego, CA) was the first nucleic acid-based amplification test (NAAT) to be cleared by the FDA for the detection, and identification of *M. tuberculosis* from direct specimens. This assay utilizes transcription-mediated amplification of a portion of the 16S rRNA gene (specific to the *M. tuberculosis* complex) for the rapid detection and identification of TB bacilli. The sensitivity and specificity of this test are found to be 58% and 96%, respectively (Dunn, Starke et al. 2016). Apart from AMTD, various other NAAT has so far clear FDA standards and are being commercialized for laboratory testing.

Some commercially available NAAT kits include: (i) Amplicor MTB, Cobas Amplicor, and Light Cycler Mycobacterium Detection kits (Roche, San Diego, CA, USA), Amplified *M. tuberculosis* Direct Test (Genprobe, San Diego, CA, USA),

BDProbeTec-ET (Becton Dickinson Diagnostic Systems, Sparks, MD, USA); (ii) Genotype MTBDR*plus* and Genotype MTBDR*sl*(Hain Lifescience, Nehren, Germany), and INNO LiPARif.TB (Innogenetics, Ghent, Belgium) line probe assays; (iii) Capilia TB-Neo (Taunus, Numazu, Japan) rapid detection and speciation assay; and (iv) Xpert MTB/RIF assay(Cepheid), TruNatMicro RT PCR device.

Amidst the tally of all the commercially NAAT kits, Xpert MTB/RIF (rifampicin) (Cepheid, Sunnyvale, CA), is the most recent one which has been approved by Food and Drug Administration (FDA). It has also been recommended as the initial diagnostic test for children suspected of having multidrug-resistant TB (MDR-TB) or HIV-associated TB in the 2013 WHO policy update (Strategic 2013). This is fully automated *M. tuberculosis* NAAT. The GeneXpert platform is a self-contained cartridge-based system that utilizes microfluidics dynamics for an automated nucleic acid extraction, amplification, detection, and identification of *M. tuberculosis* directly from clinical samples (fresh sputum samples) within two hours. The Xpert MTB/RIF assay uses specific genes to detect *M. tuberculosis* sequences amplified in a real-time PCR assay. The assay is FDA cleared for use on smear-positive and smear-negative respiratory specimens, however, a specific clearance for extra pulmonary specimens is yet to be availed. Apart from *M. tuberculosis* detection, the Xpert MTB/RIF assay can also perform drug susceptibility testing (DST) for RIF resistance. It can detect defined mutations within the core region of the RNA polymerase b (rpoB) gene, and predict RIF resistance of MDR-TB (Theron, Peter et al. 2011).

Xpert MTB/RIF is suggested to hold considerable promise in the diagnosis of TB. This can be attributed to its innate attributes *viz.* accuracy, non-requirement of a dedicated testing facility (the procedure does not generate any aerosolized viable bacilli thus it can be performed in the absence of a biosafety cabinet) (Banada, Sivasubramani et al. 2010, Dheda, Ruhwald et al. 2013). However, Xpert MTB/RIF is more costly than smear microscopy, and its negative predictive value is significantly diminished in HIV-infected patients. Therefore, it is suggested to be useful in combination with other diagnostic tests. Recent work has shown that performing Xpert MTB/RIF only in smear-negative individuals can substantially reduce the overall cost of diagnosis (Vassall, van Kampen et al. 2011).

## 2.3. Advancements in the Diagnosis of TB from Direct Specimens: Serological detection

The tests, which monitor the humoral antibody immune responses to antigens of *M. tuberculosis* in serum/blood sample of patients is known as serological tests. Most of the bacteria or viruses-borne diseases are currently diagnosed *via* the detection of antibodies or antigens in the blood, but diagnosing TB serologically remains a challenge. The most commonly used serological tests are rapid immune-chromatographic assays (ICAs) or enzyme-linked immunosorbent assay (ELISA).

Serological tests of TB are based on either detection of antibodies or detection of antigen in serum/blood. For antibody detection, the immune response to specific antigens of *M. tuberculosis* (such as Mpt64, the 6-kDa early secreted antigenic target (Esat6), the 10-kDa culture filtrate protein (Cfp10), and the antigen 85 complex (Ag85)) (Ngeow, Wong et al. 2011) are identified in serum. The immuno-chromatography assays (ICAs) such as lateral flow assay or ELISA-based methods are coated with these antigens, and their corresponding antibodies produced in TB patients are detected.

Mpt64 is a secreted, immunogenic protein that is highly specific for the *M. tuberculosis* complex including *M. tuberculosis*, *M. africanum,* and virulent *M. Bovis*. Most importantly, it is not found in Mycobacteria other than tuberculosis (MOTT) (García, Restrepo et al. 2010). The Ag85 complex of proteins are secreted and can be found in the external layer of the bacterium's cell wall, and the blood (Ronning, Klabunde et al. 2000, Anderson, Harth et al. 2001). In blood, the Ag85 complex binds to plasma fibronectin or immunoglobulin G, and thus alters the host's immune response. This interaction appears to reduce phagocytosis of *M. tuberculosis,* thereby promoting infection (Bentley-Hibbert, Quan et al. 1999, Ronning, Vissa et al. 2004). Cfp10 and Esat6 are early secreted proteins with molecular weights of 10- and 6-kDa, respectively. In culture, they are some of the most abundant *M. tuberculosis* antigens. They are not found in *M. Bovis* BCG or most MOTT (Nguyen, Ma et al. 2012). Since one-third of all TB patients are seronegative for any single antigen mentioned above, therefore combinations of all above proteins, either separately or fused, have been used for diagnosis of TB (Raja, Ranganathan et al. 2008). Commercially available ICAs detect IgG and IgM antibodies against a combination of the above-mentioned TB antigens.

At present, detecting antibodies in blood for TB diagnosis doesn't provide enough sensitivity, and specificity to be used as a first-line screening tool. This can be attributed to the fact that these tests lead to false-positive results since antibodies against environmental Mycobacteria are often present in the general population. Therefore major challenge in the use of serology is the ability to distinguish between infection by *M. tuberculosis*, and infection by Mycobacteria other than tuberculosis. Moreover, serological methods cannot reliably distinguish between active, and latent TB.

Moreover, in 2011, WHO banned the use of serological tests for the detection of antibodies against *M. tuberculosis* infection in TB patients. The experts of WHO analyzed published data indicating specificity, sensitivity, and deciphered that numerous false-positive results were given by serological tests, which ultimately compromised the validity of these assays. Whilst, many scientists criticized this ban, and suggested that serological diagnosis is a very useful tool that is inexpensive, rapid, and an easy approach to complement diagnosis, and prognosis of *M. tuberculosis* infection (Dowdy, Steingart et al. 2011, Maes 2016).

On the brighter side, antigen detection is still a very useful tool for the diagnosis of TB patients. LAM (lipoarabinomannan) is a 17.3-kDa immunogenic glycolipid component of the mycobacterial cell wall. LAM is released from metabolically active, and degrading mycobacteria. The previous study shows that LAM in urine and pulmonary specimen showed a sensitivity of 93% and a specificity of 95%, respectively (Shah, Martinson et al. 2010). It has been the most extensively studied antigen and offers potential clinical utility in HIV-infected patients with advanced immunosuppression in both inpatient and outpatient (antiretroviral clinic) settings (Dheda, Davids et al. 2010, Minion, Leung et al. 2011).

TB LAM ELISA by Alere had an overall sensitivity of ~50%, which increased to 67% and 85% in HIV-infected patients with CD4 count <50 cells/mL from outpatient and inpatient settings, respectively. The aforementioned assay also depicted an overall specificity of 83–100% (Dheda, Davids et al. 2010, Dheda, Ruhwald et al. 2013). However, apart from LAM, other known antigens could not deliver any promising diagnostic test. So far, TB antigen test sensitivity was as low as 2%, and specificity was suboptimal. Most of the ICAs manufacturing companies detecting antigens of *M. tuberculosis* now use their tests for identification of *M. tuberculosis* in condensation fluid of the solid and/or liquid culture media incubated with patient samples.

## 2.4 Molecular Techniques for TB Detection

When it comes to the detection of TB using molecular techniques, a majority of these WHO endorsed methods rely upon the detection of bacilli (dead and alive) using Deoxyribose nucleic acid (Oommen and Banaji 2017). These methods have proven to be a souvenir and have inherently revolutionized the field of diagnostic sciences. The reduced cost, sensitivity, specificity are some of the advantageous aspects which have placed these diagnostic methods par apart from the traditional methods *viz.* culture and phenotypic DST (Oommen and Banaji 2017). Diminished contamination is another vital attribute which is offered by these diagnostic methods. Unlike their predecessors, NAATs these methods don't involve the amplification of target DNA, hence probable chances of contamination from amplicons, and inhibitors are circumvented (Parsons, Somoskövi et al. 2011, Oommen and Banaji 2017) thereby resulting in enhanced and accurate detection of TB. Lately, three molecular methods have been devised for TB detection, and they are as follows;

## Cartridge Based NAAT (CB-NAAT)

CB-NAAT is a real-time PCR-based nested method that was endorsed by WHO in 2010. This test, in particular, aids in the semi-quantitative detection of MTB, and RIF resistant TB bacilli from clinical specimens (Oommen and Banaji 2017). Amidst, inception, the aforementioned method has been largely employed for the detection of pulmonary and extra pulmonary TB, pediatric TB, and HIV patients with TV. With the advent of time, CB-NAAT has been further utilized for the presumptive diagnosis of drug-resistant-TB (DR-TB) as well.

The method is highly versatile and steady, as it works within an analytical limit of 131 CU/mL within a stipulated period of 2-3 h. Additionally, the corresponding method portrays a respectively high sensitivity of < 99%. All this can be attributed to the self-contained attribute of the cartridges, which play an intricate role in eluding the chances of cross-contamination amidst the given samples(Oommen and Banaji 2017). When it comes to the standard operating protocol, the working comprises of seven stages;

1$^{st}$ stage: Liquefaction of sample (sputum).
2$^{nd}$ stage: Inactivation of the liquefied sample using a sample reagent (killing, and immobilization of TB bacilli).
3$^{rd}$ stage: Transferring of inactivated sample to the cartridge.
4$^{th}$ stage: Insertion of the cartridge into the MTB-RIF test platform.
5$^{th}$ stage: Capturing, followed by washing and lysis of TB bacilli using ultrasonic waves.
6$^{th}$ stage: Amplification of obtained DNA (*rpoB* gene 81 bp) using specific primers, and molecular probes.
7$^{th}$ stage: Detection.

## Line Probe Assay (LPA)

For decades, LPA has been endorsed by WHO as a *"gold"* standard method for the diagnosis of TB. This can be attributed to its ability to detect MDR-TB against first-line anti-TB drugs (*viz.* isoniazid (INH), and RIF) (World Health Organization the UNICEF/UNDP/World Bank/WHO Special Programme for Research Training in Tropical Diseases . World Health Organization 2008, MacLean, Kohli et al. 2020). These assays are capable of detecting TB bacilli possessing MDR against fluoroquinolones (FLQs), and second line anti-TB drugs (*viz.* kanamycin, amikacin, and capreomycin) (Diagnostics , Tortoli, Cichero et al. 1999, Tortoli, Benedetti et al. 2002, QING, QIN et al. 2004, Richter, Rüsch-Gerdes et al. 2009, Barnard, Warren et al. 2012, Organization 2017, MacLean, Kohli et al. 2020). Speaking on a broader term, LPA is basically a hybridization strip assay which allows for a qualitative detection of both the TB DNA, and genetically mutated MDR-TB bacilli in a given sample specimen (Oommen and Banaji 2017). In addition to this, differentiation of variegated type of *Mycobacterium* species is another advantageous perk which is offered by LPA (Oommen and Banaji 2017). Amidst all these, Genotype MTBDRplus (Hain Lifesciences-Bruker, Nehren, Germany), Genotype MTBDRsI version 2.0 (Hain Lifesciences-Bruker, Nehren, Germany), and Nipro NTM+MDRTB II (Osaka, Japan), InnoLiPA assay-Innogenetics are some of the LPA which have been devised on a commercial level (Organization 2016, Maningi, Malinga et al. 2017, Oommen and Banaji 2017, MacLean, Kohli et al. 2020).

A typical LPA assay strip comprises of 27 reaction bands also commonly referred to as reaction zones. Distinctive from these there are six controls and twenty probes. Unlike its predecessor, CB-NAAT, the corresponding assay offers numerous *viz.* reduced time of operation (72 h), enhanced sensitivity (97.9 - 99.7% (RIF resistance)), and specificity, (94.2 – 99.7% (RIF resistance)), respectively (Oommen and Banaji 2017).

## Loop-Mediated Isothermal Amplification (LAMP)

On a distinctive note, molecular amplification methods have proven their worth when it comes to the detection of TB. Whilst, owing to factors such as high cost, and complex working protocols the full-scale potential of these expedient techniques is yet to be explored (Oommen and Banaji 2017, Shete, Farr et al. 2019, Phetsuksiri, Rudeeaneksin et al. 2020, Sreedeep, Sethi et al. 2020). In this context, a potential breakthrough was achieved by Eiken Chemical Co., Ltd, Japan (Iwamoto, Sonobe et al. 2003). They devised a highly cost-effective, swift (detection time ((35 (solid medium) - 60 (liquid medium)) and sensitive nucleic acid amplification (5-50 copies) based TB-LAMP assay (MacLean, Kohli et al. 2020).

Speaking in general terms, LAMP is an isothermal PCR amplification-based diagnostic method that aids in the visual identification of TB bacilli. This assay bestows a cutting edge over the traditional diagnostic methodology as it can be performed in a peripheral health care setting, and the requirement for a sophisticated laboratory setup (*viz.* thermocycler, detection system) is circumvented (Organization 2016, Kim, Cho et al. 2018, MacLean, Kohli et al. 2020, Toonkomdang, Phinyo et al. 2020). In addition to this, enhanced diagnostic performance is another technical benefit which is offered by LAMP-based TB assay. Keeping in mind, the variegated advantages presented by the aforementioned assay WHO endorsed this method since 2016 (MacLean, Kohli et al. 2020).

In this method, TB bacillus DNA is amplified using four different types of primers. These primers have an innate tendency to recognize six characteristic regions (*viz. gyrB*, and 16S rRNA) on the target gene. These distinct regions are targeted by carrying out an auto-cycling strand displacement reaction at a constant temperature (65 °C). Further, an incessant stride has been made in this field which has significantly resulted in the development of a single-step reaction for the amplification and detection of the target genes. Herein, the sample specimen, along with DNA polymerase, strand displacement activity, and substrates are incubated altogether and the reaction process is carried out at a constant temperature. The aforementioned process offers enhanced amplification of target genes (109-1010 times) within a short period (15-60 min) (MacLean, Kohli et al. 2020).

Apart from the aforementioned assays, recently several incessant strides have been made in the field of diagnostic methods. This has ultimately resulted in the innovation of novel diagnostic methods. Some of these methods have been depicted in Table 1, and 2, respectively.

## 3. CULTURING MYCOBACTERIUM FROM SPECIMEN FOR DIAGNOSIS

Despite many available POC diagnosis methods of TB detection, culturing bacilli, and then identifying it is a suboptimal "*gold standard*". However, a grave predicament is possessed in many TB patients, where an adequate amount of biological specimen is not readily available. Most importantly the sample specimen has a very low concentration of *M. tuberculosis* bacteria, which gives smear-negative results in microscopy. Therefore, WHO has recommended culturing of the specimen in media, which are specific for *M. tuberculosis* growth. This in turn will result in a proper diagnosis of TB disease.

Traditionally, *M. tuberculosis* culture is done on an egg-based solid media, known as Lowenstein-Jensen (LJ) medium or a liquid medium known as Middlebrook 7H9. On a comparative note, Middlebrook 7H9 liquid medium is more commonly used as the bacilli isolated from it is more as compared to LJ media (Pfyffer 2015). The growth of *M. tuberculosis* is very slow, and it may take up to as long as 4–6 weeks to actually visualize bacterial colonies. Nowadays, the traditional culturing methods are replaced by automated detection systems with liquid media such as BACTEC TB-460 (Becton Dickinson, Sparks, MD), BACTEC MGIT 960, or Bact/Allert 3D (bioMerieux, Durham, NC)]. These automated culturing methods reduce the duration for the detection of *M. tuberculosis* by about 1–2 weeks.

## 3.1. M. Tuberculosis Culturing in Automated Detection System

Since *M. tuberculosis* is slow-growing bacilli, therefore systems in which its growth can be sensed early were developed. The salient feature of these systems is the selection of *M. tuberculosis* from the specimen by its selective media, and then quantifying the rate of its growth based on the consumption of the media ingredients. BACTEC 460 is one such system that was the first of its kind (Kirihara, Hillier et al. 1985). It is a radiometric system for TB identification. In this system, *M. tuberculosis* is inoculated in a Middlebrook 7 H12 broth with carbon 14 ($^{14}C$) labeled palmitic acid.

The $^{14}C$ labeled palmitic acid acts as a substrate for *M. tuberculosis* to feed upon. The system estimates $^{14}CO_2$ liberated during the decarboxylation of $^{14}C$ labeled substrates in the growth medium (Lakshmi, Patil et al. 2006). Although this method is still used in various laboratories, still this system has several drawbacks. The most important drawback of this methodology is the use of radioactive materials in this system. This system requires a labor-intensive workflow and has a potential risk of cross-contamination (Tortoli, Cichero et al. 1999). To overcome the drawbacks of the BACTEC 460 system Becton Dickinson has developed the BACTEC MGIT 960 system. Mycobacteria Growth Indicator Tube (MGIT™), is a non-radiometric method of testing *M. tuberculosis* (Kanchana, Cheke et al. 2000).

It consists of a liquid broth medium that is known to yield better recovery and faster growth of mycobacteria. The MGIT contains 7.0 ml of modified Middlebrook 7H9 broth base. This medium is terminally sterilized by autoclaving. An enrichment, MGIT OADC (Oleic acid, Albumin, Dextrose, and Catalase) or MGIT 960 Growth Supplement, is added to make the medium complete. This Growth Supplement is essential for the growth of many Mycobacteria, especially those belonging to the *M. tuberculosis* complex. The addition of PANTA is necessary to suppress contamination of other bacterias. The MGIT tube contains an oxygen-quenched fluorochrome, tris 4, 7-diphenyl-1, 10-phenanthroline ruthenium chloride pentahydrate, embedded in silicone at the bottom of the tube.

During bacterial growth within the tube, the free oxygen is utilized and is replaced with carbon dioxide. With the depletion of free oxygen, the fluorochrome is no longer inhibited, resulting in fluorescence within the MGIT tube. The intensity of fluorescence is directly proportional to the extent of oxygen depletion. The growth of bacteria, as well as Mycobacteria, increases the fluorescence (Hanna, Ebrahimzadeh et al. 1999, Somoskövi, Ködmön et al. 2000). Currently, the WHO and the Strategic and Technical Advisory Group on TB recommends phased implementation of these automated liquid culture systems where feasible including low-income countries .

### 3.2. Differentiation of Mycobacterium

Once the culture has confirmed the presence of *M. tuberculosis* in the specimen, it is verified for the type of Mycobacterium. Classically, these differentiations are done by biochemical tests, which were not very accurate, and were time-consuming (Neonakis, Gitti et al. 2008). At present, various other molecular methods are used for identifying the type of Mycobacterium. PCR-based sequencing is considered the "*gold*" standard for the identification of Mycobacterium. In this method, Mycobacterium specific gene is amplified in an automated thermocycler *via* polymerase chain-based method. There are currently several specific genes of Mycobacterium known to date, however, the most commonly used is the 16S rRNA gene (Kox, Van Leeuwen et al. 1995, Kirschner, Rosenau et al. 1996).

Similarly, DNA probe technology also helps in differentiating variegated clinically important strains of Mycobacterium species *viz. M. tuberculosis complex, M. avium complex, M. avium, M. kansasii,* and *M. gordonae*. The DNA probes are single-stranded DNA oligonucleotides that are labeled with acridinium ester. These probes are complementary to the rRNA of specific Mycobacterium. In this method, the sample specimen is initially sonicated, and later on, the probes are added to the broken Mycobacterial cells, to form a stable DNA-RNA complex. This hybridization is then detected by light emission in a luminometer (Badak, Goksel et al. 1999, Tortoli, Cichero et al. 1999).

Moreover, there are many other molecular techniques available like PCR that can be used for the identification of Mycobacterium from culture. One such method is restriction enzyme analysis. In this method, a 65 kDa heat shock protein is initially amplified, and then it is subsequently cleaved by restriction enzymes. Further, the analysis is carried out on an agarose gel (Telenti, Imboden et al. 1993, Telenti, Marchesi et al. 1993). Following similar footsteps, another novel method by the name of pyrosequencing was innovated. The corresponding method derives its base from the phenomenon of nucleic acid sequencing-by-synthesis. In this method, the pyrophosphate (PPi) released during DNA synthesis is chiefly detected which further gives an affirmation about the target organism (Tuohy, Hall et al. 2005).

## 4. DRUG SUSCEPTIBILITY TESTING (DST)

Mycobacterium is evolving rapidly and has developed drug resistance. At present, the biggest challenge of TB management is multidrug resistance (MDR) and extensively drug-resistant (XDR) Mycobacterium species. Therefore, after isolation of bacilli in culture for definitive diagnosis, DST determination also becomes a prerequisite.

Moreover, culturing Mycobacterium is also vital, since it helps in developing a plan of action for drug administration.

Lately, two different approaches (*viz.* phenotypic and genotypic methods) have been frequently employed for determining the drug susceptibility for *M. tuberculosis*. Phenotypic methods assess the inhibition of *M. tuberculosis* growth in the presence of antibiotics and define resistance based on the response of the organism when exposed to the drug. Whilst, genotypic methods, on the other hand, are based on the detection of genes or mutations known to be associated with resistance.

Conventional DST utilizes phenotypic methods and depends on a variety of factors. The first of which is the definitive microbiologic diagnosis of *M. tuberculosis* with the isolation of the organism. The classical phenotypic method of DST is done by measuring the growth rate of *M. tuberculosis* in a culture media (LJ or Middlebrook 7 H12 broth) with antibiotics added to it. In phenotypic methods such as commercial liquid culture DST, generally, the WHO-approved radiometric BACTEC 460 TB system and BACTEC MGIT 960 system are extensively utilized (Wallis, Pai et al. 2010). Apart from these, microscopic observation drug susceptibility, colorimetric redox indicator, and nitrate reductase assay are some of the other methods which are also comprehensively used for phenotypic DST (Farnia, Masjedi et al. 2008, Coronel, Roper et al. 2010, Gupta, Singh et al. 2010).

MDR is generally defined as the resistance evoked in Mycobacterium in response to two vital first-line antibiotics *i.e.* RIF and isoniazid (INH). RIF, basically a rifamycin, acts by inhibiting the DNA-dependent RNA polymerase which further hampers the multiplication of Mycobacterium multiplication . RIF resistance is predominantly generated due to the mutations occurring in the RIF resistance determining region (RRDR) of the *rpoB* gene (Telenti, Imboden et al. 1993). Whilst, INH inhibits the pathways of mycolic acid synthesis by ceasing the activity of the Mycobacterial catalase-peroxidase enzyme (Timmins and Deretic 2006). Identification of these altered genes in the Mycobacterium can easily and rapidly identify drug resistance.

Coming onto the genotypic DST, NAAT assays are the invigorated tools of choice. Amidst all the NAAT, LPA is the kind of NAAT that simultaneously detects infection of Mycobacterium and amplifies regions of drug resistance for both RIF and INH, respectively. This system employs DNA strip technology, whereby amplified DNA is applied to strips containing probes specific for *M. tuberculosis* identification and RIF, INH resistance regions. The DNA in the specimen hybridizes with these probes and thus identifies the resistance (Nyendak, Lewinsohn et al. 2009).

An alternative to conventional liquid and solid media culture is the microscopic observation drug susceptibility (MODS) assay. This is a low-cost alternative to the detection of drug resistance. This system involves direct inoculation of a decontaminated sample into the wells of a tissue culture plate containing a liquid growth medium. Some wells include RIF and INH at critical concentrations, allowing simultaneous detection of drug resistance. Growth is determined by visual inspection using an inverted microscope to detect the presence of Mycobacteria in cords. MODS detection of MDR TB gave a sensitivity and specificity of 95 and 100% respectively when compared with the MGIT 960 system. Moreover, the time for detection was found to be 7 days.

## CONCLUSION

Despite the advancements and investments in the field of TB diagnosis, several shortcomings and obstacles are being faced by the scientific community taking into consideration the geographical aspects, cost-related constraints which critically hampers the hope of low-income nations but with a higher burden of TB, the sensitivity of assays and time consumption as well as delivery of results. A quantum leap is still needed in the world to provide an easily accessible-affordable, sensitive, and promising diagnosis. The current scenario demands the need to discover and flourish the novel diagnostic therapies in the otherwise perishable environment.

*Table 1. WHO and CFDA endorsed test for TB diagnosis (MacLean, Kohli et al. 2020)*

| Test | Technique Based Upon | Sensitivity | Specificity | Diagnosis | Ref |
|---|---|---|---|---|---|
| Xpert MTB/RIF ultra | qPCR/melting temperature analysis | 90-94 | 96-98 | MTB/RIF resistance | (Organization 2020) |
| GenoType MTBDRsl | PCR/hybridization | 86-87 | ~ 99 | FLQ and SLID resistance | (Organization 2016) |
| MTBC assay | LAMP | ~ 78 | ~ 98 | MTB | (Organization 2016) |
| Truenat MTB plus | Micro RT-PCR | ~ 80 | ~ 96 | MTB | (Organization 2020) |
| Truenat MTB-RIF Dx | Micro RT-PCR | ~ 84 | ~ 97 | RIF resistance | (Organization 2020) |
| EasyNAT | Cross priming amplification | ~ 87 | ~ 97 | MTB | (Deng, Sun et al. 2019) |
| MeltPro TB | PCR/melting curve analysis | 64-98 | 97-99 | DST | (Sun, Gao et al. 2019) |
| GeneChip MDR | PCR/hybridization | 79-89 | 97-98 | MDR-TB/ INH and RIF resistance | (Sun, Gao et al. 2019) |

RT-PCR: reverse transcriptase PCR; SLID: second-line injectable drugs; CFDA: China Food and Drug Administration.

*Table 2. Diagnostic methods for detection of TB*

| Diagnostic Methods | Instrumentation/ Peripheral/ Staining | Principle/Method | Advantages | Load Required | Detection Limit | Place/ Firm of origin/ Innovator | Ref |
|---|---|---|---|---|---|---|---|
| **Direct Detection (Microscopy)** | | | | | | | |
| (i) Smear microscopy | Acid fast staining (carbol fuschin, fluorochrome dye (auramine, and rhodamine)) | | (a) Rapid<br>(b) Inexpensive<br>(c) Specific | 10,000 bacilli/ ml | 20-80% | | (Behr, Warren et al. 1999, Mase, Ramsay et al. 2007, Organization 2008, Parsons, Somoskövi et al. 2011) |
| (ii) Fluorescent microscopy (LED) | Royal blue colored LED | | a) Swift<br>(b) Robust<br>(c) User friendly<br>(d) Sustainable<br>(e) Cost effective<br>(f) No hi-tech infrastructure/ instrumentation required<br>(g) Sensitivity/ specificity | | | | (Minion, Shenai et al. 2011, Organization 2011, Shenai, Minion et al. 2011, UNITAID 2012) |
| (iii) Sodium hypochlorite (Bleach) microscopy | Mathare sodium hypochlorite (MaSH) method | | Detection of TB bacilli was increased by 15% | | | Medecins Sans Frontieres (Mathare, Nairobi) | (Organization 2008, Schramm, Hewison et al. 2012) |
| (iv) Vital fluorescent staining | Fluorescent viability marker (fluorescein diacetate (FDA)) | | Detection of live bacteria | | | | (Al-Moamary, Black et al. 1999, Schramm, Hewison et al. 2012) |
| Automated microscopic technology coupled with CellScope (portable digital FM) | | (a) Automated loading, and high-resolution digital imaging of stained slides<br>(b) Swift analysis (200 slides can be loaded at the same time) | | | | TBDx (Signature Mapping Medical Sciences, USA) | (Lewis, Chihota et al. 2012, Tapley, Switz et al. 2013, Oommen and Banaji 2017) |
| **Indirect Detection** | | | | | | | |
| (v) Tuberculin skin testing (Mantoux test) | | Initially, the purified protein derivative (PPD) of MTB is administered intradermally in the forearm. Thereafter, the inference is made after 48-72 h of administration (Oommen and Banaji 2017). | Aids in the screening of infection (detection of potential exposure to MTB or latent MTB)(Oommen and Banaji 2017) | | | Charles Mantoux | (Reichman and Hershfield 2000, Nayak and Acharjya 2012, Oommen and Banaji 2017) |

*continues on following page*

*Table 2. Continued*

| Diagnostic Methods | Instrumentation/ Peripheral/ Staining | Principle/Method | Advantages | Load Required | Detection Limit | Place/ Firm of origin/ Innovator | Ref |
|---|---|---|---|---|---|---|---|
| Modified tuberculin skin test | | | | | | | |
| (a) PPD-S PPD<br>(b) PPD-RT 23 (1-TU, 2-TU)<br>(c)International Standard Tuberculins | | | | | | (a) Siebert (MTB)<br>(b) BCG Vaccine Laboratory, Guindy, Chennai<br>(c) Laboratory of Biological Standards, Staten, Serum Institute, Copenhagen, Denmark | (Oommen and Banaji 2017) |
| (vi) Interferon-gamma release (*in vitro*) assay | | On encountering TB bacilli antigen (ESAT-6, 10 (CFP-10), and TB 7.7) the T-cells sensitized MTB releases interferon gamma (TH 1 cytokine) which is detected and affirms the presence of TB. | (a) Enhanced sensitivity<br>(b) Escalated specificity (recognize T-cells of TB patient)<br>(c) Swift assay (result can be viewed within 24 h) | | | | (Mazurek 2010) |
| (a) QuantiFeron-TB Gold (QFT-G), and QuantiFeron-TB Gold in Tube (QFT-GIT) | | | | | | Cellestis, Australia | |
| (b) T-SPOT TB | | | | | | Immunotec, UK | |
| (vii)Beta-Lactamase detection | LED filter and mobile phone camera. | Detects metabolic signature of MTB *viz.* BlaC (ref online bookmark). | (a) Enhanced specificity<br>(b) Enhanced sensitivity (10 CFU (unprocessed sputum)) | | | Global Biodiagnostics, Temple, TX, USA, | (Xie, Mire et al. 2012) |

## REFERENCES

Al-Moamary, M. S., Black, W., Bessuille, E., Elwood, R. K., & Vedal, S. (1999). The significance of the persistent presence of acid-fast bacilli in sputum smears in pulmonary tuberculosis. *Chest*, *116*(3), 726–731.

Anderson, D. H., Harth, G., Horwitz, M. A., & Eisenberg, D. (2001). An interfacial mechanism and a class of inhibitors inferred from two crystal structures of the Mycobacterium tuberculosis 30 kda major secretory protein (antigen 85B), a mycolyl transferase1. *Journal of Molecular Biology*, *307*(2), 671–681.

Badak, F. Z., Goksel, S., Sertoz, R., Nafile, B., Ermertcan, S., Cavusoglu, C., & Bilgic, A. (1999). Use of nucleic acid probes for identification of Mycobacterium tuberculosis directly from MB/BacT bottles. *Journal of Clinical Microbiology*, *37*(5), 1602–1605.

Banada, P. P., Sivasubramani, S. K., Blakemore, R., Boehme, C., Perkins, M. D., Fennelly, K., & Alland, D. (2010). Containment of bioaerosol infection risk by the Xpert MTB/RIF assay and its applicability to point-of-care settings. *Journal of Clinical Microbiology*, *48*(10), 3551–3557.

Barnard, Warren, Van Pittius, van Helden, Bosman, Streicher, Coetzee, & O'Brien. (2012). *Genotype MTBDR sl line probe assay shortens time to diagnosis of extensively drug-resistant tuberculosis in a high-throughput diagnostic laboratory*. Academic Press.

Behr, M., Warren, S., Salamon, H., Hopewell, P., De Leon, A. P., Daley, C., & Small, P. (1999). Transmission of Mycobacterium tuberculosis from patients smear-negative for acid-fast bacilli. *Lancet*, *353*(9151), 444–449.

Bentley-Hibbert, S. I., Quan, X., Newman, T., Huygen, K., & Godfrey, H. P. (1999). Pathophysiology of antigen 85 in patients with active tuberculosis: Antigen 85 circulates as complexes with fibronectin and immunoglobulin G. *Infection and Immunity*, *67*(2), 581–588.

Central TB Division. (2010). *Revised National Tuberculosis Control Programme status report. Overview*. Author.

Coronel, J., Roper, M., Mitchell, S., Castillo, E., Gamarra, N., Drobniewski, F., Luna, G., Mendoza, A., & Moore, D. (2010). MODS accreditation process for regional reference laboratories in Peru: Validation by GenoType® MTBDRplus. *The International Journal of Tuberculosis and Lung Disease*, *14*(11), 1475–1480.

Deng, S., Sun, Y., Xia, H., Liu, Z., Gao, L., Yang, J., Zhao, Y., Huang, F., Feng, J., & Wang, L. (2019). Accuracy of commercial molecular diagnostics for the detection of pulmonary tuberculosis in china: A systematic review. *Scientific Reports*, *9*(1), 1–11.

Dheda, K., Davids, V., Lenders, L., Roberts, T., Meldau, R., Ling, D., Brunet, L., van Zyl Smit, R., Peter, J., & Green, C. (2010). Clinical utility of a commercial LAM-ELISA assay for TB diagnosis in HIV-infected patients using urine and sputum samples. *PLoS One*, *5*(3), e9848.

Dheda, K., Ruhwald, M., Theron, G., Peter, J., & Yam, W. C. (2013). Point-of-care diagnosis of tuberculosis: Past, present and future. *Respirology (Carlton, Vic.)*, *18*(2), 217–232.

Diagnostics, T. (2016). *Molecular Line-probe Assay for the Detection of Resistance to Second-line Anti-TB Drugs (SL-LPA)*. WHO Recommendations on the Use of the SL-LPA.

Doctor, T. H., Mehra, N., & Vaidya, P. C. (2019). Tuberculosis Burden among Household Pediatric Contacts of Adult Tuberculosis Patients: Correspondence. *Indian Journal of Pediatrics*, *86*(7), 667–668.

Dowdy, D. W., Steingart, K. R., & Pai, M. (2011). Serological testing versus other strategies for diagnosis of active tuberculosis in India: A cost-effectiveness analysis. *PLoS Medicine*, *8*(8), e1001074.

Dunn, Starke, & Revell. (2016). Laboratory diagnosis of Mycobacterium tuberculosis infection and disease in children. *Journal of Clinical Microbiology*.

Ejigu, G., Woldeamanuel, Y., Shah, N., Gebyehu, M., Selassie, A., & Lemma, E. (2008). Microscopic-observation drug susceptibility assay provides rapid and reliable identification of MDR-TB. *The International Journal of Tuberculosis and Lung Disease*, *12*(3), 332–337.

Farnia, P., Masjedi, M. R., Mohammadi, F., Tabarsei, P., Farnia, P., Mohammadzadeh, A. R., Baghei, P., Varahram, M., Hoffner, S., & Velayati, A. A. (2008). Colorimetric detection of multidrug-resistant or extensively drug-resistant tuberculosis by use of malachite green indicator dye. *Journal of Clinical Microbiology*, *46*(2), 796–799.

García, J. C., Restrepo, S., & Zambrano, M. M. (2010). Comparative analysis of six Mycobacterium tuberculosis complex genomes. *Biomedica*, *30*(1), 23–31.

Gondil, V. S., & Chhibber, S. (2018). Exploring potential of phage therapy for tuberculosis using model organism. *Biomedical Biotechnology Research Journal*, *2*(1), 9.

Gupta, M., Singh, N., & Kaur, I. (2010). Evaluation of nitrate reductase assay for direct detection of drug resistance in Mycobacterium tuberculosis: Rapid and inexpensive method for low-resource settings. *Indian Journal of Medical Microbiology*, *28*(4), 363.

Hanna, B. A., Ebrahimzadeh, A., Elliott, L. B., Morgan, M. A., Novak, S. M., Rusch-Gerdes, S., Acio, M., Dunbar, D. F., Holmes, T. M., & Rexer, C. H. (1999). Multicenter evaluation of the BACTEC MGIT 960 system for recovery of mycobacteria. *Journal of Clinical Microbiology*, *37*(3), 748–752.

Iwamoto, T., Sonobe, T., & Hayashi, K. (2003). Loop-mediated isothermal amplification for direct detection of Mycobacterium tuberculosis complex, M. avium, and M. intracellulare in sputum samples. *Journal of Clinical Microbiology*, *41*(6), 2616–2622.

Kanchana, M., Cheke, D., Natyshak, I., Connor, B., Warner, A., & Martin, T. (2000). Evaluation of the BACTEC™ MGIT™ 960 system for the recovery of mycobacteria. *Diagnostic Microbiology and Infectious Disease*, *37*(1), 31–36.

Kim, C.-K., Cho, E. A., Shin, D. M., Choi, S. W., & Shin, S. Y. (2018). Comparative evaluation of the loop-mediated isothermal amplification assay for detecting pulmonary tuberculosis. *Annals of Laboratory Medicine*, *38*(2), 119.

Kirihara, J., Hillier, S., & Coyle, M. (1985). Improved detection times for Mycobacterium avium complex and Mycobacterium tuberculosis with the BACTEC radiometric system. *Journal of Clinical Microbiology*, *22*(5), 841–845.

Kirschner, P., Rosenau, J., Springer, B., Teschner, K., Feldmann, K., & Böttger, E. (1996). Diagnosis of mycobacterial infections by nucleic acid amplification: 18-month prospective study. *Journal of Clinical Microbiology*, *34*(2), 304–312.

Kox, L., Van Leeuwen, J., Knijper, S., Jansen, H., & Kolk, A. (1995). PCR assay based on DNA coding for 16S rRNA for detection and identification of mycobacteria in clinical samples. *Journal of Clinical Microbiology*, *33*(12), 3225–3233.

Lakshmi, V., Patil, M., Subhadha, K., & Himabindu, V. (2006). Isolation of mycobacteria by Bactec 460 TB system from clinical specimens. *Indian Journal of Medical Microbiology*, *24*(2), 124.

Lalvani, A. (2007). Diagnosing tuberculosis infection in the 21st century: New tools to tackle an old enemy. *Chest*, *131*(6), 1898–1906.

Lewis, J. J., Chihota, V. N., Van Der Meulen, M., Fourie, P. B., Fielding, K. L., Grant, A. D., Dorman, S. E., & Churchyard, G. J. (2012). "Proof-of-concept" evaluation of an automated sputum smear microscopy system for tuberculosis diagnosis. *PLoS One*, *7*(11), e50173.

MacLean, E., Kohli, M., Weber, S. F., Suresh, A., Schumacher, S. G., Denkinger, C. M., & Pai, M. (2020). Advances in molecular diagnosis of tuberculosis. *Journal of Clinical Microbiology*, *58*(10).

Maes, R. (2016). A Critical Appraisal of the Ban on Serological Tests for Tuberculosis Usefulness of Serological Monitoring of Tuberculosis Antibodies during the Treatment of TB Patients. *Clinical Microbiology (Los Angeles, Calif.)*, *5*(3).

Mandell & Dolin. (Eds.). (2005). Principles and Practice of Infectious Diseases (6th ed.). Elsevier.

Maningi, N. E., Malinga, L. A., Antiabong, J. F., Lekalakala, R. M., & Mbelle, N. M. (2017). Comparison of line probe assay to BACTEC MGIT 960 system for susceptibility testing of first and second-line anti-tuberculosis drugs in a referral laboratory in South Africa. *BMC Infectious Diseases*, *17*(1), 1–8.

Mase, S., Ramsay, A., Ng, V., Henry, M., Hopewell, P., Cunningham, J., Urbanczik, R., Perkins, M., Aziz, M., & Pai, M. (2007). Yield of serial sputum specimen examinations in the diagnosis of pulmonary tuberculosis: A systematic review. *The International Journal of Tuberculosis and Lung Disease*, *11*(5), 485–495.

Mazurek, G. H. (2010). IGRA Expert Committee; Centers for Disease Control and Prevention (CDC): Updated guidelines for using Interferon Gamma Release Assays to detect Mycobacterium tuberculosis infection-United States, 2010. *MMWR. Recommendations and Reports*, *59*, 1–25.

Minion, J., Leung, E., Talbot, E., Dheda, K., Pai, M., & Menzies, D. (2011). Diagnosing tuberculosis with urine lipoarabinomannan: Systematic review and meta-analysis. *The European Respiratory Journal*, erj00257–erj02011.

Minion, J., Shenai, S., Vadwai, V., Tipnis, T., Greenaway, C., Menzies, D., Ramsay, A., Rodrigues, C., & Pai, M. (2011). Fading of auramine-stained mycobacterial smears and implications for external quality assurance. *Journal of Clinical Microbiology*, *49*(5), 2024–2026.

Moore, D. A., Evans, C. A., Gilman, R. H., Caviedes, L., Coronel, J., Vivar, A., Sanchez, E., Piñedo, Y., Saravia, J. C., & Salazar, C. (2006). Microscopic-observation drug-susceptibility assay for the diagnosis of TB. *The New England Journal of Medicine*, *355*(15), 1539–1550.

Nayak, S., & Acharjya, B. (2012). Mantoux test and its interpretation. *Indian Dermatology Online Journal*, *3*(1), 2.

Neonakis, I. K., Gitti, Z., Krambovitis, E., & Spandidos, D. A. (2008). Molecular diagnostic tools in mycobacteriology. *Journal of Microbiological Methods*, *75*(1), 1–11.

Ngeow, Y. F., Wong, Y. L., Ng, K. P., Ong, C. S., & Aung, W. W. (2011). Rapid, cost-effective application of Tibilia TB Rapid Test for culture confirmation of live and heat-killed Mycobacterium tuberculosis. *Journal of Clinical Microbiology*, *49*(7), 2776–2777.

Nguyen, Y. H., Ma, X., & Qin, L. (2012). Rapid identification and drug susceptibility screening of ESAT-6 secreting Mycobacteria by a NanoELIwell assay. *Scientific Reports*, *2*, 635.

Nyendak, M. R., Lewinsohn, D. A., & Lewinsohn, D. M. (2009). New diagnostic methods for tuberculosis. *Current Opinion in Infectious Diseases*, *22*(2), 174.

Oommen, S., & Banaji, N. (2017). Laboratory diagnosis of tuberculosis: Advances in technology and drug susceptibility testing. *Indian Journal of Medical Microbiology*, *35*(3), 323–331.

Organization, W. H. (2008). *New laboratory diagnostic tools for tuberculosis control*. WHO.

Organization, W. H. (2011). *Fluorescent light-emitting diode (LED) microscopy for diagnosis of tuberculosis: policy statement*. World Health Organization.

Organization, W. H. (2016). *The use of loop-mediated isothermal amplification (TB-LAMP) for the diagnosis of pulmonary tuberculosis: policy guidance*. WHO.

Organization, W. H. (2016). *The use of molecular line probe assays for the detection of resistance to second-line anti-tuberculosis drugs: policy guidance*. World Health Organization.

Organization, W. H. (2017). *South-East Asia Regional response framework for DR-TB, 2017-2021*. World Health Organization. Regional Office for South-East Asia.

Organization, W. H. (2020). *WHO consolidated guidelines on tuberculosis: module 3: diagnosis–rapid diagnostics for tuberculosis detection: web annex 4: evidence synthesis and analysis*. WHO.

Parsons, L. M., Somoskövi, Á., Gutierrez, C., Lee, E., Paramasivan, C., Abimiku, A. l., Spector, S., Roscigno, G., & Nkengasong, J. (2011). Laboratory diagnosis of tuberculosis in resource-poor countries: Challenges and opportunities. *Clinical Microbiology Reviews*, *24*(2), 314–350.

Pfyffer, G. E. (2015). *Mycobacterium: general characteristics, laboratory detection, and staining procedures. In Manual of Clinical Microbiology* (11th ed.). American Society of Microbiology.

Phetsuksiri, B., Rudeeaneksin, J., Srisungngam, S., Bunchoo, S., Klayut, W., Nakajima, C., Hamada, S., & Suzuki, Y. (2020). Comparison of Loop-Mediated Isothermal Amplification, Microscopy, Culture, and PCR for Diagnosis of Pulmonary Tuberculosis. *Japanese Journal of Infectious Diseases*, *73*(4), 272–277.

Qing, Qin, & Zhou. (2004). Evaluation of Automated BACTEC MGIT 960 System for Testing Susceptibility of Mycobacterium Tuberculosis. *West China Medical Journal, 2*.

Raja, A., Ranganathan, U. D., & Bethunaickan, R. (2008). Improved diagnosis of pulmonary tuberculosis by detection of antibodies against multiple Mycobacterium tuberculosis antigens. *Diagnostic Microbiology and Infectious Disease*, *60*(4), 361–368.

Reichman, L. B., & Hershfield, E. S. (2000). *Tuberculosis: a comprehensive international approach*. Marcel Dekker.

Richter, E., Rüsch-Gerdes, S., & Hillemann, D. (2009). Drug-susceptibility testing in TB: Current status and future prospects. *Expert Review of Respiratory Medicine*, *3*(5), 497–510.

Ronning, D. R., Klabunde, T., Besra, G. S., Vissa, V. D., Belisle, J. T., & Sacchettini, J. C. (2000). Crystal structure of the secreted form of antigen 85C reveals potential targets for mycobacterial drugs and vaccines. *Nature Structural & Molecular Biology*, *7*(2), 141.

Ronning, D. R., Vissa, V., Besra, G. S., Belisle, J. T., & Sacchettini, J. C. (2004). Mycobacterium tuberculosis antigen 85A and 85C structures confirm binding orientation and conserved substrate specificity. *The Journal of Biological Chemistry*, *279*(35), 36771–36777.

Schramm, B., Hewison, C., Bonte, L., Jones, W., Camélique, O., Ruangweerayut, R., Swaddiwudhipong, W., & Bonnet, M. (2012). Field evaluation of a simple fluorescence method for detection of viable Mycobacterium tuberculosis in sputum specimens during treatment follow-up. *Journal of Clinical Microbiology*, *50*(8), 2788–2790.

Shah, M., Martinson, N. A., Chaisson, R. E., Martin, D. J., Variava, E., & Dorman, S. E. (2010). Quantitative analysis of a urine-based assay for detection of lipoarabinomannan in patients with tuberculosis. *Journal of Clinical Microbiology*, *48*(8), 2972–2974.

Shenai, S., Minion, J., Vadwai, V., Tipnis, T., Shetty, S., Salvi, A., Udwadia, Z., Pai, M., & Rodrigues, C. (2011). Evaluation of light emitting diode-based fluorescence microscopy for the detection of mycobacteria in a tuberculosis-endemic region. *The International Journal of Tuberculosis and Lung Disease*, *15*(4), 483–488.

Shete, P. B., Farr, K., Strnad, L., Gray, C. M., & Cattamanchi, A. (2019). Diagnostic accuracy of TB-LAMP for pulmonary tuberculosis: A systematic review and meta-analysis. *BMC Infectious Diseases*, *19*(1), 1–11.

Somoskövi, Á., Ködmön, C., Lantos, Á., Bártfai, Z., Tamási, L., Füzy, J., & Magyar, P. (2000). Comparison of recoveries of Mycobacterium tuberculosis using the automated BACTEC MGIT 960 system, the BACTEC 460 TB system, and Löwenstein-Jensen medium. *Journal of Clinical Microbiology*, *38*(6), 2395–2397.

Sreedeep, K., Sethi, S., Yadav, R., Vaidya, P. C., Angurana, S. K., Saini, A., Mehra, N., & Singh, M. (2020). Loop-mediated isothermal amplification (LAMP) in the respiratory specimens for the diagnosis of pediatric pulmonary tuberculosis: A pilot study. *Journal of Infection and Chemotherapy*, *26*(8), 823–830.

Steingart, K. R., Henry, M., Ng, V., Hopewell, P. C., Ramsay, A., Cunningham, J., Urbanczik, R., Perkins, M., Aziz, M. A., & Pai, M. (2006). Fluorescence versus conventional sputum smear microscopy for tuberculosis: A systematic review. *The Lancet. Infectious Diseases*, *6*(9), 570–581.

Strategic, W. (2013). Technical Advisory Group for Tuberculosis (STAG-TB). Report on Conclusions and Recommendations.

Sun, Y., Gao, L., Xia, H., Yang, Z., Deng, S., Yang, J., Zhao, Y., Wang, L., Feng, J., & Huang, F. (2019). Accuracy of molecular diagnostic tests for drug-resistant tuberculosis detection in China: A systematic review. *The International Journal of Tuberculosis and Lung Disease*, *23*(8), 931–942.

Tapley, A., Switz, N., Reber, C., Davis, J. L., Miller, C., Matovu, J. B., Worodria, W., Huang, L., Fletcher, D. A., & Cattamanchi, A. (2013). Mobile digital fluorescence microscopy for diagnosis of tuberculosis. *Journal of Clinical Microbiology*, *51*(6), 1774–1778.

Telenti, A., Imboden, P., Marchesi, F., Matter, L., Schopfer, K., Bodmer, T., Lowrie, D., Colston, M., & Cole, S. (1993). Detection of rifampicin-resistance mutations in Mycobacterium tuberculosis. *Lancet*, *341*(8846), 647–651.

Telenti, A., Marchesi, F., Balz, M., Bally, F., Böttger, E., & Bodmer, T. (1993). Rapid identification of mycobacteria to the species level by polymerase chain reaction and restriction enzyme analysis. *Journal of Clinical Microbiology, 31*(2), 175–178.

Theron, G., Peter, J., & Dheda, K. (2011). Xpert MTB/RIF test for tuberculosis. *Lancet, 378*(9790), 481.

Timmins, G. S., & Deretic, V. (2006). Mechanisms of action of isoniazid. *Molecular Microbiology, 62*(5), 1220–1227.

Toonkomdang, S., Phinyo, P., Phetsuksiri, B., Patumanond, J., Rudeeaneksin, J., & Klayut, W. (2020). Pragmatic accuracy of an in-house loop-mediated isothermal amplification (LAMP) for diagnosis of pulmonary tuberculosis in a Thai community hospital. *PLoS One, 15*(7), e0236496.

Tortoli, E., Benedetti, M., Fontanelli, A., & Simonetti, M. T. (2002). Evaluation of automated BACTEC MGIT 960 system for testing susceptibility of Mycobacterium tuberculosis to four major antituberculous drugs: Comparison with the radiometric BACTEC 460TB method and the agar plate method of proportion. *Journal of Clinical Microbiology, 40*(2), 607–610.

Tortoli, E., Cichero, P., Piersimoni, C., Simonetti, M. T., Gesu, G., & Nista, D. (1999). Use of BACTEC MGIT 960 for recovery of mycobacteria from clinical specimens: Multicenter study. *Journal of Clinical Microbiology, 37*(11), 3578–3582.

Tuohy, M. J., Hall, G. S., Sholtis, M., & Procop, G. W. (2005). Pyrosequencing™ as a tool for the identification of common isolates of Mycobacterium sp. *Diagnostic Microbiology and Infectious Disease, 51*(4), 245–250.

UNITAID. (2012). Diagnostic Technology Landscape. World Health Organization.

Vassall, A., van Kampen, S., Sohn, H., Michael, J. S., John, K., den Boon, S., Davis, J. L., Whitelaw, A., Nicol, M. P., & Gler, M. T. (2011). Rapid diagnosis of tuberculosis with the Xpert MTB/RIF assay in high burden countries: A cost-effectiveness analysis. *PLoS Medicine, 8*(11), e1001120.

Wallis, R. S., Pai, M., Menzies, D., Doherty, T. M., Walzl, G., Perkins, M. D., & Zumla, A. (2010). Biomarkers and diagnostics for tuberculosis: Progress, needs, and translation into practice. *Lancet, 375*(9729), 1920–1937.

WHO. (2007). *Strategic and Technical Advisory Group For Tuberculosis*. Report on Conclusions and Recommendations. WHO.

World Health Organization the UNICEF/UNDP/World Bank/WHO Special Programme for Research Training in Tropical Diseases. World Health Organization. (2008). Molecular line probe assays for rapid screening of patients at risk of multi-drug resistant tuberculosis (MDR TB). Expert Group Report. World Health Organization.

Xie, H., Mire, J., Kong, Y., Chang, M., Hassounah, H. A., Thornton, C. N., Sacchettini, J. C., Cirillo, J. D., & Rao, J. (2012). Rapid point-of-care detection of the tuberculosis pathogen using a BlaC-specific fluorogenic probe. *Nature Chemistry*, *4*(10), 802.

# Chapter 5
# Extra-Pulmonary TB:
## Changing Paradigm in Diagnosis

**Sarita Rani**
*Central University of Rajasthan, India*

**Ankur Kaul**
*Institute of Nuclear Medicine and Allied Sciences, India*

**Anil Kumar Mishra**
*Institute of Nuclear Medicine and Allied Sciences, India*

**Umesh Gupta**
*Central University of Rajasthan, India*

## ABSTRACT

*Tuberculosis is considered a fatal respiratory disease commonly seen in developing countries. This chapter includes the global scenario of TB patients and brief description of TB history, its pathogenesis, types, diagnosis tests, emergence of MDR (multi drug resistance) and XDR (extensively drug resistance). The traditional chemotherapy of TB includes first and second line drug therapy. These lines of therapies face many difficulties such as low solubility, low bioavailability, and stability issues. Therefore, some new drugs were introduced in the market that showed effective results to the patients. Nanoparticulate drug delivery gained much focus in recent years due to its advantages and ideal characteristics. Numerous nanoparticles, liposomal formulations, and polymeric micelles were reported by the researchers with significant and considerable results. Inhalable formulations were also prepared by scientists that showed effective and remarkable anti-tuberculosis action on TB patients. Many efforts are awaited to completely eradicate TB from the planet.*

DOI: 10.4018/978-1-7998-0307-2.ch005

# 1. INTRODUCTION

## 1.1 Global Scenario

Tuberculosis (TB) is an infectious respiratory disease caused by the bacteria, *mycobacterium tuberculosis* (MTB). The slow-growing acid-fast bacilli, MTB infect lungs, spleen, brain, kidneys and other organs (Sosnik, Carcaboso, Glisoni, Moretton, & Chiappetta, 2010). TB is mostly detected in people of developing countries and increasing the mortality rate rapidly. TB exists as a dreadful disease and comes in the list of top 10 death causing diseases globally. Understanding the pathophysiology and immune response of TB is an immense challenge to the scientists and medical supervisors (Sandhu, 2011). This disease becomes a burden on public health with the high death toll rate only second to deadly after HIV infections. To combat this disease, it is obligatory to understand the pathogenesis of TB in the hope of better management and treatment. According to World Health Organization (WHO) 28 million cases of TB were reported in India, which was about one fourth of the world. Out of this enormous ratio, about 4,23,000 death occurred due to TB excluding HIV patients and 87,000 cases were of HIV patients. Multi drug resistant (MDR) TB patient's population includes 1,47,000 cases in India as per the global TB statistics data 2017. In 2016, 95% deaths occurred due to TB in middle- and lower-income countries such as India, Pakistan, China, Philippines, South Africa, Indonesia and Nigeria. About 10.4 million TB patients were detected and 7.1 million people were died due to TB. About 53 million lives have been saved through TB diagnosis and treatment since year (WHO TB report 2017). In case of drug resistant TB, more than 0.6 million resistant cases were examined and screened for rifampicin (RIF), the first line drug of TB treatment and 0.49 million cases were of MDR-TB in India, China and Russia. These alarming facts gained the attention of world towards TB treatment and eradication to achieve superior public health. Still, we lack approaches to TB therapy as only 54% TB cases were cured, and this cure rate is not satisfactory. Therefore, it is not wrong to recognize it as a disease without boundaries leaving powerful impact on Indian population, as highest TB patient count was reported in our country. WHO has documented the fact that skillful efforts, newer diagnosis tools and funding can completely abolish TB from our planet by 2035.

Recently, newer advancement in bacteria genomics and molecular determinants are carried out to understand the Immunopathophysiology of this disease. Despite these steps such as knowledge of immunology and pathophysiology and recent discoveries at molecular level, TB exists as a vast and challenging burden to developing countries. The sources gap and lack of funding are the the barriers to new inventions and diagnosis tests in this disease eradication pathway (Philips & Ernst, 2012). This book chapter focuses on TB, extra pulmonary tuberculosis (EPTB), knowledge of causative agent, pathogenesis, TB chemotherapy and some recent advancement (in terms of treatment) done in tuberculosis treatment.

## 1.2 History

Historically, TB was called, as a "devils' disease" as the diagnosis, cure and treatment for TB was not available. TB was first noticed in the fossil of an extinct bison; (Pleistocene bison) and it was traced by radiocarbon activity as about ten thousand-year-old (Rothschild et al., 2001). In the Eastern Mediterranean part 9000 years ago, first time TB of human origin was identified in mummies of Egypt. In 1689, Dr. Richard Morton first stated that the TB is a disease with miscellaneous symptoms. Prior to 1800's the researchers thought that TB is a not a single disease and named as "tuberculosis" by J. L. Schonlein (Hershkovitz et al., 2008). In 1865, a French surgeon, Jean-Antoine Villemin officially said that TB is a communicable disease. In 1882, a German physiologist Robert Koch discovered the causative agent of TB "*mycobacterium tuberculosis*" and he was awarded the noble prize (physiology or medicine) of 1905 for this discovery (New medical net 2010). Edward Livingston Trudeau, was the first victim of TB who also, become its first survivor in the United States in 1884. An American bacteriologist "Theobald Smith" in 1896 demonstrated that *M. bovis* and some another species also caused *M. tuberculosis*. After 12 years, two scientists Albert Calmette and Camille Guerin both jointly isolated the various form of bovine and grew separately in a culture of ox bile. Later, tuberculin was discovered in 1890 and the Bacillus-Calmette Guerin (BCG) vaccine in 1908. This was an endeavor of 13 years culture laboratory experimentation to develop oral vaccine. ATB victim child was immunized, whose mother died during delivery of baby. Currently known as BCG, the intradermal vaccine is used commonly to combat TB. TB chemotherapy was started in 1960s.

## 1.3 Pathogenesis of TB

TB is a communicable and air borne disease and the pathogenesis of TB begin with the droplet infection through inhaled air. Other than *mycobacterium tuberculosis, M. bovis, M. africanum. And M. microtia* are also responsible for the TB as causative bacteria. However, these other type of bacteria's are responsible for non-human TB (Prasad et al., 2005). TB affects most of the human body organs, but lungs are the targeted organ for TB bacteria that leads to pulmonary TB. More than 80% of the cases are of pulmonary TB (Pandey & Khuller, 2005). The lungs are the primary and foremost portal for entry of *Mycobacterium tuberculosis* and are main site of disease manifestation, which further spreads to other body organs (Flynn & Chan, 2001).

*M. tuberculosis* is an intracellular pathogen that targets and inhibits the proficient antigen presenting cells (APC's), the alveolar macrophage (AM) and the dendritic cells during the early stages of infection (Wolf et al., 2007). Whenever, *Mycobacterium* reaches to the alveoli, it starts multiplication intracellularly in macrophage cells and spread through the lymph nodes, and then moves in the bloodstream to infect other organs (Hass 2000). *M. tuberculosis* continuously grows in the lungs for 2 to 12 weeks, and develops an immunocompetent host, cell-mediated immunity (CMI) and can be identified by tuberculin skin test and other diagnosis tests. The mechanism of TB pathogenesis along with its types is depicted in **Figure 1**.

*Figure 1. Pathogenesis of TB and its type (active and latent)*

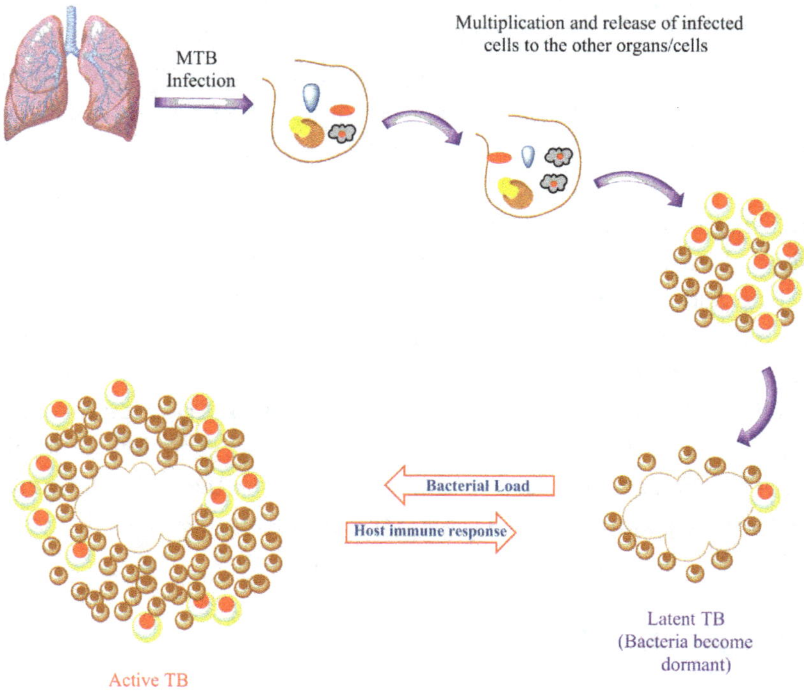


## 1.4 Types of TB

When *M. tuberculosis* multiplies and invades rapidly leading to the condition of illness is known as active TB. Cough, phlegm deposition, chest pain, weakness, weight loss, fever, chills and sweating at night are the main symptoms of active TB. LTBI patients (Latent tuberculosis infection) have *M. tuberculosis* in their bodies without TB disease condition, but they are unable to spread the TB infections to other people. About 5-10% of total LTBI cases are at risk to progressing from infection to active (primary) TB.

### 1.4.1 Pulmonary TB (PTB)

TB disease can affect many human body organs such as lungs, brain, uterus, bones etc. When TB bacteria infect mainly lungs, it is considered as pulmonary TB. For example, miliary TB is a type of PTB because the lesions are found predominately in the lungs.

### 1.4.2 Extra Pulmonary TB (EPTB)

If *M. tuberculosis* affects other organs such as larynx, lymph nodes, brain, pleura, bones, kidneys and joints then it is termed as extra pulmonary TB. TB in thoracic lymphadenitis or pleural effusion, without radiographic abnormalities in the lungs, constitutes a case of EPTB.

#### 1) LYMPHADENTIS TB (LTB)

LTB frequently occurs in children in the peak age of 15-20 years and most commonly noticed in women of America. The disease condition is described with discrete and non-tender lymph nodes and a firm mass of matted nodes were observed that can be clearly seen during examination of patient. Sometimes, severe inflammation may leads to drainage impulsively from patient nodes with sinus tract formation. In anti-tubercular therapy, some new nodes formation may occur that is a sign of an immune response to kill the pathogen bacteria (Shafer, Kim, Weiss, & Quale, 1991).

#### 2) PLEURAL TB

It mainly affects pleural fluid of lungs and characterized by pleuratic chest pain, cough, fever or dyspnea. It is the most frequently occurring TB form about 5% among the TB cases in the United states. In the chest radiography examination, more than 20% patients diagnosed with pulmonary lesions (Valdes et al., 1998). Pleural fluid of this kind of TB patients associated with lymphocyte predominance effusions within two weeks of diagnosis and initially same pattern can be observed in neutrophils.

### 3) SKELETAL TB

More than 35% of EPTB cases are related to bone and joint. Spine, weight-bearing joints and extra-spinal osteomyelitis are mainly involved in skeletal TB (Lifeso, Weaver, & Harder, 1985). Thoracic spine is the majorly affected part of spine. While in articular TB, arthritis of the hip or knee slowly progressed followed by pain, joint swelling, and reduced mobility. Systemic symptoms are absent usually but in severe cases, sinuses drainage and abscesses can be seen.

### 4) CENTRAL NERVOUS SYSTEM TB

Tuberculosis meningitis is the most common form of central nervous system TB. After that intracranial tuberculoma, and spinal tuberculous arachnoiditis are other commonly diagnosed forms of EPTB. Meningitis occurs by rupturing of subependymal tubercle into the subarachnoid space at high intensity. Symptoms such as unconsciousness, seizures, and high intracranial pressure are noticed in the patients. Even on the onset of TB manifestation, the patient suffers from headache, fever, or personality change, meningismus, vomiting, confusion, and focusing problems. Sometimes if not treated, mental status worsens and converted into stupor or coma (Kennedy & Fallon, 1979). A young adult is frequently affected by this TB in comparison to older (above 50 years of age).

### 5) ABDOMINAL TB

Abdominal TB may involve the gastrointestinal tract, peritoneum, mesenteric lymph nodes, or genital-urinary tract. Other organs such as liver, spleen, adrenal glands are the target sites for miliary TB. Swallowing of infected sputum, ingestion of contaminated food are the main reasons for initiation of this TB. Abdominal pain, diarrhea, weight loss, and fever are the symptoms noticed in the patient. Melena, rectal bleeding, and abdominal tenderness also can be present. Anti-TB chemotherapy is recommended for more than six months but in chronic conditions surgery may be advised.

## 1.5 Emergence of Multi Drug Resistant (MDR) and Extensively Drug Resistant (XDR)

The continuous spread of tuberculosis pathogens become a threat on the global public health. TB exists as a burden due to the failure of response by patients towards drugs.

Despite the immune response pathway understanding and conceptual knowledge, the resistant strain is a challenging task to treat and known as the main reason of MDR and XDR-TB form. TB chemotherapy includes first line drugs (INH, RIF etc.), second line drugs and some fluoroquinolones-based drugs. Patients who are drug susceptible towards chemotherapy are difficult to cure. TB patient who is resistance to INH and RIF, fall in the category of MDR-TB patient (Seung K.J. 2015). MDR-TB is mainly reported in Eastern Europe and Central Asian population and exists as a challenge for physicians to defeat. Early diagnosis is the only suitable way to identify TB patients.

XDR TB term was coined for those MDR TB patients who are resistant towards second line drugs and fluoroquinolones both. WHO have recommended some guidelines for MDR and XDR TB treatment that is included separately here. Both these TB forms are responsible for the increased mortality rate due to TB since last few years. Drug resistance is a biological phenomenon as the bacterial body changes can be observed through clinical and molecular approaches (Seung, Keshavjee, & Rich, 2015). Since the discoveries of RIF in 1947, drug resistance can be controlled in combinatorial approaches with other first line drugs such as INH, PYZ, some injectable drugs and fluoroquinolones. In 1993, DOTS (directly observed therapy, short course) was introduced that become a standard cure for TB even in the limited resources.

## 1.6 HIV and TB

HIV positive patients are more susceptible to TB infection as, compared to HIV negative person. It is a leading cause of death in HIV patients. According to WHO Indian TB, report more than one quarter of HIV patients with TB died since 2000 (Manosuthi, Chottanapand, Thongyen, Chaovavanich, & Sungkanuparph, 2006). The reason for this is the accelerated progression of HIV disease due to infected TB bacteria in patient body (Seung et al., 2015). HIV and TB exhibited a synergistic and dual direction interaction of infection inside body. The chance of MDR TB development from TB is higher due to co-infection of HIV and TB as the HIV infection favors the resistance TB strains. HIV-AIDS weakens the immune system of patient hence, enhances the probability of getting TB by 30%. This mainly examined with a higher rate in the age range from 15-60 years old. HIV infection drastically changes the nature of cell wall of *mycobacterium tuberculi* and, further the epidemiology that more worsen this disease. In Africa population, HIV positive patients with TB were reported in a large number than other countries in last decade. The diagnosis test sometimes fails to detect TB with HIV positive patients. The diagnosis test such as tuberculin skin, smear-sputum test, atypical X-ray gave negative results and therefore, it is difficult to detect TB. The longer time consumed in detection and diagnosis can be a reason for resistant TB form (Fatkenheuer, Taelman, Lepage, Schwenk, & Wenzel, 1999). In India, National AIDS Control Organization and the Central TB Division, function as an organization and take some directions and actions for TB treatment.

## 1.7 Diagnosis of TB

Most people with TB have latent infection, the development of new diagnostic and screening tools and standards has become necessary in order to control the disease (Diel, Loddenkemper, & Nienhaus, 2012). Drug susceptibility testing (DST) is strongly recommended by WHO for previously treated patients at predetermine time intervals.

### i. IMMUNOLOGICAL TESTS

Tuberculin skin test (TST) and IFN-γ releasing assay (IGRA) are widely used for diagnosing EPTB, but it has some limitations. IGRA diagnoses LTBI but it is costly, while TST is cost effective test (Thillai, Pollock, Pareek, & Lalvani, 2014). Both, TST and IGRA measures the response of T cells to TB antigens. The results of TST reactivity can be complicated by cross-reactivity with previous bacillus Calmette-Guerin vaccination or latent TB infection in countries where TB is prevalent. Various factors like HIV infection, poor nutritional status, recent viral or bacterial infections, or vaccination with live virus can reduce response to the TST.

### ii. TUBERCULIN SKIN TEST (TST)

In this test, a tuberculin purified protein derivative (PPD) mix of proteins is injected into a person under the top layer of skin *via* intradermal route. It causes a type IV delayed hypersensitivity skin reaction to examine whether the individual was either previously exposed to the mycobacterial infection (Thillai et al., 2014). Size of the dermal reaction is measured to determine the TB infection. However, the TST is known to lead to false-positive responses in those who are BCG vaccinated and to false-negative responses in immunosuppressed individuals.

## iii. INTERFERON-GAMMA RELEASE ASSAYS (IGRAS)

IGRAs is most reliable, sensitive, and specific TB diagnosis test (Diel et al., 2012) but costly and more technical (Goldman and Schafer 2011). In this test, the release cytokine IFN-g from T cells is detected from the collected blood sample of person and react with antigens that are absent in the BCG vaccine. As per the guidelines in some European countries, IGRAs and the TST are quietly used together to diagnose LTBI, while these tests are not definitive.

## iv. T-SPOT TB TEST

This test is performed as described by the manufacturer included in the assay kit. About 10 mL of blood was withdrawn from the subject by veinpuncture and transported to the clinical testing laboratory. After 4 h of venipuncture, the peripheral blood mononuclear cells (PBMs) were isolated through centrifugation. PBMs were washed thrice with serum free culture media and cell number was determined. Then an adequate cell count was adjusted per mL of media, transferred into the wells of 96 well plates with monoclonal IFN-γ antibodies, and seeded in to each well. The positive and negative control is decided and rest will be treated as per designed experiment. The 96-well plate is incubated at 5% $CO_2$ and 37 °C temperature was maintained. Washed the plates with phosphate-buffered saline (PBS)and reincubated with 50 μL of alkaline-phosphatase conjugated anti-IFN-γ monoclonal antibody. The results will be evaluated by the expert based on reaction (Meier, Eulenbruch, Wrighton-Smith, Enders, & Regnath, 2005).

## v. SPUTUM AND SMEAR TEST

Diagnosis of TB should not be time consuming, as the bacteria grow rapidly. TB can be examined by sputum and smear test in clinical laboratories. This is one of the most frequently over looked test for diagnosis of *mycobacterium tuberculi*. TB suspicious person sputum and smear is collected and tested. The diagnosis yield of TB bacteria is evaluated by the experts (Conde et al., 2003).

## vi. PLEURAL FLUID ASSAY

TB patient's pleural fluid level is found to be ≥5 g/dL (Light 2007). In majority of TB patients 50-90% small lymphocytes are present in their pleural fluid. In the starting phase (less than 2 weeks), pleural fluid of some patient may have neutrophils. Glucose level in pleural fluid of TB patient may decrease sometimes but showed similar serum level. The pleural fluid pH in TB patient is $\geq 7.3$ but it may be lower in some cases. Mesothelial cells are rarely observed in TB patients as most of them coated in the visceral and parietal pleura (Levine, Szanto, & Cugell, 1968).

## 2. CHEMOTHERAPY

After the unsatisfactory results of BCG vaccination, some therapeutics comes in the list of tuberculosis chemotherapy such as INH, RIF, PYZ, antibiotics and become the only hope for TB treatment. The goal of chemotherapy includes a high cure rate without relapse, to decrease mortality rate and to prevent drug resistance. Treatment of active TB should not be possible by a single drug that is why drugs are given in combination.

### 2.1 WHO Guidelines

WHO recommended TB chemotherapy should include first and second-line drugs. Drugs available for anti-TB therapy can be categorized into two types:

- First line anti-TB drugs: Rifampicin (RIF), Isoniazid (INH), Streptomycin (SM), Pyrazinamide (PZA), Ethambutol (EMB).
- Second line anti-TB drugs: Para Amino Salicylate (PAS), Kanamycin, Cycloserine (CS), Ethionamide (ETA), Amikacin, Capreomycin, Thiacetazone, Fluoroquinolones.

The daily doses are required in different phases of treatment. TB chemotherapy consists of two phases (intensive and continuous). In the intensive phase, INH, RMP, PYZ, and ETB are administered to the patients in different doses, while in the continuous phase, INH and RMP both are given to the patients in combination. INH initially is responsible for killing 95% of TB bacteria, and then, RMP and PYZ replace its role in the intensive phase. When either INH or RMP is not used, then chemotherapy duration is prolonged by 12–18 months. Hence, INH and RMP plays a pivotal role in the TB chemotherapy (Mitchison, 2000). The Most preferred therapeutic regimen for the treatment of pulmonary tuberculosis includes a combination therapy of three drugs, i.e., INH, RMP, and PYZ (Joshi, 2011) (WHO 2000).

However, it is suggested that the addition of Fluoroquinolones along with the appropriate dosage of first-line TB drugs, the total duration of drug-regimen in the case of EPTB may be cut down to 12 from 15 months after sputum-culture negative confirmation. Certain conditions require special attention before the start of anti-TB therapy: liver and renal disorders, visual and auditory conditions, drug-allergy, and co-administered medications and more caution is also advised in case of administration of second-line anti-TB drugs. Other special considerations for EPTB include identifying best possible therapy regimen for patients with AIDS, particularly noting adverse effects of drug/drug interactions for those receiving anti-retroviral therapy (Iseman, 2000). The severity of the disease is determined by bacillary load, the anatomical site of infection and the extent of disease and resultantly, the appropriate therapy is planned.

TB chemotherapy dose regimen is divided further in to different categories. Category 1 includes four drugs ETB, INH, RIF, PYZ for 2 months and INH, RIF are given for 4 months in combination phase. Category 2 use streptomycin (ST) along with category 1 drugs for a longer period of 8 months and considered for relapse and retreatment cases (WHO Geneva 2008). In category 3, ETB is omitted for children, patients of smear-negative pulmonary or EPTB and for HIV-negative and fully drug-susceptible patients. Category 4, includes combination of second-line drugs and the initial phase five drugs PYZ, kanamycin (Km), ofloxacin (OFX), ethionamide (ETD) and cycloserine (CS) for 6-8 months and use of OFX, ETD and CS in continuous phase for 12 months to treat drug-resistant TB (Caminero, Sotgiu, Zumla, & Migliori, 2010). In case of XDR-TB treatment, capreomycin (CPM), moxifloxacin (MFX), para amino salicylic acid (PAS) +/- cycloserine (CS) are used in addition of two or three agents **(Table 1).**

## 2.2 Current Drugs used in TB treatment and their Major Issues

At present first line drugs, second line drugs, antibiotics, some injectable and fluoroquinolones are also used for TB therapy. The longer duration of chemotherapy is responsible for some major issues of this chemotherapy in patients. As in case of MDR-TB and XDR-TB patient needs higher amount of chemotherapeutics for treatment and it enhances the chances of toxicity incidences. Some major issues and side effects of these drugs are discussed in this paragraph. INH eradicates most of bacilli in first 2 weeks of TB and it has a great distribution in combination therapy with RIF in longer duration TB therapy. The continuous administration of these drugs in maximum dose as 450 mg for INH and 600 mg for RIF causes many toxic effects in patient's body (Keshavjee and Farmer 2012).

TB chemotherapy causes some major issues related to different organs. Some of them are headache, drowsiness, fatigue, ataxia, dizziness, inability to concentrate, mental confusion, visual disturbances, muscular weakness, pain in extremities and generalized numbness. In case of gastrointestinal disturbances, heartburn, epigastric distress, anorexia, nausea, vomiting, gas, cramps, and diarrhea are common.

Liver dysfunctions, thrombocytopenia, transient leukopenia, hemolytic anaemia, eosinophilia, and decreased hemoglobin can be seen. Thrombocytopenia occurs after administration of RIF and ETB combination for twice weekly.

*Table 1. Chemotherapy of TB*

| Drugs used in TB treatment |
|---|
| Fist line drugs |
| Isoniazid (INH) |
| Rifampicin (RIF) |
| Ethambutol (ETB) |
| Pyrazinamide (PYZ) |
| Streptomycin (S) |
| Second line drugs |
| Kanamycin |
| Amikacin |
| Capreomycin |
| Ofloxacin |
| Ciprofloxacin |
| Levofloxacin |
| Capreofloxacin |
| Moxifloxacin |
| Gatifloxacin |
| Ethionamide |
| Prothionamide |
| Cycloserine |
| p-aminosalicylic acid |
| Terizidone |
| Other drugs |
| Clofazimine |
| Linezolid |
| Thioacetazone |
| Amoxycillin-clavulanate |
| Imipenem |
| Clarithromycin |
| Thioridazone |

## 3. CHALLENGES IN TB CHEMOTHERAPY

The conventional anti-TB Therapy has some limitations:

- Patient non-compliance due to lengthy therapy duration
- Toxic adverse effects of anti-TB drugs to non-target organs

- Co-existing HIV infections make TB more difficult to treat
- Degradation of active drug molecules before reaching the target-site
- Lower intracellular permeability of drugs to unique structural aspect of mycobacterial cell-wall

Other causes for the failed chemotherapy could be poor water solubility of drug, inadequate bioavailability, unpredictable plasma drug-levels or high dependency upon food (Paramasivan & Venkataraman, 2004).

However, it has become clear recently that the development of new anti-TB drugs alone is not enough to ensure improvement in anti-TB chemotherapy (Gaspar et al., 2008; Kumar, Asthana, Dutta, & Jain, 2006).

Although it may be imperative to search for new drugs and new drug targets to treat TB infection of multi-drug resistant TB (MDRTB), it is also essential to utilize currently available drug-therapy to treat initial mycobacterial infection such that MDRTB does not occur and to overcome the inadequate and failed anti-TB chemotherapy (Pandit and Choudhary, 2006).

However, to minimize drug-induced toxicity and improve compliance of the patient towards the therapy regimen, widespread aggressive efforts are being made to develop drug delivery systems to either target the mycobacterial infection more efficaciously or lessen the drug dosing frequency to improve patient compliance for achieving the therapeutic goal.

## 3.1 RIF Solubility, Bioavailability Concern

RIF is a water insoluble drug with solubility profile of 0.014 mg/mL. RIF's bioavailability (BA) is the major concern to think for TB therapy. As the drug is required in fixed dose combination (FDC) but due to hydrophobic issues, it is hard to achieve the required BA for RIF.RIF is well absorbed following oral administration. Its absorption from the gut is almost complete but is impaired or delayed by food. About 10 µg/mL peak plasma level is reached within 3 h after a single oral dose of 600 mg and provides effective blood level for more than 8 h. The $t_{1/2}$ is 2 to 5 h, and renal insufficiency does not significantly raise the plasma levels. The particle size and crystalline form of the drug, manufacturing process and the excipients are the factors responsible for the variation of RIF bioavailability (Laing et al. 1999). RIF in combination with INH showed better results, as INH accelerates the degradation rate of RIF in the acidic pH. In presence of INH, RIF undergoes higher decomposition rate as compare to single drug administration (Shishoo, Shah, Rathod, Savale, & Vora, 2001). The longer duration if TB therapy leads to further worsening of patient non-compliance. Drug resistance is also become a major issue to treat as it is a natural tendency of bacterial body and natural phenomenon too. Nanotherapeutic approach can deal with this and improve the TB chemotherapy via controlled or sustained drug release effect in short time. Therefore, it is an urgency to take some powerful steps to improve the chemotherapy problems for the sake of people life globally. The collaborative affords of scientists and public can resolve this problem in a systematic way and quickly too.

## 4. NEWER DRUGS INCLUDED IN TB TREATMENT

Unfortunately, the TB chemotherapeutics are still facing some solubility, bioavailability and multi-drug resistant issues. Therefore, the limited sources led clinicians to add some toxic old drugs (PAS, ETH and CS) to improve the patient health and survival rate. New TB drugs with their clinical phases are included in (**Table 2**).

*Table 2. List of anti-TB drugs in clinical trials*

| Newer drugs for TB treatment | Clinical trials phase |
|---|---|
| Gatifloxacin, Moxifloxacin | III |
| Rifapentine | II |
| Bedaquiline (TMC207) | II |
| Oxazolidinones (linezolid and PNU-100480) | I |
| Substituted ethylenediamines (SQ 109) | I |
| Benzothiazinones and Dinitrobenzamides | Preclinical phase |
| Fluroquinolones | II |
| Nitroimidazoles (PA-824 and OPC-67683) | II |

### 4.1 Clinical Development of Existing Drug: RIF

Several known drugs and their derivatives are improved, modified, identified, and added in the current TB regimen list. These include rifampicin and fluoroquinolones.

- **Rifapentine, (rifamycin, rifampin):** Rifampin is the most widely used drug for tuberculosis treatment. In clinical studies, it has been seen that higher dose of rifampin showed higher bactericidal activity, so the high dose of rifampin is included in treatment of TB. Rifapentine, in short time produced higher bactericidal activity against *mycobacterium tuberculi* than rifampin and become an attractive chemotherapeutic for TB regimen. Combination of rifapentine along with INH, PYZ, ETB reduced the time duration of TB treatment according to the previous research studies (Rosenthal et al., 2006; Rosenthal et al., 2007).

- **Fluoroquinolones (FQR):** These are broad-spectrum antimicrobial agents (DNA gyrase inhibition) and have potency against TB bacteria in the *in vitro* and *in vivo* studies. FQR are enlisted in the second line drugs for MDR-TB regimen. The new FQRs methoxy fluoroquinolones, gatifloxacin and moxifloxacin have been shown great potential against *M. tuberculosis* than the previously used, OFX and ciprofloxacin (Gillespie & Kennedy, 1998; Moadebi, Harder, Fitzgerald, Elwood, & Marra, 2007). The newer compounds are in clinical phase trials III and may shorten the TB chemotherapy.

## 4.2 Newer Anti-TB Regimen

- **Delamanid (OPC-67683) and pretomanid:** These are nitroimidazoles derivatives of antibiotics and shown potential against mycolic acid synthesis inhibitions against drug resistant strains of *M. tuberculi.* Both drugs have shown higher bactericidal activity in case of pulmonary TB. Recently, undergoes in phase clinical trials (Matsumoto et al., 2006).
- **Bedaquiline (TMC207):** It is a diarylquinoline class of antibiotics. The mechanism of action is to target the proton pump of adenosine triphosphate (ATP) synthase, causing inadequate synthesis of ATP, which is required for bacterial metabolism. The minimal inhibitory concentration (MIC) of bedaquiline against *M. tuberculi* is very low, but it has shown more bactericidal activity than INH and RIF. Based on clinical trials (phase II) results, WHO and the US Centers for Disease Control and Prevention suggested that bedaquiline can be prescribed to the patient for2 weeks with 400 mg daily dose and for longer duration the dose should be 200 mg thrice a week for 4.2 months (Diacon et al., 2012).
- **PA-824:** It is anitroimidazole oxazine derivative of metronidazole and act by inhibiting the synthesis of ketomycolates, an essential component of the *mycobacterium tuberculi* cell wall and by providing nitric oxide during enzymatic reduction.
- **Substituted Ethylenediamines:** SQ109 is a 1,2-ethylenediamine ethamubutol. The MOA is to inhibit the cell wall formation by interfering with the function of enzyme named as trehalose monophosphate transferase.

It is free from cross-resistance with ETB. Recently, it is under safety and efficacy studies in humans (Grosset, Singer, & Bishai, 2012).

- **Oxazolidinones:** They act by enzymatic competitive inhibition and causing interference with the ribonucleic acid (RNA) and inhibit translation. Oxazolidinedione, linezolid are given for MDR-TB patients as optimized background observations (OBR). PNU-100480 (sutezolid) has been well tolerated at a dose of 1200 mg by TB patient. AZD-5847 (posizolid) is in phase II trials at 500 mg single or double dose daily, 800 mg twice daily and 1,200 mg in a single dose daily (Swindells 2012).

## 5. ROLE OF NANOMEDICINE IN EPTB CHEMOTHERAPY

In the last decades, many liposomes, NPs and PM based formulation were developed and investigated for TB treatment. The newer drug delivery therapeutics can be administered via intravenous (*iv*), oral, intradermal and inhalable routes to identify a suitable and effective regimen. These new approaches can reduce the duration therapy ultimately reduce the burden of TB. Currently in TB therapy, there has been extensive progressive efforts in nanotechnology to develop nano-based drug delivery systems to target the *M. tuberculosis* infection and improve therapeutic index resulting in the more positive outcome in the TB therapy (Gelperina, Kisich, Iseman, & Heifets, 2005; Huh & Kwon, 2011; Loudos, Kagadis, & Psimadas, 2011). The high surface-area to volume ratio and unique physicochemical properties have made NPs potential carrier in the novel delivery system.

### 5.1 Liposomal Drug Delivery

Liposomal systems are self-closed and spherical structures that are created by one or many concentric lipid bilayers as outer shell with an aqueous phase as inner core where the polar head groups interact with the aqueous phase and the fatty acids form the hydrophobic core of the bilayers that are shielded from the aqueous core. The diameter of liposomes may range from 20 nm to several hundreds of nanometers. However, the dual character of liposomes with both hydrophilic and lipophilic regions, their application has expanded from purely bio-membrane mimetic systems to pharmaceutical carriers for both hydrophobic and hydrophilic drugs.

Therefore, the application of liposomes as nano-based drug delivery started in the early 70s (Gregoriadis, 1993) and these liposomal systems became the first type of nanocarriers approved by the US FDA (1995) for application in therapy of Kaposi's sarcoma related with acquired immunodeficiency syndrome (Krown, Northfelt, Osoba, & Stewart, 2004).

Among various nano particles formulations available for design, liposomes have shown enormous potential in the intravenous delivery of therapeutic drugs because of being biodegradable, biocompatible, non-immunogenic in nature and capable of incorporating both hydrophilic and lipophilic drugs and discharging them in a controlled mode (Jia, Joly, & Omri, 2008). Generally, liposomes can be categorised as unilamellar or multilamellar. Unilamellar liposomes are further classified into small size (SUV, 50-100 nm) or large size (LUV, 100-250 nm). SUV and LUV have a single lipid bilayer with a large aqueous central core, apt for loading of hydrophilic drugs; whereas multilamellar liposomes (MLV), generally with a diameter of 1-5 mm, have several lipid bilayers with a limited aqueous space, appropriate for loading of hydrophobic drugs.

Biological properties of liposomes make them widely accepted delivery systems for administration of drug both locally as well as systemically. Some of them are listed below:

- Biocompatible and non-immunogenic
- Amphiphilic carrier of both hydrophilic as well as hydrophobic pharmaceuticals
- Unique opportunity to deliver drug into cell or even inside cellular components
- Their surface properties, size and charge can be easily altered as per specific need
- Reduction in undesirable side-effects to non-target organs when drug is incorporated into these carrier systems. Gaspar and group developed and characterized liposome-based preparations of rifabutin and delivered the formulations via *iv* route. Multilamellar micelles were prepared by a remote loading method using ammonium sulphate. The *in vivo* studies shown great distribution of antibiotic through liposome approach in the liver, spleen and lungs in pathopysiological results up to 24 h after administration as compare to free rifabutin. In the lungs, the concentration of bacteria was lower than free antibiotic. The constituents dipalmitoyl phosphatidylcholine, dipalmitoyl phosphatidylglycerol (DPPC:DPPG) were the most effective lipids used in this formulation and exhibited great potency for bactericidal activity (Gaspar et al., 2008). The liposome preparation has shown better bactericidal activity against *mycobacterium tuberculi.* The developed liposomes were effective against EPTB treatment as well. Kaul et al. 2016 developed $^{99m}$Tc radiolabeled folate targeted PEGylated liposomes of rifampicin and ofloxacin as a theranostic against tuberculosis. PEGylation enhances the circulation time via surface modification approach and it allowed more interaction between cell receptors and targeted moiety to facilitate internalization. The *in vitro* release profile indicated that rifampicin release and ofloxacin released

28.33% and 30.09% respectively upto 6 h from the liposomal formulation. In contrast the pure rifampicin and ofloxacin released upto 84,2% and 96.1% respectively within 1 h (Kaul et al. 2016).

## 5.2 Nanoparticles (NPS) Based Drug Delivery

Various polymers such as PLGA, PEG, PLA has been utilized to develop nanoparticulate anti-TB formulations. Different respirable formulation was evaluated in the past for the better cure and treatment of this dreadful disease TB.

## 5.3 Inhalable or Respiratory Formulations as Nanocarriers

Hwang and colleagues prepared hyaluronan (HN) microspheres of OFX. They delivered the formulation to the lungs and alveolar macrophages via microsphere drug delivery via inhalation route. HN is the sodium salt of hyaluronic acid (HA) and preferred for its great mucoadhesive properties. Microspheres were prepared by co-spray drying method. The *in vitro* studies results revealed that the microspheres showed more AUC concentration in the lungs and plasma. The *in vitro* uptake of OFX from microsphere formulation in the cultured cells line RAW 264.7 was 2.1 and 1.7 folds more than the free OFX and without HN-OFX microsphere formulation (Hwang, Kim, Chung, & Shim, 2008).

## 5.4 Chitosan and Gelatin NPs as Nanocarriers

A research group formulated and characterized CS and gelatin (G) based PLA-PEG-NPs for controlled drug delivery of RIF for TB treatment. They prepared CS-PLA NPs loaded with RIF (CS-PLA-RIF). RIF release pattern from CS-PLA was more than the CS–PLA–PEG and CS–PLA–PEG–G (gelatin) NPs. In the initial phase, RIF released rapidly and then showed sustained release pattern for few hours. CS–PLA–PEG–G composite exhibited strong electrostatic interaction with drug than CS–PLA and CS–PLA–PEG composite. Conclusively, it was shown that the alkaline environment is more favorable for the release of drug from their respective composites, while acidic environment release the drug rapidly due to the ease of interaction breakage (Rajan & Raj, 2013). CS can be used in combination with PEG and G for controlled and targeted drug delivery system. Saraogi et al. prepared gelatin loaded NPs of RIF by desolvation method and evaluated them for anti-tubercular activity. The spherical shaped NPs were observed in TEM. AFM morphology analysis of NPs. The size of NPs were found to be 264±11.2 nm with a low PDI suggesting the narrow particle size distribution with homogeneity. The cytotoxicity study revealed that the gelatin NPs were safe and non-toxic. Bio-distribution studied gave an idea about the localization of NPs in lungs and spleen of infected mice. In cultured cell lines studies, the bacterial count of gelatin loaded RIF NPs were less than the blank NPs. Therefore, from the prepared formulation results we can conclude that the NPs can reduce the dose frequency via controlled release effect and helpful in the better management of TB (Saraogi, Gupta, Gupta, Jain, & Agrawal, 2010).

## 5.5 Polymeric Micelles (Pm) Based Drug Delivery

Gupta and coworkers reported the polymeric micelles of N-2-hydroxypropylmethacrylamide (HPMA) and PLA recently for delivery of anti-tubercular drugs rifampicin (RIF) and isoniazid (INH). Firstly, HPMA was conjugated with a hydrophobic polymer (PLA) and the INH was conjugated as a result, a hydrazone linkage was formed between INH and PLA. The synthesized block conjugates was characterized by thin layer chromatography (TLC), NMR and FT-IR spectroscopy. The CMC value of polymers was determined by Iodine method and UV-Visible spectrophotometric analyzer took measurements of polymers at different concentrations. RIF was physically encapsulated and resultant micelles were prepared by the hydrophobic interaction. The size of PMCs was found to be approximately 87 nm that is sufficing to exclude the renal filtration and but small enough to protect the micelles from macrophages attack (phagocytosis). The anti-microbial/anti-bacterial assays (MABA) were performed to determine the anti-microbial activity. The prepared PMs exhibited a greater anti-bacterial activity than the pure drugs INH, RIF with less minimum inhibitory concentration (MIC). Hence, a combination therapy of both first line drugs was prepared successfully for better efficacy and treatment (Upadhyay et al., 2017). Tyloxapol is a non-ionic polymer consists of alkyl and aryl alcohol chains. It acts as a surfactant and aid to liquefy the mucus and remove the pulmonary secretions or mucus from the respiratory tract. Mehta and Jindal prepared a formulation of mixed micelles that may be use for treatment and better management of EPTB. Lecithin act as a hydrophobic core and tyloxapol is utilized as surfactant. Lecithin and tyloxapol both solubilize the anti-TB drug. The prepared mixed micelles were characterized by surface tension measurement and conductivity both. The interaction between the mixed micelles formulation and drug RIF was synergistic. The morphology analysis suggested that the micelles showed homogeneity with good particle size distribution. The *in vitro* release data revealed that the drug loaded mixed micelles formulation exhibited sustained drug release effect. Kinetic models result also supported the *in vitro* release (Mehta & Jindal, 2013). Recently, Rani and colleagues reported the dual drug delivery favor TB via conjugations of polymers and INH. The author first prepared the polymeric conjugates of PEG-PLA and activated the conjugates. Further, INH was conjugated with activated PEG-PLA conjugate via hydrazine linkage and PEG-PLA-INH

(PPI) conjugate was developed. PM of both PPI conjugate and blank PEG-PLA was prepared. RIF was loaded in prepared polymeric micelles and characterized via size analysis and morphology. The critical micelles concentration of PEG-PLA PM was found to be 8.9 ± 0.96 mg/L with size range of 187.9 ± 2.68 nm. The PM formulation showed less hemolytic toxicity than free RIF. Minimum inhibitory concentration (MIC) of polymeric micelles was 8 times less than the RIF. PPI conjugate's MIC was also less than the pure INH as shown in the MABA results. Microscopic images showed the action of prepared pm formulation on bacterial cell wall. The *in vitro* release studies showed controlled release of RIF from smartly engineered PM. Microplate Alamar blue assay (MABA) assay results revealed that the nano structured PM exhibited better anti-bacterial activity than INH and RIF. Therefore, the PM may be established for *in vivo* studies in future to achieve and confirm better anti-TB therapy (Rani et al., 2018).

## 6. FUTURE PROSPECTS

Increasing morality rates due to TB rift the different public health organization of developing countries to take some major steps to eradicate this disease. Extra pulmonary TB patient population is causing more harm to patients in different parts of body such as bone, joints, abdominal parts and brain etc. EPTB should be treated with more attention as the last two decades data become an alarming factor for this. Some new diagnostic test must be used at primary basis as sometimes the symptoms are absent in few TB population but after some days the bacterial load increases in the macrophages and become more virulent. WHO threatening data of co-infected TB and HIV patients is terrible. WHO and some health agencies, must join hands together to help these developing countries and make the world population free from TB. The newer treatment strategies with the use of nanotechnology tools can further boost the chemotherapeutic approaches for the betterment of RB as well as EPTB.

However, the development of newer multidrug carrier systems for TB therapy, still poses a great challenge to the global scientific community.

## REFERENCES

Caminero, J. A., Sotgiu, G., Zumla, A., & Migliori, G. B. (2010). Best drug treatment for multidrug-resistant and extensively drug-resistant tuberculosis. *The Lancet. Infectious Diseases*, *10*(9), 621–629. doi:10.1016/S1473-3099(10)70139-0 PMID:20797644

Conde, M. B., Loivos, A. C., Rezende, V. M., Soares, S. L., Mello, F. C., Reingold, A. L., Daley, C. L., & Kritski, A. L. (2003). Yield of sputum induction in the diagnosis of pleural tuberculosis. *American Journal of Respiratory and Critical Care Medicine*, *167*(5), 723–725. doi:10.1164/rccm.2111019 PMID:12598215

Diacon, A. H., Donald, P. R., Pym, A., Grobusch, M., Patientia, R. F., Mahanyele, R., Bantubani, N., Narasimooloo, R., De Marez, T., van Heeswijk, R., Lounis, N., Meyvisch, P., Andries, K., & McNeeley, D. F. (2012). Randomized Pilot Trial of Eight Weeks of Bedaquiline (TMC207) Treatment for Multidrug-Resistant Tuberculosis: Long-Term Outcome, Tolerability, and Effect on Emergence of Drug Resistance. *Antimicrobial Agents and Chemotherapy*, *56*(6), 3271–3276. doi:10.1128/AAC.06126-11 PMID:22391540

Diel, R., Loddenkemper, R., & Nienhaus, A. (2012). Predictive value of interferon-gamma release assays and tuberculin skin testing for progression from latent TB infection to disease state: A meta-analysis. *Chest*, *142*(1), 63–75. doi:10.1378/chest.11-3157 PMID:22490872

Fatkenheuer, G., Taelman, H., Lepage, P., Schwenk, A., & Wenzel, R. (1999). The return of tuberculosis. *Diagnostic Microbiology and Infectious Disease*, *34*(2), 139–146. doi:10.1016/S0732-8893(99)00006-1 PMID:10354864

Flynn, J. L., & Chan, J. (2001). Immunology of tuberculosis. *Annual Review of Immunology*, *19*(1), 93–129. doi:10.1146/annurev.immunol.19.1.93 PMID:11244032

Gaspar, M. M., Cruz, A., Penha, A. F., Reymao, J., Sousa, A. C., Eleuterio, C. V., Domingues, S. A., Fraga, A. G., Filho, A. L., Cruz, M. E. M., & Pedrosa, J. (2008). Rifabutin encapsulated in liposomes exhibits increased therapeutic activity in a model of disseminated tuberculosis. *International Journal of Antimicrobial Agents*, *31*(1), 37–45. doi:10.1016/j.ijantimicag.2007.08.008 PMID:18006283

Gelperina, S., Kisich, K., Iseman, M. D., & Heifets, L. (2005). The potential advantages of nanoparticle drug delivery systems in chemotherapy of tuberculosis. *American Journal of Respiratory and Critical Care Medicine*, *172*(12), 1487–1490. doi:10.1164/rccm.200504-613PP PMID:16151040

Gillespie, S. H., & Kennedy, N. (1998). Fluoroquinolones: A new treatment for tuberculosis? *The International Journal of Tuberculosis and Lung Disease*, *2*(4), 265–271. PMID:9559396

Gregoriadis, G. (1993). Liposomes, a tale of drug targeting. *Journal of Drug Targeting*, *1*(1), 3–6. doi:10.3109/10611869308998758 PMID:8069542

Grosset, J. H., Singer, T. G., & Bishai, W. R. (2012). New drugs for the treatment of tuberculosis: Hope and reality. *The International Journal of Tuberculosis and Lung Disease*, *16*(8), 1005–1014. doi:10.5588/ijtld.12.0277 PMID:22762423

Hershkovitz, I., Donoghue, H. D., Minnikin, D. E., Besra, G. S., Lee, O. Y., Gernaey, A. M., Galili, E., Eshed, V., Greenblatt, C. L., Lemma, E., Bar-Gal, G. K., & Spigelman, M. (2008). Detection and molecular characterization of 9,000-year-old Mycobacterium tuberculosis from a Neolithic settlement in the Eastern Mediterranean. *PLoS One*, *3*(10), e3426. doi:10.1371/journal.pone.0003426 PMID:18923677

Huh, A. J., & Kwon, Y. J. (2011). "Nanoantibiotics": A new paradigm for treating infectious diseases using nanomaterials in the antibiotics resistant era. *Journal of Controlled Release*, *156*(2), 128–145. doi:10.1016/j.jconrel.2011.07.002 PMID:21763369

Hwang, S. M., Kim, D. D., Chung, S. J., & Shim, C. K. (2008). Delivery of ofloxacin to the lung and alveolar macrophages via hyaluronan microspheres for the treatment of tuberculosis. *Journal of Controlled Release*, *129*(2), 100–106. doi:10.1016/j.jconrel.2008.04.009 PMID:18538437

Jia, Y., Joly, H., & Omri, A. (2008). Liposomes as a carrier for gentamicin delivery: Development and evaluation of the physicochemical properties. *International Journal of Pharmaceutics*, *359*(1-2), 254–263. doi:10.1016/j.ijpharm.2008.03.035 PMID:18485632

Joshi, J. (2011). Tuberculosis chemotherapy in the 21 century: Back to the basics. *Lung India: Official Organ of Indian Chest Society*, *28*(3), 193–200. doi:10.4103/0970-2113.83977 PMID:21886955

Kennedy, D. H., & Fallon, R. J. (1979). Tuberculous meningitis. *Journal of the American Medical Association*, *241*(3), 264–268. doi:10.1001/jama.1979.03290290032021 PMID:102806

Keshavjee, S., & Farmer, P. (2012). Tuberculosis drug, resistance, and the history of modern medicine. *The New England Journal of Medicine*, *10*(10), 931–936. doi:10.1056/NEJMra1205429 PMID:22931261

Krown, S. E., Northfelt, D. W., Osoba, D., & Stewart, J. S. (2004). Use of liposomal anthracyclines in Kaposi's sarcoma. *Seminars in Oncology*, *31*(6, Suppl 13), 36–52. doi:10.1053/j.seminoncol.2004.08.003 PMID:15717737

Kumar, P. V., Asthana, A., Dutta, T., & Jain, N. K. (2006). Intracellular macrophage uptake of rifampicin loaded mannosylated dendrimers. *Journal of Drug Targeting*, *14*(8), 546–556. doi:10.1080/10611860600825159 PMID:17050121

Laing, R., Fourie, B., & Ellard, G. (1999). Fixed-dose combination tablets for the treatment of tuberculosis. Report of an informal meeting held in Geneva, Tuesday, 27April 1999 World Health Organization, Geneva, WHO/CDS/CPC/TB/99.267.

Levine, H., Szanto, P. B., & Cugell, D. W. (1968). Tuberculous pleurisy. An acute illness. *Archives of Internal Medicine*, *122*(4), 329–332. doi:10.1001/archinte.1968.00300090039009 PMID:5676418

Lifeso, R. M., Weaver, P., & Harder, E. H. (1985). Tuberculous spondylitis in adults. *The Journal of Bone and Joint Surgery. American Volume*, *67*(9), 1405–1413. doi:10.2106/00004623-198567090-00017 PMID:4077912

Loudos, G., Kagadis, G. C., & Psimadas, D. (2011). Current status and future perspectives of in vivo small animal imaging using radiolabeled nanoparticles. *European Journal of Radiology*, *78*(2), 287–295. doi:10.1016/j.ejrad.2010.06.025 PMID:20637553

Manosuthi, W., Chottanapand, S., Thongyen, S., Chaovavanich, A., & Sungkanuparph, S. (2006). Survival rate and risk factors of mortality among HIV/tuberculosis-coinfected patients with and without antiretroviral therapy. *Journal of Acquired Immune Deficiency Syndromes*, *43*(1), 42–46. doi:10.1097/01.qai.0000230521.86964.86 PMID:16885778

Matsumoto, M., Hashizume, H., Tomishige, T., Kawasaki, M., Tsubouchi, H., Sasaki, H., Shimokawa, Y., & Komatsu, M. (2006). OPC-67683, a nitro-dihydro-imidazooxazole derivative with promising action against tuberculosis in vitro and in mice. *PLoS Medicine*, *3*(11), e466. doi:10.1371/journal.pmed.0030466 PMID:17132069

Mehta, S. K., & Jindal, N. (2013). Mixed micelles of Lecithin-Tyloxapol as pharmaceutical nanocarriers for anti-tubercular drug delivery. *Colloids and Surfaces. B, Biointerfaces*, *110*, 419–425. doi:10.1016/j.colsurfb.2013.05.015 PMID:23751420

Meier, T., Eulenbruch, H. P., Wrighton-Smith, P., Enders, G., & Regnath, T. (2005). Sensitivity of a new commercial enzyme-linked immunospot assay (T SPOT-TB) for diagnosis of tuberculosis in clinical practice. *European Journal of Clinical Microbiology & Infectious Diseases*, *24*(8), 529–536. doi:10.100710096-005-1377-8 PMID:16133410

Mitchison, D. A. (2000). Role of individual drugs in the chemotherapy of tuberculosis. *The International Journal of Tuberculosis and Lung Disease*, *4*(9), 796–806. PMID:10985648

Moadebi, S., Harder, C. K., Fitzgerald, M. J., Elwood, K. R., & Marra, F. (2007). Fluoroquinolones for the treatment of pulmonary tuberculosis. *Drugs, 67*(14), 2077–2099. doi:10.2165/00003495-200767140-00007 PMID:17883288

Pandey, R., & Khuller, G. K. (2005). Antitubercular inhaled therapy: Opportunities, progress and challenges. *The Journal of Antimicrobial Chemotherapy, 55*(4), 430–435. doi:10.1093/jac/dki027 PMID:15761077

Pandit, N., & Choudhary, S. K. (2006). A study of treatment compliance in directly observed therapy for tuberculosis. *Indian Journal of Community Medicine, 31*, 241–243.

Paramasivan, C. N., & Venkataraman, P. (2004). Drug resistance in tuberculosis in India. *The Indian Journal of Medical Research, 120*(4), 377–386. PMID:15520487

Philips, J. A., & Ernst, J. D. (2012). Tuberculosis pathogenesis and immunity. *Annual Review of Pathology, 7*(1), 353–384. doi:10.1146/annurev-pathol-011811-132458 PMID:22054143

Prasad, H. K., Singhal, A., Mishra, A., Shah, N. P., Katoch, V. M., Thakral, S. S., Singh, D. V., Chumber, S., Bal, S., Aggarwal, S., Padma, M. V., Kumar, S., Singh, M. K., & Acharya, S. K. (2005). Bovine tuberculosis in India: Potential basis for zoonosis. *Tuberculosis (Edinburgh, Scotland), 85*(5-6), 421–428. doi:10.1016/j.tube.2005.08.005 PMID:16253560

Rajan, M., & Raj, V. (2013). Formation and characterization of chitosan-polylacticacid-polyethylene glycol-gelatin nanoparticles: A novel biosystem for controlled drug delivery. *Carbohydrate Polymers, 98*(1), 951–958. doi:10.1016/j.carbpol.2013.05.025 PMID:23987433

Rani, S., Gothwal, A., Khan, I., Pachouri, P. K., Bhaskar, N., Gupta, U. D., Chauhan, D. S., & Gupta, U. (2018). Smartly Engineered PEGylated Di-Block Nanopolymeric Micelles: Duo Delivery of Isoniazid and Rifampicin Against Mycobacterium tuberculosis. *AAPS PharmSciTech, 19*(7), 3237–3248. doi:10.120812249-018-1151-8 PMID:30191379

Rosenthal, I. M., Williams, K., Tyagi, S., Peloquin, C. A., Vernon, A. A., Bishai, W. R., Grosset, J. H., & Nuermberger, E. L. (2006). Potent twice-weekly rifapentine-containing regimens in murine tuberculosis. *American Journal of Respiratory and Critical Care Medicine, 174*(1), 94–101. doi:10.1164/rccm.200602-280OC PMID:16574936

Rosenthal, I. M., Zhang, M., Williams, K. N., Peloquin, C. A., Tyagi, S., Vernon, A. A., Bishai, W. R., Chaisson, R. E., Grosset, J. H., & Nuermberger, E. L. (2007). Daily dosing of rifapentine cures tuberculosis in three months or less in the murine model. *PLoS Medicine*, *4*(12), e344. doi:10.1371/journal.pmed.0040344 PMID:18092886

Rothschild, B. M., Martin, L. D., Lev, G., Bercovier, H., Bar-Gal, G. K., Greenblatt, C., Donoghue, H., Spigelman, M., & Brittain, D. (2001). Mycobacterium tuberculosis complex DNA from an extinct bison dated 17,000 years before the present. *Clinical Infectious Diseases*, *33*(3), 305–311. doi:10.1086/321886 PMID:11438894

Sandhu, G. K. (2011). Tuberculosis: Current situation, challenges and overview of its control programs in India. *Journal of Global Infectious Diseases*, *3*(2), 143–150. doi:10.4103/0974-777X.81691 PMID:21731301

Saraogi, G. K., Gupta, P., Gupta, U. D., Jain, N. K., & Agrawal, G. P. (2010). Gelatin nanocarriers as potential vectors for effective management of tuberculosis. *International Journal of Pharmaceutics*, *385*(1-2), 143–149. doi:10.1016/j.ijpharm.2009.10.004 PMID:19819315

Seung, K. J., Keshavjee, S., & Rich, M. L. (2015). Multidrug-Resistant Tuberculosis and Extensively Drug-Resistant Tuberculosis. *Cold Spring Harbor Perspectives in Medicine*, *5*(9), a017863. doi:10.1101/cshperspect.a017863 PMID:25918181

Shafer, R. W., Kim, D. S., Weiss, J. P., & Quale, J. M. (1991). Extrapulmonary tuberculosis in patients with human immunodeficiency virus infection. *Medicine*, *70*(6), 384–397. doi:10.1097/00005792-199111000-00004 PMID:1956280

Shishoo, C. J., Shah, S. A., Rathod, I. S., Savale, S. S., & Vora, M. J. (2001). Impaired bioavailability of rifampicin in presence of isoniazid from fixed dose combination (FDC) formulation. *International Journal of Pharmaceutics*, *228*(1-2), 53–67. doi:10.1016/S0378-5173(01)00831-6 PMID:11576768

Sosnik, A., Carcaboso, A. M., Glisoni, R. J., Moretton, M. A., & Chiappetta, D. A. (2010). New old challenges in tuberculosis: Potentially effective nanotechnologies in drug delivery. *Advanced Drug Delivery Reviews*, *62*(4-5), 547–559. doi:10.1016/j.addr.2009.11.023 PMID:19914315

Thillai, M., Pollock, K., Pareek, M., & Lalvani, A. (2014). Interferon-gamma release assays for tuberculosis: Current and future applications. *Expert Review of Respiratory Medicine*, *8*(1), 67–78. doi:10.1586/17476348.2014.852471 PMID:24308653

Upadhyay, S., Khan, I., Gothwal, A., Pachouri, P. K., Bhaskar, N., Gupta, U. D., Chauhan, D. S., & Gupta, U. (2017). Conjugated and Entrapped HPMA-PLA Nano-Polymeric Micelles Based Dual Delivery of First Line Anti TB Drugs: Improved and Safe Drug Delivery against Sensitive and Resistant Mycobacterium Tuberculosis. *Pharmaceutical Research*, *34*(9), 1944–1955. doi:10.100711095-017-2206-3 PMID:28685299

Valdes, L., Alvarez, D., San Jose, E., Penela, P., Valle, J. M., Garcia-Pazos, J. M., Suárez, J., & Pose, A. (1998). Tuberculous pleurisy: A study of 254 patients. *Archives of Internal Medicine*, *158*(18), 2017–2021. doi:10.1001/archinte.158.18.2017 PMID:9778201

Wolf, A. J., Linas, B., Trevejo-Nunez, G. J., Kincaid, E., Tamura, T., Takatsu, K., & Ernst, J. D. (2007). Mycobacterium tuberculosis infects dendritic cells with high frequency and impairs their function in vivo. *Journal of Immunology (Baltimore, Md.: 1950)*, *179*(4), 2509–2519. doi:10.4049/jimmunol.179.4.2509 PMID:17675513

# Section 2

# Countermeasures and Strategies for Remediation of AMR

*This section comprises of chapters that cover the application of traditional as well as novel therapies in combating antibiotic-resistant bacterial strains. A chapter on computer-aided drug-designing strategies for COVID-19 has also been added.*

# Chapter 6
# Traditional Medicine:
## Exploring Their Potential in Overcoming Multi-Drug Resistance

**Pavitra Solanki**
https://orcid.org/0000-0001-9199-9566
*Jamia Hamdard, India*

**Yasmin Sultana**
*Jamia Hamdard, India*

**Satyavir Singh**
*Bundelkhand University, India*

## ABSTRACT

*Everybody is at risk of being infected by drug-resistant microscopic organisms. Managing with sickness has never been less demanding within the history of our species. At the current rate of antimicrobial resistance (AMR) in microbes, specialists foresee that battling infections tuberculosis, HIV, and intestinal sickness will become more complicated. Antimicrobial resistance is rendering numerous life-saving drugs useless. Antibiotic-resistant microbes, known as "superbugs," are getting to be more various and more harmful, thanks to the proceeding abuse of anti-microbials. Natural medication offers an alternative to these progressively ineffectual drugs. According to the World Health Organization (WHO), traditional medicine is a holistic term enclosing diverse health practices. Concurring to a report by the College of Maryland Therapeutic Center, turmeric's volatile oil serves as a common anti-microbial.*

DOI: 10.4018/978-1-7998-0307-2.ch006

## INTRODUCTION

Multi drug resistance (MDR) is the oversensitivity or resistance of microorganisms such as (bacteria, viruses, fungi and parasites) to the antimicrobial drugs which results in ineffective treatment and spreading of infections. In bacteria it occurs due to the over expression of bacterial genes and accumulation of multiple genes in the bacterial cells which results in multidrug and single drug resistance simultaneously (Nikaido, 2009). MDR with increased morbidity and mortality due to the infecting agents such as bacteria, virus, fungi and parasite is rendered to as "super bugs". Worldwide multidrug resistance is a serious threat to public health in life threatening disease such as tuberculosis, pneumonia, HIV, malaria, yeast infections and other diseases. At whatever point we feel a small beneath the climate, we discover out what's off-base with us and take whatever is endorsed to create our side effects for all intents and purposes vanished. It appears nearly cursory until you think almost the infection-causing bacteria's capacity to adjust. The normal determination can be a perilous amusement. Within the case of disease-causing microscopic organisms, the last mentioned is an unfavorable result of our abuse of anti-microbials. Most major wellbeing specialists fear that drug-resistant microbes, or "superbugs," maybe our following worldwide wellbeing emergency. It's been utilized to treat all sorts of skin and respiratory diseases for thousands of a long time. Lab reports appeared that Guduchi diminished or killed E. coli and upper respiratory disease, concurring to the Indian Journal of Pharmacology 2003. A 2006 study by the College of Madras appeared Triphala restrained development of common bacterial segregates from HIV patients. Risorine having piperine as normal bio-enhancer is utilized in a settled measurements composition with Rifampicin and Isoniazid for the administration of tuberculosis. This diminished the dosage of rifampicin and moved forward its bioavailability. The multidrug resistance can be classified as essential resistance, auxiliary and clinical resistance. Essential resistance is when the medicate of intrigued doesn't stand up to by the specific have micro-organism (Tanwar et al, 2014; Vranakis et al, 2013). Auxiliary resistance happens for the most part after the introduction of the drugs to the living being and classified as: Natural resistance: When a single species of all micro-organisms appeared heartlessness to certain common to begin with line drugs. Broad Resistance: It is the capacity of living beings to stand up to the inhibitory impacts of one or two antimicrobial drugs which are most viable (Lee et al, 2013; Marks et al, 2014). Clinical Resistance: Restraint of contaminating living being by a concentration of an antimicrobial specialist due to helpful disappointment or return of contaminations inside an life form (Tanwar et al, 2014). Earlier studies have shown that for the pain coupled with boils Datura stramonium has been used as a medicinal plant, and also has been used in the treatment of gout, abscesses, rheumatism, and asthma. From the previous reports, it is clear that natural remedies can be considered as a good alternative to the antibiotics which have shown microbial resistance. From the ancient time, Indian Rishi and Vaidhyas have shown to cure various dangerous diseases and further ayurvedic research is going on in many ayurvedic universities of India. Although the practice of herbal medicine in India is going to be increased, still it requires some therapeutic affectivity in terms of pharmacokinetics data and pharmacodynamic evaluation scientifically. The use of plant by a specific ethnic group in medicine is known as ethnobotanical medicine. Thus the medicinal plant can be defined as any plant having medicinal value or the components present in the plants are used as drugs. According to World Health Organization (WHO) traditional medicine is a holistic term enclosing diverse health practices, plant, animal and mineral based medicines knowledge and beliefs, manual techniques, spiritual therapies and exercises bid to maintain well-being either singularly or in combination and also to diagnose, treat and prevent illness" (Fabricant et al, 2001; Reid et al, 2018). Homegrown pharmaceutical or phytomedicine alludes to the utilisation of plants and herbs for the reason of remedy and moderation of human afflictions. Plants have been utilized for therapeutic purposes by people since long sometime recently recorded history. In spite of the fact that cutting edge pharmaceutical has taken over the lead from homegrown medications within the treatment of illnesses in people, the utilization of herbals has expanded in later a long time around the world, as they are accepted to be more secure than advanced solutions with few or no side effects. The potential of homegrown pharmaceutical in wellbeing care has been built up by different phytochemical and pharmacological studies. Homegrown drugs are commonly managed as an extricate of the total herb, as homegrown tea or new juice. Numerous times, the whole herb is expended either new or within the dried and powdered shape. The adequacy of a medicate can be significantly influenced by the strategy of sedate conveyance. The gradualness within the adequacy of a treatment proposes a growing requirement for a claim to fame approach to the conveyance of drugs to particular targets within the body. Hence a proficient sedate conveyance framework that can deliver an ideal sum of the drug to the location of activity must be created for homegrown pharmaceutical to form a fruitful offered within the treatment of different human afflictions. India is focusing on the idea of synergism for enhancing the safety and therapeutic efficacy of traditional medicines and the drugs used for the treatment of the disease which is also called 'samyoga' in Ayurveda. Synergism is termed as combining the two or more ingredients for achieving the maximum therapeutic efficacy and safety. 'Sarangdhar Samhita' the Ayurvedic literature highlighted the concept of poly-herbalism to achieve higher therapeutic efficacy. Tea tree oil has been used for combating the Mycobacterium tuberculosis infections for eradicating the methicillin-resistant Staphylococcus aureus bacterium from the wounds.

## History of Traditional Medicine

There are various medicinal plants which have been used since ancient time for the treatment of skin disease, tuberculosis, mental disorders, cancer, AIDS and other infectious disease. Countries like India, China, South America and Egypt from ancient time are using herbal medicines as treatment remedies for the cure of such diseases ("HST", 2015). The bioactive constituents present in medicinal plants act as a defense system against pathogenic micro-organism. In India the traditional medicines are being practiced for years under (AYUSH) which is Ayurveda, Yoga, Unani, Siddha and Homeopathy (Mukherjee et al, 2017). For centuries, in India Ayurveda has been the mainstream healthcare system for the treatment of various diseases. For enhancing the drug efficacy and safety the traditional medicine of India focusing on the idea of synergism which is also called 'samyoga' in Ayurveda. Synergism is termed as combining the two or more ingredients for achieving the maximum therapeutic efficacy and safety. 'Sarangdhar Samhita' the Ayurvedic literature highlighted the concept of poly-herbalism to achieve higher therapeutic efficacy (Mukherjee et al, 2018). Traditional medicines were sole medicinal system in the African continent approximately 4000 years ago. This also remains dominant in the present with about 80% of the total estimated population of Africa using traditional medicine as a primary source for their health needs. There are ~27 million individuals who use traditional medicine as their health care need in South Africa. In case of plant biodiversity 250,000-500,000 plants species worldwide are found from which South Africa has been richest center comprising 368 families of plants having 24,000 specific and infra-specific taxa and 3000 plants are medicinally recognized. China has been using herbal medicine from last 5000 years for the treatment of diseases (Fair et al, 2014; Mukherjee et al, 2018).

*Aloe ferox* a South African plant used traditionally as laxative, for conjunctivitis, hypertension eczema and arthritis. Now the plant is commercially used as laxative. *Boophone disticha* also known as 'gifbol bushman poison bulb' or leshoma have used traditionally for septic wounds, postcircumcision wounds and as dressing agents for boils. After mild décoctions the plant provides relief from headaches, eye infections, and abdominal pain while after strong decoctions it is used as hallucinogens and sedatives (Reid et al, 2018). Earlier for the treatment of diabetes *Catharanthus roseus also termed* Madagascar periwinkle or isisushlungu was used, it consist two alkaloids which are in combination with chemotherapy available commercially for the treatment of cancer. Earlier studies shown that for the pain coupled with boils *Datura stramonium* has been used as a medicinal plant, and also has been used in the treatment of gout, abscesses, rheumatism and asthma. The alkaloids present within the plant are commercially available for the treatment of Parkinsonism, motion sickness and as an eye drop. *H. perforatum* was used as antidepressant and as an antimicrobial agent which was evaluated pharmacologically. *Harpagophytum procumbens* termed as 'thornapple' was used traditionally for the treatment of rheumatism, arthritis and digestion problems. Traditionally *Cinnnamomum camphora* has been reported to be used for the treatment of heart disease, fever, cold, infectious disease, inflammatory disease and respiratory pneumonia. *Mandragora officinarum* known as Mandrake was used as surgical anesthetic in Greek medicine due to the presence of hyoscine alkaloid which steer the use of mandrake as a medicinal plant into traditional medicine practices (Petrovska, 2012).

Tea tree oil (5-10% v/v) has been used for combating the *Mycobacterium tuberculosis* infections before administration of anti-tubercular drugs (ethambutol, isoniazid, pyrazinamide, and rifampin) for eradicating the methicillin resistant staphylococcus aureus bacterium from the wounds (Sherry et al, 2004).

## Current Trends in Herbal Medicine for Drug Resistance

Agreeing to World health organization (WHO), 60% of the world's populace depends on home grown pharmaceutical and 80% of the populace in creating nations depends nearly completely on it for their essential wellbeing care needs (Modak et al, 2007). There is 15% growth rate in the trade of medicinal plants and herbal drugs annually according to WHO report. For centuries, in India Ayurveda has been the mainstream healthcare system. From last two decades researchers have intensified their research towards herbal medicines to resist the multidrug resistance. The popularity and acceptability of herbal medicines is increasing due to their safety, easy availability and low cost.

The relentless emergence and increase in resistance of antimicrobial drugs to pathogenic bacteria raised the eyes of the researchers for newer drug formulations approaches to combat the multidrug resistance. Use of traditional medicine can be a valuable approach for combating the multidrug resistance (Mukherjee et al, 2017). Piperine confined from *P Piper longum L.* a normal bio-enhancer utilized in a settled dosage composition with Rifampicin and Isoniazid for the administration of tuberculosis. This combination is promoted beneath the exchange title Risorine which decreased the dosage of rifampicin and made strides its bioavailability (10,16). Piperine moreover has shown synergistic impacts on nimesulide-induced anti-nociception which is dose-dependent. Studies uncovered that co-administration of *Carum carvi L.* which is known as ('Jeera'), when co-administered with a few anti-microbials, antifungals, antivirals, anticancer, anti-ulcers, anti-inflammatory, cardiovascular, anti-leprosy and antihistaminic drugs expanded their bioavailability (25% to 300%) (Qazi et al, 2009). It was found that the bioactivity of rifampicin, tetracycline and ampicillin were improved with the utilize of novel bioactive nitrile glycosides, Niaziridin and Niazirin individually gotten from *Moringa oleifera Adans*, which prevent tumor promoter-induced Epstein–Barr infection (Khanuja et al, 2005).

There are various bioactive constituent presents in the plants which proved their efficacy against various bacterial and fungal strains such as *Bridelia micrantha*, *Senna occidentalis* and *Clerodendrum myricoides* are the plants which have proven pharmacological activity against malarial infection. *Rauvolfia vomitoria* exhibited a strong anti-plasmodial activity. Phyto-constituents presents in natural terpenoids showed their efficacy against vancomycin resistant enterococci. Plant derived Antimicrobials substance ($PDA_m$) are the secondary metabolite of plants which had shown profound antibiotic activity against the gram positive and negative strains of bacteria without further causing resistance. $PDA_m$ showed antimicrobial activity by reducing the pH, altering the efflux pumping and reducing membrane permeability of the microbes and found to be effective in the dose range 100 - 1000 μg $ml^{-1}$ (Srivastava et al, 2014).

Curcuminoids exhibited the anti-chemoresistance activity and can be employed for overcoming the drug resistance induced by adriamycin through the down regulation of p-glycoprotein (pgp) and ABCB1 mRNA. Study reveals that the bioactive presents in the curcumin have low toxicity and the potential to suppress proliferation, metastasis, invasion and angiogenesis in human cancers (Rodrigues, et al 2016).

Network pharmacology is gaining popularity to endorse the synergistic interaction of phyto-constituents present in the botanical drugs and traditional medicines using techniques like protein interaction, genomic expression and mRNA expression data for prediction and validation of the mechanism of action of the drugs to combat the drug resistance. Researchers have explored the connectivity between the phyto-constituents and the target drug molecules using the network pharmacology technique in food plant for hyperlipidemia elucidating the relationships between bio-actives, targets and pathways. *Lagenaria siceraria* ('Lauki') containing bio-active such as p-Coumaric acid, genistin, sinapaldehyde showed high and synergistic activity against hyperlipidemia associated targets (Mukherjee et al, 2017).

## Traditional Medicine for Overcoming Drug Resistance

*Bhringraj* is one of the most potent plants found in various region of India like, Uttar Pradesh, Assam, *Gujrat* etc. This plant has inherent property of antibacterial resistance as it can survive in excessive muddy area and near water canal. Bhringraj has shown its effects since long time according Indian system of medicine and Ayurveda. It has strong effect in treating wound healing specially in case of diabetes and in toothache (Darah et al, 2013). Berberine an isoquilonine alkaloid having broad spectrum antibacterial activity against streptococcal, staphylococcal and enterococcal species found to be 10 times active against gram negative strains of bacteria (Brown D). Sharma and co-workers suggested that Manuka honey can be efficient in reversing antimicrobial resistance to antibiotics and for the treatment of chronically infected wounds or drug resistance wounds (Sharma et al, 2014). Kudos CM 9 herbal tablets are recently approved for the treatment of cancer. Kudos CM9 tablets could be an alternative approach to treat the cancer in chemo-resistant patients. Figure 1 consists of few herbal medicines with improved drug resistance.

*Figure 1.*

Table 1 consists of list of traditional medicines which possesses multidrug resistance activity.

## FUTURE PERSPECTIVES

In the present era everyone is focusing for a faster treatment of their illness. This may be a reason for higher use of antibiotics; this can lead to the MDR. In India every prescription contains antibiotics. Undoubtedly herbal/traditional drugs still are prominent options to avoid MDR as they never develop MDR. Even repeated therapy of herbal medicine may not cause MDR. The current issue of rising MDR microscopic organisms is posturing a worldwide therapeutic danger and is persistently challenging the logical community. The diminishing viability and expanding harmfulness of engineered drugs is encouraged exasperating the issue. This has driven analysts to look for plant based antimicrobials for arrangement as they are presently known to play an imperative part within the improvement of compelling therapeutics. Phytoconstituents, either alone or in combination with anti-microbials may be a successful approach to bargain with the worldwide antimicrobial resistance. The viability of herbals in treatment of infections for decades proposes that microbes, organisms and infections may have a diminished capacity to adjust to a plant based antimicrobial administration.

*Table 1. Traditional Medicine having Multi drug Resistance activity*

| Traditional Medicine | Source | Activity | References |
|---|---|---|---|
| Bhringraj (*Eclipta alba*) | Leaves, flowers, stem and root | Antibacterial agent, antifungal, wound healing, sinusitis, hair fall, toothache. | Darah et al, 2013 |
| Berberine (Isoquilonine alkaloid) | Roots stems and barks of golden seal, golden thread, Oregon grape, barberry and tree turmeric | Antibacterial activity against streptococaal, staphylococcal and enterococcal bacteria. Kill *Streptococcus pneumoniae* and *S.aureus* bacteria | Tiwari et al, 2015 |
| Curcumin | *Curcuma longa* | Antichemo-resistance activity, Potential to suppress proliferation, metastasis, invasion and angiogenesis | Rodrigues et al, 2016 |
| Piperine marketed under trade name "Risorsine" | *P Piper Longum L.* | Increase efficacy and decrease tubercular resistance of rifampicin and isoniazid. | Mukherjee et al, 2018 |
| 'Triphala' Churna | *Terminilia chebula, Terminilia belerica, Emblica officinalis* | Possess strong antibacterial activity against *Staphlococcus epidermidis* and *S. aureus* and other strains of bacteria. | Sharma et al, 2014 |
| Epigallocatechin-3-gallate | Polyphenols found in green tea | Mild antibiotic activity against gram positive and negative strains of bacteria | Tiwari et al, 2015 |
| Phenolic extracts of essential oil | *Achillea biebersteinii* | Exhibited antifungal and anticandidal activity for treating infectious diseases without conferring resistance. | Baris et al, 2006 |
| 'Kutaj' | *Holarrhena antidysenterica* | Strong antibacterial activity against *Salmonella typhi.* | Sharma et al, 2014 |

## CONCLUSION

In India ministry of AYUSH is focusing on development of herbal drugs for MDR. Currently, there's a pressing need for unused commercial models to be created to support improvement of herbal medicine to counter medicate safe organisms as well as administrative changes so that clinical improvement programs are impartial, attainable, thorough, and clinically important.

## CONFLICT OF INTEREST

Author declares no conflict of interest.

## ACKNOWLEDGMENT

I am highly thankful to Dr. Dimple Sethi Chopra for their continuous efforts and guidance and the Department of Biotechnology, New Delhi for providing financial support reference no. BT/PR32389/TRM/120/262/2019.

## REFERENCES

Baris, O., Gulluce, M., & Sahin, M. (2006). Biological Activities of the Essential Oil and Methanol Extract of Achillea biebersteinii Afan. (Asteraceae). *Turkish Journal of Biology*, *3*, 65–73.

Chawla, P.C. (2010). Resorine A novel CSIR drug curtails TB treatment. *CSIR News*, *60*, 52–4.

Darah, I., Lim, S. H., & Nithianantham, K. (2013). Effects of Methanol Extract of Wedelia chinensis Osbeck (Asteraceae) Leaves against Pathogenic Bacteria with Emphasise on Bacillus cereus. *Indian Journal of Pharmaceutical Sciences*, *75*(5), 533–539. PMID:24403653

Fabricant, D. S., & Farnsworth, N. R. (2001). The value of plants used in traditional medicine for drug discovery. *Environmental Health Perspectives*, *109*, 69–75. PMID:11250806

Fair, R. J., & Tor, Y. (2014). Antibiotics and Bacterial Resistance in the 21st Century. *Perspectives in Medicinal Chemistry*, *6*, 25–64. doi:10.4137/PMC.S14459 PMID:25232278

HST. (2015). *Economics of traditional medicine trade in South Africa*. https://www.hst.org.za/uploads/files/chap13_07.pdf

Jeon, J., Kim, J. H., Lee, C. K., Oh, C. H., & Song, H. J. (2014). The Antimicrobial Activity of (-)-Epigallocatehin-3-Gallate and Green Tea Extracts against *Pseudomonas aeruginosa* and *Escherichia coli* Isolated from skin wounds. *Annals of Dermatology, 26*(5), 564–569. doi:10.5021/ad.2014.26.5.564 PMID:25324647

Khanuja, S. P., Arya, J. S., Tiruppadiripuliyur, R. S., Saikia, D., Kaur, H., & Singh, M. (2005). *Nitrile glycoside useful as a bioenhancer of drugs and nutrients, process of its isolation from Moringa oleifera.* United States Patent 6,858,588.

Kumar, D., Kumar, A., Sondhi, S., Sharma, P., & Gupta, N. (2018). An alkaline bacterial laccase for polymerization of natural precursors for hair dye synthesis. *3 Biotech, 8*(3), 1-10.

Lee, C. R., Cho, I. H., Jeong, B. C., & Lee, S. H. (2013). Strategies to minimize antibiotic resistance. *International Journal of Environmental Research and Public Health, 10*(9), 4274–4304. doi:10.3390/ijerph10094274 PMID:24036486

Marks, S. M., Flood, J., Seaworth, B., Hirsch-Moverman, Y., Armstrong, L., Mase, S., Salcedo, K., Oh, P., Graviss, E. A., Colson, P. W., Armitige, L., Revuelta, M., & Sheeran, K. (2014). Treatment practices, outcomes, and costs of multidrug-resistant and extensively drugresistant tuberculosis, United States, 2005–2007. *Emerging Infectious Diseases, 20*(5), 812–821. doi:10.3201/eid2005.131037 PMID:24751166

Modak, M., Dixit, P., Londhe, J., Ghaskadbi, S., & Devasagayam, T. P. A. (2007). Indian Herbs and Herbal Drugs Used for the Treatment of Diabetes. *Journal of Clinical Biochemistry and Nutrition, 40*(3), 163–173. doi:10.3164/jcbn.40.163 PMID:18398493

Mukherjee, P. K., Bahadur, S., Harwansh, R. K., Biswas, S., & Banerjee, S. (2017). Paradigm shift in natural product research: Traditional medicine inspired approaches. *Phytochemistry Reviews, 16*(5), 1–24. doi:10.100711101-016-9489-6

Mukherjee, P.K., Banerjee, S., & Kar, A. (2018). Exploring synergy in Ayurveda and traditional Indian systems of medicine. *Int J Synergy Res in Life Sci.,* 1-10.

Mukherjee, P. K., Harwansh, R. K., Bahadur, S., Banerjee, S., Kar, A., Chanda, J., Biswas, S., Ahmmed, S. M., & Katiyar, C. K. (2017). Development of Ayurveda Tradition to trend. *Journal of Ethnopharmacology, 197,* 10–24. doi:10.1016/j.jep.2016.09.024 PMID:27633405

Nikaido, H. (2009). Multidrug resistance in bacteria. *Annual Review of Biochemistry, 78*(1), 119–146. doi:10.1146/annurev.biochem.78.082907.145923 PMID:19231985

Petrovska, B. B. (2012). Historical review of medicinal plants' usage. *Pharmacognosy Reviews*, *6*(11), 1–6. doi:10.4103/0973-7847.95849 PMID:22654398

Ponnusankar, S., Pandit, S., Babu, R., Bandyopadhyay, A., & Mukherjee, P. K. (2011). Cytochrome P450 inhibitory potential of Triphala – A Rasayana from Ayurveda. *Journal of Ethnopharmacology*, *133*(1), 120–125. doi:10.1016/j.jep.2010.09.022 PMID:20883765

Qazi, G. N., Bedi, K. L., Johri, R., Tikoo, M. K., Tikoo, A. K., & Sharma, S. C. (2009). *Bioavailability/bioefficacy enhancing activity of Cuminum cyminum and extracts and fractions thereof*. United States Patent 7514105.

Reid, A. M., Oosthuizen, C. B., & Fibrich, B. D. (2018). *Traditional medicine: the ancient roots of modern practice*. doi:10.1016/B978-0-12-812475-8.00001-9

Rodrigues, V. L., Sousa, E., & Vasconcelos, M. H. (2016). Curcumin as a Modulator of P-Glycoprotein in Cancer. *Challenges and Perspectives.*, *9*(4), 1–11.

Sharma, C., Singh, C., & Sharma, L. N. (2014). Antibiotic resistant organism: An emerging public health problem and role of ayurveda (an overview). *International Journal of Ayurveda and Pharma Research*, *2*(1), 17–29.

Sherry, E., Reynolds, M., Sivananthan, S., Mainawalala, S., & Warnke, P. H. (2004). Inhalational phytochemicals as possible treatment for pulmonary tuberculosis, two case reports. *American Journal of Infection Control*, *32*(6), 36–70. doi:10.1016/j.ajic.2004.01.001 PMID:15454898

Srivastava, J., Chandra, H., Nautiyal, A.R., & Klara, S.J.S. (2014). Antimicrobial resistance (AMR) and plant-derived antimicrobials ($PDA_m$s) as an alternative drug line to control infections. *3Biotech.*

Subramani, R., Narayanasamy, M., & Feussner, K.D. (2017). Plant-derived antimicrobials to fight against multi-drug-resistant human pathogens. *3Biotech*, *7*(3), 1-15.

Tambekar, D. H., & Dahikar, S. B. (2011). Antibacterial activity of some Indian Ayurvedic preparations against enteric bacterial pathogens. *Journal of Advanced Pharmaceutical Technology & Research*, *2*(1), 24–29. doi:10.4103/2231-4040.79801 PMID:22171288

Tanwar, J., Das, S., Fatima, Z., & Hameed, S. (2014). Multidrug Resistance: An Emerging Crisis. *Interdiscip Prespect Infect Dis.*, 1-7. . doi:10.1155/2014/541340

Tiwari, V., Roy, R., & Tiwari, M. (2015). Antimicrobial active herbal compounds against *Acinetobacter baumannii*and other pathogens. *Frontiers in Microbiology*, *6*(618), 1–11. PMID:26150810

Vranakis, I., Goniotakis, I., Psaroulaki, A., & Sandalakis, V. (2013). Proteome studies of bacterial antibiotic resistance mechanisms. *Journal of Proteomics*, 1–5. PMID:24184230

# Chapter 7
# Natural Products in the Fight Against Multi-Drug-Resistant Bacteria:
## Natural Antibiotics and Resistance

**Mümtaz Güran**
*Eastern Mediterranean University, Turkey*

## ABSTRACT

*It is well recognised that the antimicrobial resistance crisis has approached critical levels, and current treatment options are very limited, especially in the treatment of infections caused by resistant bacteria. Thus, ongoing research is focused on the development of new molecules which have broader antimicrobial activity. However, the advancements in drug development studies using synthetic compounds has led to a lack of success. Also, economic and regulatory issues have formed a challenge as well. Therefore, research has focused again on natural products. A large number of natural products and natural product-derived compounds are still in various stages of clinical development. Here, current research on the potential uses of natural products or their templates as viable sources of new drug candidates have been discussed to construct an understanding towards the goal of development of new antimicrobials to overcome resistant pathogens.*

DOI: 10.4018/978-1-7998-0307-2.ch007

## INTRODUCTION

The life before antibiotics was dark and the discovery of antibiotics has shined upon the darkness of humanity and saved many lives. Average lifespan of a human being has increased significantly after the introduction of antibiotics. But, after a century of gains in the war against infectious diseases, the humanity is losing ground again. Common bacteria, parasites and fungi are developing widespread immunity to our best weapons.

Since its development, the humanity thought that penicillin will be a magic bullet for the infectious diseases. It has been used extensively and irresponsibly together with other following antimicrobials. Since many decades, there is a tremendous effort throughout the world to overcome the global antibiotic resistance. But now, medical authorities are raising concerns that treatment options in certain infectious diseases are narrowing because of alarming global antibiotic resistance problem. As new antimicrobial drugs are developing, the need for newer ones grow as well.

The looming public health crisis of antimicrobial-resistance have been evaluated by Centers for Disease Control and Prevention (CDC) and a report on antimicrobial resistance and its impact on various sectors in United States of America (USA) which highlights the scope of the problem have been published (CDC, 2013). CDC estimates that in the United States, more than two million people are infected every year with antibiotic-resistant bacteria, ending with at least 23,000 deaths as a result. From an economic standpoint, an economic burden of $18,000 -- 29,000 of medical costs per patient in a single year have been reported for USA in the year of 2000 which is similar in many geographies (Thabit et al., 2015). For example, high burden of antimicrobial drug resistance in Asia have been reported for various pathogens, namely; Methicillin-Resistant *Staphylococcus aureus* (MRSA), Vancomycin Resistant *S. aureus* (VISA), Vancomycin resistant *Enterococcus* (VRE), macrolide and penicillin resistant *Streptococcus pneumoniae*, Extended Spectrum Beta- Lactamase Producer *Escherichia coli* and *Klebsiella pneumoniae*, Carbapenem Resistant Enterococci (CRE), Multi drug resistant (MDR) *Pseudomonas aeruginosa* and *Acinetobacter* spp. are mentioned as superbugs of the 21st century (Lai et al., 2014).

As a result, the emergence of pathogens with different mechanisms of resistance has intensified the challenges associated with infection control and treatment strategies. Therefore, as prudent use of antimicrobials is not at the intended level, development of novel antimicrobial molecules remains as the main strategy to overcome the present narrow pass. In this context, many products have been studied to develop antimicrobials and the major sources of chemical diversity for starting materials were mainly natural products because of their abundant scaffold diversity (Mishra & Tiwari, 2011). Before, the pharmaceutical companies were using crude plant extracts to produce relatively simple therapeutic formulations but nowadays purified compounds and their chemically modified derivates have been included in many drug formulations. Despite there are plenty of possibilities coming from pure natural compounds, sometimes modification of formulations of natural compounds can be an option in developing antimicrobial molecules too. On the other hand, discovery of antimicrobial effectiveness of a molecule is not enough to use it as a therapeutic agent. For this, a drug approval process has to be carried out. To register a novel molecule as a drug, an Investigational New Drug (IND) application must be submitted to the Food and Drug Administration (FDA) in USA or European Medicines Agency (EMA) in Europe. Afterwards, the process continues with successful clinical trials ending with another application for marketing licenses namely, New Drug Application (NDA) in the USA or a Marketing Authorization Application (MAA) in Europe. This though period rarely ends with the introduction of novel, safe antimicrobials into the market. Nevertheless, many natural products and derived compounds are still in various stages of clinical development. So, the use of natural products or their templates is a viable source of new drug candidates.

Here, various natural products and their applications have been discussed to construct an understanding towards the goal of development of new strategies to overcome infectious diseases caused by antimicrobial resistant bacteria.

## HISTORICAL DEVELOPMENT OF NATURAL ANTIMICROBIALS IN DRUG INDUSTRY

The significant history of antimicrobial drug development starts with the fight against African Sleep Sickness which is also called as African trypanosomiasis; a disease spread by infected fly vectors. In 1859, Antoine Bechamp has discovered a drug named as “Atoxyl” with an aim to be used against African Sleep Sickness, which was unsuccessful (Burke, 1925). Despite the discovery of Atoxyl was as unsuccessful attempt, it was a good starting point for Paul Ehrlich and colleagues as it inspired them a lot. In 1909, Ehrlich and colleagues have showed the efficiency of this compound in the treatment of syphilis-infected rabbits. Subsequently, despite its adverse effects, the drug was marketed under the name of “Salvarsan”, which was a great success. A less toxic and more soluble form was also introduced afterwards which was called as “Neosalvarsan”(Aminov, 2017). The need for a drug effective against infectious diseases was so huge in that days. For this reason, Neosalvarsan quickly became the most frequently prescribed drug in the world until its replacement by penicillin in the 1940s (Mahoney et al., 1943).

Many antibiotic molecules have been developed from natural compounds after the “first natural antimicrobial approved for use in humans”, which is Penicillin. Penicillin was discovered in 1928 when Alexander Fleming noticed a halo of inhibited bacterial growth around mould colonies which led Fleming to conclude that the mould released a substance that repressed the growth of the bacteria (Lax et al., 2005). The discovery of penicillin along with other antibiotics such as sulphonamides or streptomycin have led to a period of 15 years, from 1938 to 1952, with highest rate of decline in infectious disease mortality (a decrease by 8.2% per year in the USA) (G. L. Armstrong et al., 1999). As penicillin was discovered it did not immediately take off as a clinically useful antibiotic. Despite the advantages, penicillin had many drawbacks in the first decade of introduction, such as low yield, instability, purification, and other problems. Actually, the World War 2 has stimulated the development of penicillin into a valuable treatment option for previously untreatable infectious diseases (Aminov, 2017). Afterwards, penicillin has been studied widely and extensively for a better understanding of the molecule structure and action mechanism.

Identification of 6-aminopenicillanic acid as the core of penicillin allowed the synthesis and production of numerous semisynthetic penicillin such as penicillinase-resistant penicillin (namely, methicillin, oxacillin, and nafcillin), the aminopenicillins (namely, ampicillin, amoxicillin, and bacampicillin), the carboxypenicillins (namely, carbenicillin and ticarcillin), and the ureidopenicillins (namely, mezlocillin, azlocillin, and piperacillin) (Aminov, 2017; Batchelor et al., 1959; A. J. Wright, 1999).

Notably, in his Nobel Speech Fleming has warned about the misuse of the drug. He stated that resistant bacteria can be selected evolutionary among the susceptible ones after misuse. Throughout the history, words of Fleming have been approved and in the 21st century, humanity still struggles to find a solution for this problem. Many antibiotic molecules have been developed after Penicillin which can be classified as purely natural, natural derived or synthetic molecules.

## TYPES AND CLASSIFICATION OF NATURAL ANTIBIOTIC MOLECULES

Antibiotics can be classified in many ways according to purpose. To summarise, the classifications can occur as; (i) According to mode of action, (ii) According to chemical structure, (iii) According to infections to be used, and (iv) According to sources.

### Sources of Antimicrobials

Antibiotics sources are so versatile. For instance, molecules that can have antimicrobial activity can be obtained from marine environments, to fungal agents. Obviously, the role of microbes in antimicrobial discovery has been always important and it seems like they have a fight by themselves as well.

The majority of antimicrobials have been discovered from prokaryotes, on the other hand, there are many molecules derived from eukaryotic sources such as fungi or plants (Ali et al., 2018). As mentioned before, Penicillin molecule was derived from Penicillium fungi. The other major natural products obtained from fungi are, Cephalosporins from *Acremonium* spp., Gliotoxins from *Aspergillus* spp. and *Trichoderma* spp., Indanonaftol A from *Aureobasidium* spp., Ascochital from *Kirschsteiniothelia maritimais*, Pestalone from *Pestalotia* spp., Pestalachloride D from *Pestalotiopsis* spp., *Aminolipopeptides trichoderins* A, A1, and B from *Trichoderma* spp. (Abraham et al., 1953; Biabani & Laatsch, 1998; Bugni & Ireland, 2004; Cueto et al., 2001; Jones & Hancock, 1988; Masuma et al., 2001; Newton & Abraham, 1955; Okutani, 1977; Pruksakorn et al., 2010; Silber et al., 2016; Wei et al., 2013).

Another eukaryotic group are plants which include some other natural products with antimicrobial activity. Some major natural products obtained from plants are; Sesquiterpenes from terrestrial medicinal plants and marine algae, Halogenated compounds from marine algae *Bonnemaisoniaceae*, Ar-turmerone, turmerone, curlene from turmeric and procyanidins from cranberry (Ahmad et al., 1994; Fenical & Mcconnell, 1977; Negi et al., 1999; Pappas & Schaich, 2009).

Last eukaryotic group of animals are potential sources for natural antimicrobials too. Some major natural products obtained from animals are mainly antimicrobial peptides such as, ll-37 from humans, sulphonamides, furanones and flavanones and some other bioactive molecules from Cockroach and Phenylacetaldehyde from *Musca domestica* (housefly)(Ali et al., 2017; Arora et al., 2011; Bonn, 2000; Srakaew et al., 2014).

The prokaryotic group which are Bacteria, are known to reserve a wide variety of natural antibiotic products which holds the majority of portion in antibiotics industry today. For example, Actinomycin, Streptomycin, Streptothricin, Erythromycin, Tetracyclines, Aminoglycosides and Lincomycin are the antibiotics which are isolated from the genus *Streptomyces* spp. (Darken et al., 1960; Majer et al., 1977; Schatz et al., 1944; Wagman, 1980; Waksman & Woodruff, 1941, 1942; J. L. C. Wright, 1983).

## Antibiotics According to their Mode of Action

One of the best approaches to classify antibiotics is to classify them according to their mode of action. As every antibiotic molecule can have its unique mechanism of action, Antibiotics mainly act on three primary targets within bacterial cells. These targets are; the inhibition of (i) cell wall synthesis, (ii) protein synthesis, and (iii) DNA or RNA synthesis (Rossiter et al., 2017).

As one of the most complex and critical organelles in the bacterial cell is the cell wall, most of the antibiotics aim to disrupt the bacterial cell wall structure or to inhibit the synthesis of it. For instance, one of the largest group of antibiotics; Beta-lactams act by disrupting the cell wall synthesis by its active site (beta-lactam ring). Beta-lactams including Penicillin or Cephalosporins inhibit the cross-linking step of peptidoglycan synthesis. A Glycopeptide, Vancomycin targets peptidoglycan synthesis as well but with a different mechanism of action which relies on inhibiting the transpeptidases from cross-linking the terminal D-Ala-D-Ala tail. Lipoglycopeptides such as Oritavancin or Dalbavancin Inhibit bacterial cell wall synthesis similarly but in addition to that, it disrupts the membrane potential, and change cell permeability due to the presence of a lipophilic side chain.

The second group of antibiotics are the group which includes drugs inhibiting the bacterial protein synthesis in the bacterial ribosome. Macrolides such as Clarithromycin or Erithromycin are in this group which targets the larger 50S subunit of bacterial ribosome. The tetracyclines, another group of natural product derived antibiotic, bind to smaller 30S subunit of the ribosome. Natural product derived, Aminoglycosides like Gentamicin, Amikacin and Tobramycin bind both 30S and/or 50S subunit of the bacterial ribosome which leads to death of bacterial cell. Chloramphenicol, Lincosamide and Oxazolidimine are other antibiotics which inhibits the bacterial protein synthesis process.

A third major target for antibiotics is nucleic acid replication and related repair mechanisms. For instance, DNA gyrase which is the enzyme taking role in the unwinding of replicated DNA is a target for the synthetic antibiotic class of quinolones.

Lastly, there are antibiotics which inhibits various steps in folic acid metabolism of bacteria. Some examples of these drugs include Sulphonamides and trimethoprim.

Antibiotics can be classified according to infections that they are used, and this type of classification can be broadened depending on the purpose of use as well, but there will be some consequences. One can classify the antibiotics as "Antibiotics to be used in Urinary Tract Infections" or "Antibiotics to be used in Upper Respiratory Infections" but this type of classification is not reliable and may lead to misuse as every pathogen requires a special attention when determining the treatment regime. For example, a group of antibiotics can be generalised as "Antibiotics to be used in Lower Respiratory Infections" but one of the most severe and hard to treat Lower Respiratory Tract infection which is "Tuberculosis" caused by *Mycobacterium tuberculosis* needs a special attention to be treated successfully.

# NATURAL PRODUCTS WITH ANTIMICROBIAL ACTIVITY AND THEIR POTENTIAL FOR NEW DRUGS AND VERSATILE APPLICATIONS

An antibiotic molecule, either natural or synthetic must have many features together and must overcome many procedures to become available for human use. There are many antibiotics discovered or developed for different purposes. Some of them have accomplished to be introduced for human use some are not. So the other antibiotic molecules that have a certain antimicrobial activity but not suitable for systemic human use can be good candidates for many other applications such as food preservation, wound dressing, antibacterial surface coating, and disinfection of pools (Carter & Joll, 2017; Juneja et al., 2012; Kana & Meimandipour, 2017).

## Antimicrobial Peptides

An important group which are good candidates for antimicrobial drug development are small antimicrobial peptides. Antimicrobial peptides are the newest group of natural products to have antimicrobial activity and these peptides are produced naturally by organisms internally to contribute the immunity of individual in the fight against infectious diseases. Moreover, they can have different roles throughout the body such as programming apoptosis or stimulating chemokine synthesis. To date, more than 2800, various Antimicrobial peptides have been shown from different organisms such as mammals, amphibians, fishes, plants, insects, echinoderms, crustaceans, fungi, and bacteria. When compared with traditional antibiotics, these Antimicrobial peptides have some advantages. For example, spectrum of activity is wide and at the same time resistance to Antimicrobial peptides is not common. Furthermore, Antimicrobial peptides killing time is quicker than conventional antibiotics where antagonism is not present (Moravej et al., 2018). The Antimicrobial Peptide Database contains 3003 antimicrobial peptides from six kingdoms (335 bacteriocins/peptide antibiotics from bacteria, 4 from archaea, 8 from protists, 14 from fungi, 343 from plants, and 2213 from animals, including some synthetic peptides) with the following activity: Antibacterial peptides; Antibiofilm peptides, Antiviral peptides; Anti-HIV peptides, Antifungal peptides, Antiparasitic peptides; Antimalarial peptides, Anti-protist peptides, Anticancer peptides, Antioxidant peptides, Chemotactic peptides, Insecticidal peptides, Protease inhibitors, Spermicidal peptides, Surface immobilized peptides and Wound healing peptides(Wang et al., 2016). The majority of Antimicrobial peptides such as LL-37 (Human cathelidins), Brilacidin (Human Defensin Analog), PXL01 (Human Lactoferrin Analog) are currently undergoing pre-clinical or clinical trials which is an indicator of their potential as promising antimicrobial candidates for human use in various infections (Crowther et al., 2013;

Malanovic et al., 2015; Maria et al., 2017). In depth detail for the promising studies on Antimicrobial peptides can be found on Moravej *et.* al.`s excellent review and AMP database (Moravej et al., 2018; Wang et al., 2016).

## Biopolymers

Biopolymers are a versatile group of natural compounds which have wide range of application capacity and have good antimicrobial activity. Biopolymers are haemostatic and anticoagulant, and they can be used as thickeners, emulsifiers or stabilizers, anti-fouling coating materials, also they can be used in wastewater management. A well-studied natural molecule, Chitosan (derived from Chitin) is a biopolymer that is found in the exoskeletons of crustaceans and arthropods with a variety of applications. Chitosan has good synergy with some other antibiotics and it can be used for a wide range of environmental and/or biomedical applications such as surface coating, tissue engineering, drug delivery and wound healing (Assaad et al., 2015; Devlieghere et al., 2004; Lee et al., 2014; Oryan & Sahvieh, 2017). Action mechanism of chitosan is mainly associated with the positive charge density of the polymer. Where negative charged bacterial cell wall and positive charged chitosan molecule interacts, this interaction leads to the lysis of the bacterial cell. The other important natural biopolymer which have antimicrobial applications is "epsilon-poly-L-lysine" which is produced by *Streptomycetaceae* and ergot fungi species (Nishikawa & Ogawa, 2002). It is an edible, non-toxic polymer that acts by electrostatic interactions between the molecule and the bacterial cell membrane resulting with physiological damage to bacterial cells which leads to death of bacteria.

## Others

In the last half-century natural antimicrobial discovery and their subsequent clinical use have peaked mainly after the developments and standardizations in antimicrobial susceptibility testing protocols. Accordingly, many interesting natural products have undergone antimicrobial susceptibility testing to detect at least meaningful antimicrobial activity. One good example for this; Animal venoms have been reviewed recently as rich sources of antimicrobials. Snake, spider, wasp, honey bee venoms have been reported to have antimicrobial activities (Perumal Samy et al., 2017). Similarly, many plant-derived compounds have been showed to possess antimicrobial activity. Flavonoids have good activity against many bacteria and they mostly have a synergistic effect in combined use with other antimicrobials (Savoia, 2012). For instance, Quercetin, a flavonoid which can be found in many fruits and vegetables have shown to possess good antibacterial and antiviral activity (Ganesan et al., 2012; Gopu et al., 2015). Apart from Flavonoids, Alkaloids like Diterpenes and Berberine had a broad range of antimicrobial effectiveness as well (Atta-ur-Rahman & Choudhary, 1999; S. H. Kim et al., 2002). Several Phenolic compounds which have the function to protect the plants from microbial infections are studied and Tannins, Quinones and Flavones showed good antibacterial activity (Savoia, 2012). Notably, Coumarin class of phenolic compounds have gained an attention for their antimicrobial activity. Furthermore, it has been reported that Coumarins can have inhibitory effect on microbial quorum sensing and biofilm formation as well (Gutierrez-Barranquero et al., 2015).

Lactoferrin, a glycoprotein included in milk is a good example of natural products which have environmental applications. It has been used as a food preservative frequently and recently it had got the approval in USA to be used as a preservative on beef (Del Olmo et al., 2009). A similar natural compound that is used in food industry is Ovotransferrin (a glycoprotein of egg white albumin) which is used as a food additive to increase the lifetime of various food products (J. Kim et al., 2012).

## MAJOR NATURAL ANTIBIOTIC DRUG CLASSES

Antibiotics that are drugs are mainly developed in 3 ways; (i)directly from natural products, (ii)by modifying natural products (natural product derived/semi-synthetic molecules) and (iii)by designing synthetic products. In early ages of antibiotic drug development, many of the agents were discovered in fermentation products without further modifications such as aminoglycosides, vancomycin and tetracycline. But the pipeline supplying new drugs from fermentation products has been drained and new antibiotic drugs could not be discovered like at the beginning in the last decade. The only exception is a drug developed for the treatment of *Clostridium difficile* related diarrhoea; Fidaxomicin which got the approval in 2011 (Hardesty & Juang, 2012). The other most common method in the discovery of antibiotics is to modify the known natural products. Many antibiotics such as daptomycin have been developed with this method (Debono et al., 1987). Sulphonamides, Diaminopyrimidines, Fluoroquinolones, Oxazolidinones, Nitroimidazoles are some of the synthetic antibiotics which will not be covered within the scope of this article.

### Beta-lactams

Antibiotics which its active site is the Beta-lactam ring are one of the most widely used antibiotic classes as they are efficient and safe. But the resistance is emerging for this group of antibiotics especially because of Extended Spectrum Beta-Lactamase producer microorganisms such as *E. coli* and *K. pneumoniae*. Beta-lactamases are enzymes produced by certain bacterial species which hydrolyse the beta-lactam ring of the compound which is the active site resulting with the ineffectiveness of the drug. Therefore, biggest challenge in overcoming the beta-lactam resistance is to develop newer beta-lactamase inhibitors (Bush & Bradford, 2016). Combination of beta-lactamase inhibitors are also suggested as a promising strategy to inhibit bacterial beta-lactamases. Ceftazidime-avibactam, Ceftaroline-avibactam, Ceftolozane-tazobactam are some of the promising beta-lactam and beta-lactamase inhibitor combinations (D. T. King et al., 2016).

*Table 1. Some antibiotics in clinical use*

| Group Name | Discovery Year | Major Representative antibiotics | Source | Target |
|---|---|---|---|---|
| Beta-lactams | 1929 | Penicillin | N/ND* | Cell wall |
| | | Cephalosporins | N/ND | Cell wall |
| | | Carbapenems | N/ND | Cell wall |
| Glycopeptides | 1953 | Vancomycin | N/ND | Cell wall |
| Macrolides | 1950 | Erythromycin | N/ND | Protein synthesis |
| | | Clarithromycin | N/ND | Protein synthesis |
| | | Azithromycin | N/ND | Protein synthesis |
| Lincosamide | 1963 | Clindamycin | N/ND | Protein synthesis |
| Aminoglycosides | 1943 | Gentamicin | N/ND | Protein synthesis |
| | | Tobramycin | N/ND | Protein synthesis |
| | | Amikacin | N/ND | Protein synthesis |
| Streptogramin | 1953 | Quinupristin/Dalfopristin | N/ND | Protein synthesis |
| Tetracycline | 1945 | Doxycycline | N/ND | Protein synthesis |
| | | Minocycline | N/ND | Protein synthesis |
| Rifamycin (Anzamycins) | 1957 | Rifampin | N/ND | DNA/RNA synthesis |
| Lipopeptide | 1947 | Daptomycin | N/ND | Cell wall |
| | | Colistin | N/ND | Cell wall |
| Amphenicols | 1947 | Chloroamphenicol | NS*** | Protein synthesis |
| Sulfonamides | 1935 | Sulfomethoxazole | S** | Folate biosynthesis |
| Diaminoprymidines | | Trimethoprim | S | Folate biosynthesis |
| Quinolones | 1962 | Ciprofloxacin | S | DNA/RNA synthesis |
| | | Levofloxacin | S | DNA/RNA synthesis |
| | | Moxifloxacin | S | DNA/RNA synthesis |
| Oxazolidinone | 1952 | Linezolid | S | Protein synthesis |
| Nitroimidazole | 1955 | Metronidazole | S | DNA/RNA synthesis |

*N: Natural/Natural derived Product, **Synthetic Product, *** Naturally discovered but produced synthetically

## Glycopeptides

Glycopeptides are natural antibiotics isolated from *Amycolatopsis orientalis* a soil bacterium. Their main action mechanism is the inhibition of cell wall synthesis. Glycopeptides are thought to be a last option in the treatment of severe, complicated infections, but global resistance problem has affected this class of antibiotics as well. First antibiotic resistant bacteria was reported in a Vancomycin resistant Enterococci strain in 1988 (Uttley et al., 1988). Afterwards, the resistance started to disseminate. Resistance is mainly related with the ability of bacteria to produce cell wall precursors that have low affinity to vancomycin due to genetic modifications of *van* genes. Nowadays, Vancomycin resistant Enterococci and Vancomycin resistant *S. aureus* strains are globally threating the healthcare facilities. In addition to misuse, use of Avoparcin (which is a growth promoting agricultural agent) is reported to be responsible from the increased resistance rates (Bager et al., 1997).

Studies working on modifications on vancomycin molecule mainly aim to increase the affinity of the molecule to the cell wall of the bacterial cell. Crowley and Boger have suggested a re-engineered vancomycin molecule which can bind to an additional cell wall target (Crowley & Boger, 2006). On the other hand, resistance levels to other Glycopeptide antibiotics are still in lower levels in contrast to Vancomycin.

Glycopeptides other than Vancomycin are Teicoplanin, Telavancin, Oritavancin and Dalbavancin. Teicoplanin is a drug which can be used against infections caused by Vancomycin resistant strains excluding VanA resistant strains (Aminov, 2017). Comparing to first generation Glycopeptides (which are Vancomycin and Teicoplanin), Telavancin, Oritavancin and Dalbavancin have better affinity features, and more than one mode of action (Klinker & Borgert, 2015; Van Bambeke, 2006; G G Zhanel et al., 2010).

## Macrolides

Azithromycin, Clarithromycin, Erythromycin, Fidaxomicin and Telithromycin are the antibiotics of Macrolides class which get the Federal Drug Administration approval. First of the macrolides, Erythromycin was isolated from an actinomycete bacteria; *Saccharopolyspora erythraea*. This group of antibiotics serve as the second mostly prescribed antibiotic group after beta-lactams (or in conditions when beta-lactams are not applicable) and mainly make action by inhibiting the protein synthesis.

Current studies aiming to modify macrolides focus mainly to improve the affinity of the molecule to bind the ribosome better and/or to find different binding sites for the molecule. There are 2 molecules currently being developed which are Ketolides (Telithromycin) and Flouroketholides (Solithromycin). Despite Solithromycin is still in the last phase of clinical trials it can be concluded that both new modifications have fulfilled the purposes as they possess a better affinity and a wider spectrum.

## Lincosamides

Clindamycine is the drug of choice among Lincosamides, whereas the original formulation; Lincomycin is not used commonly due to its low level of spectrum and serious adverse effects. Clindamycin which is a far better choice has been developed by simply adding a Chlorine atom with the inversion of chirality of the older Lincomycin molecule. The main action mechanism of lincosamides is to stop the protein synthesis thus bacteria can gain resistance by modifying the binding site (namely through modifying the *erm* gene).

## Aminoglycosides

Aminoglycosides are a wide group of antibiotics which are mainly active against gram negative bacteria and tuberculosis bacilli. Aminoglycosides which act by inhibiting the protein synthesis, are active against most gram-negative aerobic and facultative anaerobic bacilli but not active against gram-negative anaerobes and most gram-positive bacteria. There are more than 10 modifications of aminoglycosides namely, Kanamycin A, Amikacin, Tobramycin, Dibekacin, Gentamicin, Sisomicin, Netilmicin, Neomycins B, Neomycin C, Neomycin E (paromomycin) and Streptomycin.

After the discovery of the first Aminoglycoside, Streptomycin there has been a dramatic deadline especially in Tuberculosis cases. Good activity of Streptomycin in tuberculosis cases has put the drug in one of the most life-changing drugs ever as mortality of tuberculosis has decreased by 47% between 1990 and 2015 (Aminov, 2017). But there are resistant *M. tuberculosis* strains being reported from different regions in the world in the last decades. After multi drug resistant strains, extremely drug resistance strains started to appear during 2000s and in 2009 a totally drug resistant strain has been reported as well (Migliori et al., 2007; Plikaytis et al., 1994; Velayati et al., 2009). Apart from Streptomycin, Gentamicin and Amikacin have gained good popularity in clinical use as well especially in severe intensive care infections. Synergism studies exploring the interactions between Aminoglycosides and other natural antibiotics could result in newer combinations in the future.

## Streptogramins

Streptogramins which are produced by many strains of *Streptomyces* spp. are have good activity against a wide range of bacteria especially to Methycillin Resistant *S. aureus* and Vancomycin Resistant Enterococci, therefore they are thought to be one of the last-resort antibiotics. Characteristically, two different types of chemical substances; the group A streptogramins, (which are polyunsaturated macrolactones) and the group B streptogramins are combined into one drug (Mast & Wohlleben, 2014). Main action mechanism is the inhibition of protein synthesis. One of the well-known Streptogramins is Virginiamycin. The common adverse effects has prevented Virginiamycin for human use, but it has been used as a food animal growth promoting agent very popularly for a long time (Yates & Schaible, 1962). New considerable Streptogramins which are eligible for human use are Pristinamycin and Quinupristin/Dalfopristin which were introduced in 2001 (Allington & Rivey, 2001; Hamilton-miller, 1991).

The similar action mechanism of Streptogramins, Macrolides and Lincosamides lead to a resistance overlap. Once a bacterium develops resistance upon drug contact it become resistant to other two as well (Tenson et al., 2003). Future studies investigating the development of chimeric streptogramin-tyrocidine combinations can improve understanding regarding to streptogramin resistance (Mukhtar et al., 2005).

## Tetracyclines

Tetracycline group of antibiotics were firstly introduced mainly as growth promoting agents for food animals. A study reported that, Tetracyclines were most common antibiotics sold for food producing animal industry (Administration & Services, 2015).

The first Tetracycline, which was chlortetracycline, was discovered in 1945 and afterwards Aureomycin was introduced to market. Source of Aureomycin was a soil bacterium named *Streptomyces aureofaciens* (Aminov, 2017). After a long period of usage in food producing animals industry, chemical structure of natural Tetracycline molecule has been solved and this progress has led to development of new Tetracycline molecules with the aim of fighting growing antimicrobial resistance problem (Nelson & Levy, 2011). These molecules were minocycline and doxycycline which were discovered in 1966 and 1967 respectively. However, bacteria quickly have developed resistance to these antibiotics too. Therefore, studies focusing on the development of new molecules have been in progress until 2005, that a third generation Tetracycline has appeared with the name of Tigecycline (Rose & Rybak, 2006).

Introduction of Tetracycline was an important development for the drug and infectious disease societies as it has a good activity on resistant pathogens especially Multi drug resistant *Enterobacteriaceae* strains. Luckily, the resistance rates are still low. Major challenge in this class of antibiotics is to identify and elucidate further possible resistance mechanisms of bacteria. For instance, discovery of *tetX* gene has provided a good understanding of Tetracycline resistance (Aminov, 2013).

## Rifamycins

Rifamycins which are also named as Ansamycins are natural antibiotics isolated from soil bacteria *Nocardia mediterrenai* (formerly named as *Streptomyces mediterrenai*) (Margalith & Beretta, 1960). There are many types of different molecules in this class but the one which has got most of the attention is the Rifampin. Rifampins importance comes from its effectiveness in the treatment of mycobacterial infections such as tuberculosis or leprosy (Sepkowitz et al., 1995). Other modified molecules of this class include the drugs such as Rifabutin, Rifapentin and Rifaximin where the usage is not as common as Rifampisin due to poor activity and side effects. The resistance for Rifamycins is growing as similar with other classes and it needs attention especially for Multi drug resistant Tuberculosis cases (Floss & Yu, 2005).

## Lipopeptides

Lipopeptides comprise an important group of antibiotics which are highly effective against multi drug resistant strains especially. In this group, the most important antibiotics are Colistin (Polymixin E) and Daptomycin which are isolated from *Paenibacillus polymyxa*, *Streptomyces roseosporus* respectively (Storm et al., 1977; Tally & DeBruin, 2000). Another group of molecules which have good antifungal activity are Echinocandins.

Despite their toxicity in various systems, lipopeptides are life saver in many severe infections caused by multi drug resistant bacteria including gram negatives such as *P. aeruginosa, K. pneumoniae, A. baumannii* or gram positives like Vancomycin resistant Enterococci and Methicillin resistant *S. aureus*. Being active mostly in gram negative infections, Colistin can be useful in treating the most life threating infections caused by Metallo-beta lactamase producer *Enterobacteriaceae* strains as well (Kumarasamy et al., 2010). On the other hand, Daptomycin is the drug of choice in infections caused by gram positive bacteria. However, despite the reported rate of resistant strains is rare Lipopeptide resistance has started to emerge similar with other antibiotics which requires a special attention (Aminov, 2017). Further studies elaborating the possible mechanisms of resistance to this group of antibiotics and studies which will evaluate possible modifications to decrease the toxicity of lipopeptides would be useful.

## Polypeptides

This group of antibiotics did not gain much popularity due to their limited usage capacity in human infections. The main drugs in this class are Tyrocidines and Gramicidines. Tyramicidines and Gramicidines are isolated from a soil bacterium namely, *Brevibacillus brevis* (Gause & Brazhnikova, 1944). Main limitation in this class is their adverse effects and these drugs can be extremely toxic limiting their use to a topical level. However, in some extraordinary cases systemic use can be thought when run out of options. A drug which is commonly used in the differentiation of Streptococcus species, Bacitracin is a polypeptide as well (Hallen et al., 2007).

## NOVEL NATURAL ANTIMICROBIALS IN THE FIGHT AGAINST RESISTANT BACTERIA

Development of new antibiotic drugs is the main strategy to overcome superbug infections. New drugs can be developed by the modification of older ones or from new source molecules. On the other hand, rediscovery of older drugs for drug resistant infections is becoming a popular strategy for the management of these infections (Poulakou et al., 2014). Sometimes, modification of the dosage regimens or trying new combinations can be effective as well (Daikos et al., 2014; Qureshi et al., 2012; Tumbarello et al., 2015)..

Older drugs such as Colistin, Fosfomycin and Tigecycline can be effective in antibiotic resistant gram-negative bacterial infections. But there are many adverse effects and toxicity issues to tackle. For example, sensitivity to colistin's have been reported in such carbapenem resistant gram negative bacterial infections and despite the neurotoxicity and nephrotoxicity of colistin is known, it can be a last resort of option to be used in such severe infections (Giamarellou & Poulakou, 2009; Tumbarello et al., 2015). Another older drug which is effective in resistant bacterial infections is Fosfomycin. In addition to its effectiveness in gram negatives, Fosfomycin can be effective in gram positive bacterial infections as well. Like colistin, Fosfomycin must be used with care due to its side effects (Raz & Unit, 2011). Tigecycline, an old Tetracycline have a wide range of spectrum and it can be used in certain Extended Spectrum Beta-Lactamase producer infections with the consideration of adverse effects (Pournaras et al., 2011).

Apart from older drugs, new drugs or combinational regimes are alternative strategies for drug resistant gram-negative bacterial infections. In the last decade, some new beta-lactamase inhibitors have been introduced into the market and new combinations have emerged (Sfeir et al., 2018). Ceftazidime/Avibactam, one of the new combinations has reached to 100% susceptibility in Carbapenem resistant *Enterobacteriaceae* infections. Overall research suggests Ceftazidime/Avibactam as a better alternative than older ones such as colistin (M. King et al., 2017). Ceftolozane/Tazobactam is the other natural combination of cephalosporins and beta lactamase inhibitors. This combination has been reported to be very useful in the management of urinary tract infections caused by Extended Spectrum Beta-Lactamase producer *Enterobacteriaceae* and especially *P. aeruginosa* (E. S. Armstrong et al., 2016).

Carbapenemase inhibitors developed in last 2 decades are becoming very helpful in the management of infections caused by Carbapenemase producers. Use of carbapenems together with these inhibitors reactivates the action of carbapenems (Garau et al., 2005). For instance, Imipenem/Relebactam and Meropenem/Vaborbactam combinations have been showed to be very effective against Carbapenem Resistant *Enterobacteriaceae* and to lower the Minimum Inhibitory Concentrations to carbapenems in vitro (Hackel et al., 2018; Livermore et al., 2013).

A new generation Aminoglycoside, Plazomicin is currently being developed. First step in vitro studies have demonstrated good activity against Multi Drug Resistant *Enterobacteriaceae* including Aminoglycoside resistant isolates (Walkty et al., 2014). Another new drug in development, Cefederecol is a new cephalosporin with promising activity against Carbapenem Resistant *Enterobacteriaceae* (Ito et al., 2016). Tetracycline derived new antibiotic molecules; Omadacycline and Eravacycline are the latest antibiotic molecules currently being developed with promising first step results against resistant bacteria groups (Pfaller et al., 2018; George G Zhanel et al., 2016).

Fifth generation cephalosporins, Ceftaroline and Ceftobirole are the beta-lactam antibiotics which have good activity against MRSA. To compare, Ceftaroline has a broader activity in contrast to Ceftobirole (Kelley et al., 2015; Long et al., 2014; Nicholson et al., 2012). However, Ceftaroline which reported to be very effective against MRSA infections reported to be ineffective in some parts of the world where the drug never used before (Mendes et al., 2012).

Dalbavancin and Oritavancin are Lipoglycopeptides which have been introduced to market recently. FDA and EMA recently approved Dalbavancin for the treatment of skin infections. Both of the new Lipoglycopeptides possess a good activity against Resistant Gram positive pathogens with low toxicity (Klinker & Borgert, 2015; G G Zhanel et al., 2010).

Another group of novel natural derived antibiotics are from Tetracycline class namely, Omadacyclines. Omadacyclines are currently received fast track development status from FDA and they are active against a large portion of pathogens including gram positives and negatives (Villano et al., 2016).

Lastly, Lefamulin a novel semi-synthetic drug of Pleuromutilin class has been developed to inhibit 50S subunit of bacterial ribosome. Lefamulin is strongly active against MRSA, VRE and Penicillin resistant *S. pneumoniae* infections (Paukner et al., 2013). Tedizolid, Delafloxacin, Nemonoxacin and Zabofloxacin are new synthetic drugs under development for the fight against gram positive resistant bacteria which are not covered here (Abbas et al., 2017).

*Table 2. Promising natural antibiotic drugs in the fight against resistant bacteria*

| Target infections | Antibiotics |
|---|---|
| Gram negative bacterial infections | Colistin, Ceftazidime/Avibactam, Fosfomycin, Ceftolozone/Tazobactam, Tigecycline, Imipenem/Relebactam, Meropenem/Vaborbactam |
| Gram positive bacterial infections | Ceftaroline, Oritavancin, Ceftobirole, Lefamulin, Dalbavancin, Omadacycline |

## CONCLUSION

Currently a global antibiotic crisis is ongoing. Many known pathogens are developing resistance against many known antibiotics and new resistant pathogens are emerging at the same time. Mortality and morbidity rates due to infections caused by these pathogens are increasing to an alarming level. Therefore, immediate actions combining several parallel approaches are needed in the drug development industry. These actions are focused mainly on the modifications of existing natural products/ scaffolds aiming known bacterial targets. In this regard, *in-silico* studies focusing on new targets such as energy metabolism could be a promising approach for new drug discoveries.

New discoveries of natural products can also generate new insights in the fight against antibiotic resistant bacteria. Simultaneously, the experiments focusing on the optimisation of the usage of older natural antimicrobials alone or enlightening potential synergistic interactions can still be useful. Moreover, development of synthetic molecules is another field that needs attention.

Putting it all together, the current strategy should mainly focus on the preservation of existing drugs by proper usage. The large number of natural products in the current industry and natural product derived compounds in various stages of clinical development are all indicators of their potential as candidates for the new antibiotic drug development.

## ACKNOWLEDGMENT

This research received no specific grant from any funding agency in the public, commercial, or not-for-profit sectors.

## REFERENCES

Abbas, M., Paul, M., & Huttner, A. (2017). New and improved? A review of novel antibiotics for Gram-positive bacteria. *Clinical Microbiology and Infection*, *23*(10), 697–703. doi:10.1016/j.cmi.2017.06.010 PMID:28642145

Abraham, E. P., Newton, G. G. F., Crawford, K., Burton, H. S., & Hale, C. W. (1953). Cephalosporin N: A new type of penicillin. *Nature*, *171*(4347), 343. doi:10.1038/171343a0 PMID:13036884

Administration, F. and D., & Services, D. of H. and H. (2015). *Antimicrobials Sold or Distributed for Use in Food-Producing Animals*. https://www.fda.gov/downloads/ForIndustry/UserFees/AnimalDrugUserFeeActADUFA/UCM440584.pdf

Ahmad, V. U., Ahmad, W. U., Aliya, R., Baqai, F. T., Iqbal, S., Khatoon, R., Mohammad, F. V., Noorwala, M., Perveen, S., Pervez, A., Saba, N., Shah, M. G., & Siddiqui, S. (1994). New natural products from terrestrial medicinal plants and marine algae. *Pure and Applied Chemistry*, *66*(10–11), 2311–2314. doi:10.1351/pac199466102311

Ali, S. M., Siddiqui, R., & Khan, N. A. (2018). Antimicrobial discovery from natural and unusual sources. *The Journal of Pharmacy and Pharmacology*, *70*(10), 1–14. doi:10.1111/jphp.12976 PMID:30003546

Ali, S. M., Siddiqui, R., Ong, S.-K., Shah, M. R., Anwar, A., Heard, P. J., & Khan, N. A. (2017). Identification and characterization of antibacterial compound (s) of cockroaches (Periplaneta americana). *Applied Microbiology and Biotechnology*, *101*(1), 253–286. doi:10.100700253-016-7872-2 PMID:27743045

Allington, D. R., & Rivey, M. P. (2001). Quinupristin/dalfopristin: A therapeutic review. *Clinical Therapeutics*, *23*(1), 24–44. doi:10.1016/S0149-2918(01)80028-X PMID:11219478

Aminov, R. (2013). Evolution in action: dissemination of tet(X) into pathogenic microbiota. In *Frontiers in Microbiology* (Vol. 4, p. 192). https://www.frontiersin.org/article/10.3389/fmicb.2013.00192

Aminov, R. (2017). History of antimicrobial drug discovery: Major classes and health impact. *Biochemical Pharmacology*, *133*, 4–19. doi:10.1016/j.bcp.2016.10.001 PMID:27720719

Armstrong, E. S., Farrell, D. J., Palchak, M., & Steenbergen, J. N. (2016). In Vitro Activity of Ceftolozane-Tazobactam against Anaerobic Organisms Identified during the ASPECT-cIAI Study. *Antimicrobial Agents and Chemotherapy*, *60*(1), 666–668. doi:10.1128/AAC.01964-15 PMID:26552971

Armstrong, G. L., Conn, L. A., & Pinner, R. W. (1999). Trends in infectious disease mortality in the United States during the 20th century. *Journal of the American Medical Association*, *281*(1), 61–66. doi:10.1001/jama.281.1.61 PMID:9892452

Arora, S., Baptista, C., & Lim, C. S. (2011). Maggot metabolites and their combinatory effects with antibiotic on Staphylococcus aureus. *Annals of Clinical Microbiology and Antimicrobials*, *10*(1), 6. doi:10.1186/1476-0711-10-6 PMID:21299858

Assaad, E., Maire, M., & Lerouge, S. (2015). Injectable thermosensitive chitosan hydrogels with controlled gelation kinetics and enhanced mechanical resistance. *Carbohydrate Polymers*, *130*, 87–96. doi:10.1016/j.carbpol.2015.04.063 PMID:26076604

Atta-ur-Rahman, & Iqbal Choudhary. (1999). Diterpenoid and steroidal alkaloids. *Natural Product Reports*, *16*(5), 619–635. doi:10.1039/a705715f PMID:10584334

Bager, F., Madsen, M., Christensen, J., & Aarestrup, F. M. (1997). Avoparcin used as a growth promoter is associated with the occurrence of vancomycin-resistant Enterococcus faecium on Danish poultry and pig farms. *Preventive Veterinary Medicine*, *31*(1), 95–112. doi:10.1016/S0167-5877(96)01119-1 PMID:9234429

Batchelor, F. R., Doyle, F. P., Nayler, J. H. C., & Rolinson, G. N. (1959). Synthesis of Penicillin: 6-Aminopenicillanic Acid in Penicillin Fermentations. *Nature*, *183*(4656), 257–258. doi:10.1038/183257b0 PMID:13622762

Biabani, M. A. F., & Laatsch, H. (1998). Advances in chemical studies on low-molecular weight metabolites of marine fungi. *Journal Für Praktische Chemie/Chemiker-Zeitung*, *340*(7), 589–607.

Bonn, D. (2000). Maggot therapy: An alternative for wound infection. *Lancet*, *356*(9236), 1174. doi:10.1016/S0140-6736(05)72870-1 PMID:11030307

Bugni, T. S., & Ireland, C. M. (2004). Marine-derived fungi: A chemically and biologically diverse group of microorganisms. *Natural Product Reports*, *21*(1), 143–163. doi:10.1039/b301926h PMID:15039840

Burke, E. T. T. (1925). The arseno-therapy of syphilis; stovarsol, and tryparsamide. *The British Journal of Venereal Diseases*, *1*(4), 321–338. doi:10.1136ti.1.4.321 PMID:21772505

Bush, K., & Bradford, P. A. (2016). β-Lactams and β-lactamase inhibitors: An overview. *Cold Spring Harbor Perspectives in Medicine*, 025247. PMID:27329032

Carter, R. A. A., & Joll, C. A. (2017). Occurrence and formation of disinfection by-products in the swimming pool environment: A critical review. *Journal of Environmental Sciences (China)*, *58*, 19–50. doi:10.1016/j.jes.2017.06.013 PMID:28774608

CDC. (2013). Antibiotic resistance threats in the United States, 2013. *Current*, *114*.

Crowley, B. M., & Boger, D. L. (2006). Total Synthesis and Evaluation of [Ψ[CH2NH] Tpg4]Vancomycin Aglycon: Reengineering Vancomycin for Dual d-Ala-d-Ala and d-Ala-d-Lac Binding. *Journal of the American Chemical Society*, *128*(9), 2885–2892. doi:10.1021/ja0572912 PMID:16506767

Crowther, G. S., Baines, S. D., Todhunter, S. L., Freeman, J., Chilton, C. H., & Wilcox, M. H. (2013). Evaluation of NVB302 versus vancomycin activity in an in vitro human gut model of Clostridium difficile infection. *The Journal of Antimicrobial Chemotherapy*, *68*(1), 168–176. doi:10.1093/jac/dks359 PMID:22966180

Cueto, M., Jensen, P. R., Kauffman, C., Fenical, W., Lobkovsky, E., & Clardy, J. (2001). Pestalone, a new antibiotic produced by a marine fungus in response to bacterial challenge. *Journal of Natural Products*, *64*(11), 1444–1446. doi:10.1021/np0102713 PMID:11720529

Daikos, G. L., Tsaousi, S., Tzouvelekis, L. S., Anyfantis, I., Psichogiou, M., Argyropoulou, A., Stefanou, I., Sypsa, V., Miriagou, V., Nepka, M., Georgiadou, S., Markogiannakis, A., Goukos, D., & Skoutelis, A. (2014). Carbapenemase-producing Klebsiella pneumoniae bloodstream infections: Lowering mortality by antibiotic combination schemes and the role of carbapenems. *Antimicrobial Agents and Chemotherapy*, *58*(4), 2322–2328. doi:10.1128/AAC.02166-13 PMID:24514083

Darken, M. A., Berenson, H., Shirk, R. J., & Sjolander, N. O. (1960). Production of tetracycline by Streptomyces aureofaciens in synthetic media. *Applied Microbiology*, *8*(1), 46–51. doi:10.1128/am.8.1.46-51.1960 PMID:13814119

Debono, M., Barnhart, M., Carrell, C. B., Hoffmann, J. A., Occolowitz, J. L., Abbott, B. J., Fukuda, D. S., Hamill, R. L., Biemann, K., & Herlihy, W. C. (1987). A21978C, a complex of new acidic peptide antibiotics: Isolation, chemistry, and mass spectral structure elucidation. *The Journal of Antibiotics*, *40*(6), 761–777. doi:10.7164/antibiotics.40.761 PMID:3610833 del Olmo, A., Morales, P., & Nunez, M. (2009). Bactericidal activity of lactoferrin and its amidated and pepsin-digested derivatives against Pseudomonas fluorescens in ground beef and meat fractions. *Journal of Food Protection*, *72*(4), 760–765. doi:10.4315/0362-028X-72.4.760 PMID:19435224

Devlieghere, F., Vermeulen, A., & Debevere, J. (2004). Chitosan: Antimicrobial activity, interactions with food components and applicability as a coating on fruit and vegetables. *Food Microbiology*, *21*(6), 703–714. doi:10.1016/j.fm.2004.02.008

Fenical, W. H., & Mcconnell, O. J. (1977). Antibiotic and antiseptic compounds from seaweeds of family bonnemaisoniaceae. *Journal of Phycology*, *13*, 21.

Floss, H. G., & Yu, T.-W. (2005). Rifamycin - Mode of action, resistance, and biosynthesis. *Chemical Reviews*, *105*(2), 621–632. doi:10.1021/cr030112j PMID:15700959

Ganesan, S., Faris, A. N., Comstock, A. T., Wang, Q., Nanua, S., Hershenson, M. B., & Sajjan, U. S. (2012). Quercetin inhibits rhinovirus replication in vitro and in vivo. *Antiviral Research*, *94*(3), 258–271. doi:10.1016/j.antiviral.2012.03.005 PMID:22465313

Garau, G., Bebrone, C., Anne, C., Galleni, M., Frère, J. M., & Dideberg, O. (2005). A metallo-β-lactamase enzyme in action: Crystal structures of the monozinc carbapenemase CphA and its complex with biapenem. *Journal of Molecular Biology*, *345*(4), 785–795. doi:10.1016/j.jmb.2004.10.070 PMID:15588826

Gause, G. F., & Brazhnikova, M. G. (1944). Gramicidin S and its use in the treatment of infected wounds. *Nature*, *154*(3918), 703. doi:10.1038/154703a0

Giamarellou, H., & Poulakou, G. (2009). Multidrug-resistant Gram-negative infections: What are the treatment options? *Drugs*, *69*(14), 1879–1901. doi:10.2165/11315690-000000000-00000 PMID:19747006

Gopu, V., Meena, C. K., & Shetty, P. H. (2015). Quercetin Influences Quorum Sensing in Food Borne Bacteria: In-Vitro and In-Silico Evidence. *PLoS One*, *10*(8), e0134684. doi:10.1371/journal.pone.0134684 PMID:26248208

Gutierrez-Barranquero, J. A., Reen, F. J., McCarthy, R. R., & O'Gara, F. (2015). Deciphering the role of coumarin as a novel quorum sensing inhibitor suppressing virulence phenotypes in bacterial pathogens. *Applied Microbiology and Biotechnology*, *99*(7), 3303–3316. doi:10.100700253-015-6436-1 PMID:25672848

Hackel, M. A., Lomovskaya, O., Dudley, M. N., Karlowsky, J. A., & Sahm, D. F. (2018). In Vitro Activity of Meropenem-Vaborbactam against Clinical Isolates of KPC-Positive Enterobacteriaceae. *Antimicrobial Agents and Chemotherapy*, *62*(1). Advance online publication. doi:10.1128/AAC.01904-17 PMID:29084745

Hallen, H. E., Luo, H., Scott-Craig, J. S., & Walton, J. D. (2007). Gene family encoding the major toxins of lethal Amanita mushrooms. *Proceedings of the National Academy of Sciences of the United States of America*, *104*(48), 19097–19101. doi:10.1073/pnas.0707340104 PMID:18025465

Hamilton-miller, J. M. T. (1991). From foreign pharmacopoeias: "new" antibiotics from old? *The Journal of Antimicrobial Chemotherapy*, *27*(6), 702–705. doi:10.1093/jac/27.6.702 PMID:1938680

Hardesty, J. S., & Juang, P. (2012). Fidaxomicin: A Macrocyclic Antibiotic for the Treatment of Clostridium difficile Infection. *Pharmacotherapy: The Journal of Human Pharmacology and Drug Therapy*, *31*(9), 877–886. doi:10.1592/phco.31.9.877 PMID:21923589

Ito, A., Kohira, N., Bouchillon, S. K., West, J., Rittenhouse, S., Sader, H. S., Rhomberg, P. R., Jones, R. N., Yoshizawa, H., Nakamura, R., Tsuji, M., & Yamano, Y. (2016). In vitro antimicrobial activity of S-649266, a catechol-substituted siderophore cephalosporin, when tested against non-fermenting Gram-negative bacteria. *The Journal of Antimicrobial Chemotherapy*, *71*(3), 670–677. doi:10.1093/jac/dkv402 PMID:26645269

Jones, R. W., & Hancock, J. G. (1988). Mechanism of gliotoxin action and factors mediating gliotoxin sensitivity. *Microbiology*, *134*(7), 2067–2075. doi:10.1099/00221287-134-7-2067

Juneja, V. K., Dwivedi, H. P., & Yan, X. (2012). Novel Natural Food Antimicrobials. *Annual Review of Food Science and Technology*, *3*(1), 381–403. doi:10.1146/annurev-food-022811-101241 PMID:22385168

Kana, J. R., & Meimandipour, A. (2017). Antimicrobial Activity of Chitosan Film Forming Solution Enriched with Essential Oils; an in Vitro Assay. *Iranian Journal of Biotechnology*, *15*(2), 111–119. doi:10.15171/ijb.1360 PMID:29845058

Kelley, W. L., Jousselin, A., Barras, C., Lelong, E., & Renzoni, A. (2015). Missense mutations in PBP2A Affecting ceftaroline susceptibility detected in epidemic hospital-acquired methicillin-resistant Staphylococcus aureus clonotypes ST228 and ST247 in Western Switzerland archived since 1998. *Antimicrobial Agents and Chemotherapy*, *59*(4), 1922–1930. doi:10.1128/AAC.04068-14 PMID:25583724

Kim, J., Moon, S. H., Ahn, D. U., Paik, H. D., & Park, E. (2012). Antioxidant effects of ovotransferrin and its hydrolysates. *Poultry Science*, *91*(11), 2747–2754. doi:10.3382/ps.2012-02150 PMID:23091127

Kim, S. H., Lee, S. J., Lee, J. H., Sun, W. S., & Kim, J. H. (2002). Antimicrobial activity of 9-O-acyl- and 9-O-alkylberberrubine derivatives. In Planta medica (Vol. 68, Issue 3, pp. 277–281). doi:10.1055-2002-23128

King, D. T., Sobhanifar, S., & Strynadka, N. C. J. (2016). One ring to rule them all: Current trends in combating bacterial resistance to the β-lactams. *Protein Science*, *25*(4), 787–803. doi:10.1002/pro.2889 PMID:26813250

King, M., Heil, E., Kuriakose, S., Bias, T., Huang, V., El-Beyrouty, C., McCoy, D., Hiles, J., Richards, L., Gardner, J., Harrington, N., Biason, K., & Gallagher, J. C. (2017). Multicenter Study of Outcomes with Ceftazidime-Avibactam in Patients with Carbapenem-Resistant Enterobacteriaceae Infections. *Antimicrobial Agents and Chemotherapy*, *61*(7). Advance online publication. doi:10.1128/AAC.00449-17 PMID:28483952

Klinker, K. P., & Borgert, S. J. (2015). Beyond Vancomycin: The Tail of the Lipoglycopeptides. *Clinical Therapeutics*, *37*(12), 2619–2636. doi:10.1016/j.clinthera.2015.11.007 PMID:26658277

Kumarasamy, K. K., Toleman, M. A., Walsh, T. R., Bagaria, J., Butt, F., Balakrishnan, R., Chaudhary, U., Doumith, M., Giske, C. G., Irfan, S., Krishnan, P., Kumar, A. V., Maharjan, S., Mushtaq, S., Noorie, T., Paterson, D. L., Pearson, A., Perry, C., Pike, R., ... Woodford, N. (2010). Emergence of a new antibiotic resistance mechanism in India, Pakistan, and the UK: A molecular, biological, and epidemiological study. *The Lancet. Infectious Diseases*, *10*(9), 597–602. doi:10.1016/S1473-3099(10)70143-2 PMID:20705517

Lai, C. C., Lee, K., Xiao, Y., Ahmad, N., Veeraraghavan, B., Thamlikitkul, V., Tambyah, P. A., Nelwan, R. H. H., Shibl, A. M., Wu, J. J., Seto, W. H., & Hsueh, P. R. (2014). High burden of antimicrobial drug resistance in Asia. *Journal of Global Antimicrobial Resistance*, *2*(3), 141–147. doi:10.1016/j.jgar.2014.02.007 PMID:27873720

Lax, E., Holt, H., York, C. N., York, N., & Fey, P. (2005). Book review The mold in Dr. Florey ' s coat. *The Journal of Clinical Investigation*, *115*(2), 2005.

Lee, D.-S., Eom, S.-H., Kim, Y.-M., Kim, H. S., Yim, M.-J., Lee, S.-H., Kim, D.-H., & Je, J.-Y. (2014). Antibacterial and synergic effects of gallic acid-grafted-chitosan with beta-lactams against methicillin-resistant Staphylococcus aureus (MRSA). *Canadian Journal of Microbiology*, *60*(10), 629–638. doi:10.1139/cjm-2014-0286 PMID:25216286

Livermore, D. M., Warner, M., & Mushtaq, S. (2013). Activity of MK-7655 combined with imipenem against Enterobacteriaceae and Pseudomonas aeruginosa. *The Journal of Antimicrobial Chemotherapy*, *68*(10), 2286–2290. doi:10.1093/jac/dkt178 PMID:23696619

Long, S. W., Olsen, R. J., Mehta, S. C., Palzkill, T., Cernoch, P. L., Perez, K. K., Musick, W. L., Rosato, A. E., & Musser, J. M. (2014). PBP2a mutations causing high-level Ceftaroline resistance in clinical methicillin-resistant Staphylococcus aureus isolates. *Antimicrobial Agents and Chemotherapy*, *58*(11), 6668–6674. doi:10.1128/AAC.03622-14 PMID:25155594

Mahoney, J. F., Arnold, R. C., & Harris, A. (1943). Penicillin Treatment of Early Syphilis-A Preliminary Report. *American Journal of Public Health and the Nation's Health*, *33*(12), 1387–1391. doi:10.2105/AJPH.33.12.1387 PMID:18015910

Majer, J., Martin, J. R., Egan, R. S., & Corcoran, J. W. (1977). Antibiotic glycosides. 8. Erythromycin D, a new macrolide antibiotic. *Journal of the American Chemical Society*, *99*(5), 1620–1622. doi:10.1021/ja00447a055 PMID:402407

Malanovic, N., Leber, R., Schmuck, M., Kriechbaum, M., Cordfunke, R. A., Drijfhout, J. W., de Breij, A., Nibbering, P. H., Kolb, D., & Lohner, K. (2015). Phospholipid-driven differences determine the action of the synthetic antimicrobial peptide OP-145 on Gram-positive bacterial and mammalian membrane model systems. *Biochimica et Biophysica Acta*, *1848*(10, Pt A), 2437–2447. doi:10.1016/j.bbamem.2015.07.010 PMID:26210299

Margalith, P., & Beretta, G. (1960). Rifomycin. XI. taxonomic study on streptomyces mediterranei nov. sp. *Mycopathologia et Mycologia Applicata*, *13*(4), 321–330. doi:10.1007/BF02089930

Maria, O. M., Eliopoulos, N., & Muanza, T. (2017). Radiation-Induced Oral Mucositis. *Frontiers in Oncology*, *7*(May), 89. Advance online publication. doi:10.3389/fonc.2017.00089 PMID:28589080

Mast, Y., & Wohlleben, W. (2014). Streptogramins – Two are better than one! *International Journal of Medical Microbiology*, *304*(1), 44–50. doi:10.1016/j.ijmm.2013.08.008 PMID:24119565

Masuma, R., Yamaguchi, Y., Noumi, M., Omura, S., & Namikoshi, M. (2001). Effect of sea water concentration on hyphal growth and antimicrobial metabolite production in marine fungi. *Mycoscience*, *42*(5), 455–459. doi:10.1007/BF02464342

Mendes, R. E., Tsakris, A., Sader, H. S., Jones, R. N., Biek, D., McGhee, P., Appelbaum, P. C., & Kosowska-Shick, K. (2012). Characterization of methicillin-resistant Staphylococcus aureus displaying increased MICs of ceftaroline. *The Journal of Antimicrobial Chemotherapy*, *67*(6), 1321–1324. doi:10.1093/jac/dks069 PMID:22398650

Migliori, G. B., De Iaco, G., Besozzi, G., Centis, R., & Cirillo, D. M. (2007). First tuberculosis cases in Italy resistant to all tested drugs. *Weekly Releases, 12*(20).

Mishra, B. B., & Tiwari, V. K. (2011). Natural products: An evolving role in future drug discovery. *European Journal of Medicinal Chemistry*, *46*(10), 4769–4807. doi:10.1016/j.ejmech.2011.07.057 PMID:21889825

Moravej, H., Moravej, Z., Yazdanparast, M., Heiat, M., Mirhosseini, A., Moosazadeh Moghaddam, M., & Mirnejad, R. (2018). Antimicrobial Peptides: Features, Action, and Their Resistance Mechanisms in Bacteria. *Microbial Drug Resistance*. doi:10.1089/mdr.2017.0392

Mukhtar, T. A., Koteva, K. P., & Wright, G. D. (2005). Chimeric streptogramin-tyrocidine antibiotics that overcome streptogramin resistance. *Chemistry & Biology*, *12*(2), 229–235. doi:10.1016/j.chembiol.2004.12.009 PMID:15734650

Negi, P. S., Jayaprakasha, G. K., Jagan Mohan Rao, L., & Sakariah, K. K. (1999). Antibacterial activity of turmeric oil: A byproduct from curcumin manufacture. *Journal of Agricultural and Food Chemistry*, *47*(10), 4297–4300. doi:10.1021/jf990308d PMID:10552805

Nelson, M. L., & Levy, S. B. (2011). The history of the tetracyclines. In Annals of the New York Academy of Sciences (Vol. 1241, Issue 1, pp. 17–32). doi:10.1111/j.1749-6632.2011.06354.x

Newton, G. G., & Abraham, E. P. (1955). Cephalosporin C, a new antibiotic containing sulphur and D-alpha-aminoadipic acid. *Nature*, *175*(4456), 548. doi:10.1038/175548a0 PMID:14370161

Nicholson, A., Perry, J. D., James, A. L., Stanforth, S. P., Carnell, S., Wilkinson, K., Anjam Khan, C. M., De Soyza, A., & Gould, F. K. (2012). In vitro activity of S-(3,4-dichlorobenzyl)isothiourea hydrochloride and novel structurally related compounds against multidrug-resistant bacteria, including Pseudomonas aeruginosa and Burkholderia cepacia complex. *International Journal of Antimicrobial Agents*, *39*(1), 27–32. doi:10.1016/j.ijantimicag.2011.08.015 PMID:21993484

Nishikawa, M., & Ogawa, K. (2002). Distribution of microbes producing antimicrobial ε-poly-L-lysine polymers in soil microflora determined by a novel method. *Applied and Environmental Microbiology, 68*(7), 3575–3581. doi:10.1128/AEM.68.7.3575-3581.2002 PMID:12089045

Okutani, K. (1977). Gliotoxin Produced by a strain of *Aspergillus* isolated from marine mud. *Nippon Suisan Gakkaishi, 43*(8), 995–1000. doi:10.2331uisan.43.995

Oryan, A., & Sahvieh, S. (2017). Effectiveness of chitosan scaffold in skin, bone and cartilage healing. *International Journal of Biological Macromolecules, 104*, 1003–1011. doi:10.1016/j.ijbiomac.2017.06.124 PMID:28684351

Pappas, E., & Schaich, K. M. (2009). Phytochemicals of cranberries and cranberry products: Characterization, potential health effects, and processing stability. *Critical Reviews in Food Science and Nutrition, 49*(9), 741–781. doi:10.1080/10408390802145377 PMID:20443158

Paukner, S., Sader, H. S., Ivezic-Schoenfeld, Z., & Jones, R. N. (2013). Antimicrobial activity of the pleuromutilin antibiotic BC-3781 against bacterial pathogens isolated in the SENTRY antimicrobial surveillance program in 2010. *Antimicrobial Agents and Chemotherapy, 57*(9), 4489–4495. doi:10.1128/AAC.00358-13 PMID:23836172

Perumal Samy, R., Stiles, B. G., Franco, O. L., Sethi, G., & Lim, L. H. K. (2017). Animal venoms as antimicrobial agents. *Biochemical Pharmacology, 134*, 127–138. doi:10.1016/j.bcp.2017.03.005 PMID:28288817

Pfaller, M. A., Rhomberg, P. R., Huband, M. D., & Flamm, R. K. (2018). Activity of omadacycline tested against Enterobacteriaceae causing urinary tract infections from a global surveillance program (2014). *Diagnostic Microbiology and Infectious Disease, 91*(2), 179–183. doi:10.1016/j.diagmicrobio.2018.01.019 PMID:29571839

Plikaytis, B. B., Marden, J. L., Crawford, J. T., Woodley, C. L., Butler, W. R., & Shinnick, T. M. (1994). Multiplex PCR assay specific for the multidrug-resistant strain W of Mycobacterium tuberculosis. *Journal of Clinical Microbiology, 32*(6), 1542–1546. doi:10.1128/jcm.32.6.1542-1546.1994 PMID:7915723

Poulakou, G., Bassetti, M., Righi, E., & Dimopoulos, G. (2014). Current and future treatment options for infections caused by multidrug-resistant Gram-negative pathogens. *Future Microbiology, 9*(9), 1053–1069. doi:10.2217/fmb.14.58 PMID:25340835

Pournaras, S., Vrioni, G., Neou, E., Dendrinos, J., Dimitroulia, E., Poulou, A., & Tsakris, A. (2011). Activity of tigecycline alone and in combination with colistin and meropenem against Klebsiella pneumoniae carbapenemase (KPC)-producing Enterobacteriaceae strains by time-kill assay. *International Journal of Antimicrobial Agents, 37*(3), 244–247. doi:10.1016/j.ijantimicag.2010.10.031 PMID:21236643

Pruksakorn, P., Arai, M., Kotoku, N., Vilcheze, C., Baughn, A. D., Moodley, P., Jacobs, W. R. J. Jr, & Kobayashi, M. (2010). Trichoderins, novel aminolipopeptides from a marine sponge-derived Trichoderma sp., are active against dormant mycobacteria. *Bioorganic & Medicinal Chemistry Letters, 20*(12), 3658–3663. doi:10.1016/j.bmcl.2010.04.100 PMID:20483615

Qureshi, Z. A., Paterson, D. L., Potoski, B. A., Kilayko, M. C., Sandovsky, G., Sordillo, E., Polsky, B., Adams-Haduch, J. M., & Doi, Y. (2012). Treatment outcome of bacteremia due to KPC-producing Klebsiella pneumoniae: Superiority of combination antimicrobial regimens. *Antimicrobial Agents and Chemotherapy, 56*(4), 2108–2113. doi:10.1128/AAC.06268-11 PMID:22252816

Raz, R., & Unit, I. D. (2011). *Fosfomycin : an old — new antibiotic*. Academic Press.

Rose, W. E., & Rybak, M. J. (2006). Tigecycline: First of a new class of antimicrobial agents. *Pharmacotherapy, 26*(8I), 1099–1110. doi:10.1592/phco.26.8.1099

Rossiter, S. E., Fletcher, M. H., & Wuest, W. M. (2017). Natural Products as Platforms to Overcome Antibiotic Resistance. *Chemical Reviews, 117*(19), 12415–12474. doi:10.1021/acs.chemrev.7b00283 PMID:28953368

Savoia, D. (2012). Plant-derived antimicrobial compounds: Alternatives to antibiotics. *Future Microbiology, 7*(8), 979–990. doi:10.2217/fmb.12.68 PMID:22913356

Schatz, A., Bugle, E., & Waksman, S. A. (1944). Streptomycin, a Substance Exhibiting Antibiotic Activity Against Gram-Positive and Gram-Negative Bacteria.*†. *Proceedings of the Society for Experimental Biology and Medicine, 55*(1), 66–69. doi:10.3181/00379727-55-14461

Sepkowitz, K. A., Raffalli, J., Riley, L., Kiehn, T. E., & Armstrong, D. (1995). Tuberculosis in the AIDS era. *Clinical Microbiology Reviews, 8*(2), 180–199. doi:10.1128/CMR.8.2.180 PMID:7621399

Sfeir, M., Askin, G., & Christos, P. (2018). Beta-lactam/ beta-lactamase inhibitors versus carbapenem for bloodstream infections due to extended spectrum beta-lactamase producing Enterobacteriaceae: Systematic review and meta-analysis. *International Journal of Antimicrobial Agents, 52*(5), 554–570. Advance online publication. doi:10.1016/j.ijantimicag.2018.07.021 PMID:30081138

Silber, J., Kramer, A., Labes, A., & Tasdemir, D. (2016). From discovery to production: Biotechnology of marine fungi for the production of new antibiotics. *Marine Drugs*, *14*(7), 137. doi:10.3390/md14070137 PMID:27455283

Srakaew, N., Young, C. D., Sae-wu, A., Xu, H., Quesnel, K. L., di Brisco, R., Kongmanas, K., Fongmoon, D., Hommalai, G., Weerachatyanukul, W., Hall, S. H., Zhang, Y.-L., Panza, L., Franchini, L., Compostella, F., Pearson, T. W., Hancock, R. E., Oko, R. J., Hermo, L. S., & Tanphaichitr, N. (2014). Antimicrobial host defence peptide, LL-37, as a potential vaginal contraceptive. *Human Reproduction (Oxford, England)*, *29*(4), 683–696. doi:10.1093/humrep/deu018 PMID:24549217

Storm, D. R., Rosenthal, K. S., & Swanson, P. E. (1977). Polymyxin and related peptide antibiotics. *Annual Review of Biochemistry*, *46*(1), 723–763. doi:10.1146/annurev.bi.46.070177.003451 PMID:197881

Tally, F. P., & DeBruin, M. F. (2000). Development of daptomycin for Gram-positive infections. *The Journal of Antimicrobial Chemotherapy*, *46*(4), 523–526. doi:10.1093/jac/46.4.523 PMID:11020247

Tenson, T., Lovmar, M., & Ehrenberg, M. (2003). The mechanism of action of macrolides, lincosamides and streptogramin B reveals the nascent peptide exit path in the ribosome. *Journal of Molecular Biology*, *330*(5), 1005–1014. doi:10.1016/S0022-2836(03)00662-4 PMID:12860123

Thabit, A. K., Crandon, J. L., & Nicolau, D. P. (2015). Antimicrobial resistance: Impact on clinical and economic outcomes and the need for new antimicrobials. *Expert Opinion on Pharmacotherapy*, *16*(2), 159–177. doi:10.1517/14656566.2015.993381 PMID:25496207

Tumbarello, M., Trecarichi, E. M., De Rosa, F. G., Giannella, M., Giacobbe, D. R., Bassetti, M., Losito, A. R., Bartoletti, M., Del Bono, V., Corcione, S., Maiuro, G., Tedeschi, S., Celani, L., Cardellino, C. S., Spanu, T., Marchese, A., Ambretti, S., Cauda, R., Viscoli, C., & Viale, P. (2015). Infections caused by KPC-producing Klebsiella pneumoniae: Differences in therapy and mortality in a multicentre study. *The Journal of Antimicrobial Chemotherapy*, *70*(7), 2133–2143. doi:10.1093/jac/dkv086 PMID:25900159

Uttley, A. H., Collins, C. H., Naidoo, J., & George, R. C. (1988). Vancomycin-resistant enterococci. *Lancet*, *8575–6*, 57–58.

Van Bambeke, F. (2006). Glycopeptides and glycodepsipeptides in clinical development: A comparative review of their antibacterial spectrum, pharmacokinetics and clinical efficacy. *Current Opinion in Investigational Drugs (London, England)*, *7*(8), 740–749. PMID:16955686

Velayati, A. A., Masjedi, M. R., Farnia, P., Tabarsi, P., Ghanavi, J., ZiaZarifi, A. H., & Hoffner, S. E. (2009). Emergence of New Forms of Totally Drug-Resistant Tuberculosis Bacilli: Super Extensively Drug-Resistant Tuberculosis or Totally Drug-Resistant Strains in Iran. *Chest, 136*(2), 420–425.

Villano, S., Steenbergen, J., & Loh, E. (2016). Omadacycline: Development of a novel aminomethylcycline antibiotic for treating drug-resistant bacterial infections. *Future Microbiology, 11*(11), 1421–1434. doi:10.2217/fmb-2016-0100 PMID:27539442

Wagman, G. H. (1980). Antibiotics from micromonospora. *Annual Review of Microbiology, 34*(1), 537–558. doi:10.1146/annurev.mi.34.100180.002541 PMID:6159822

Waksman, S. A., & Woodruff, H. B. (1941). Actinomyces antibioticus, a new soil organism antagonistic to pathogenic and non-pathogenic bacteria. *Journal of Bacteriology, 42*(2), 231–249. doi:10.1128/jb.42.2.231-249.1941 PMID:16560451

Waksman, S. A., & Woodruff, H. B. (1942). Streptothricin, a New Selective Bacteriostatic and Bactericidal Agent, Particularly Active Against Gram-Negative Bacteria. *Proceedings of the Society for Experimental Biology and Medicine, 49*(2), 207–210. doi:10.3181/00379727-49-13515

Walkty, A., Adam, H., Baxter, M., Denisuik, A., Lagace-Wiens, P., Karlowsky, J. A., Hoban, D. J., & Zhanel, G. G. (2014). In vitro activity of plazomicin against 5,015 gram-negative and gram-positive clinical isolates obtained from patients in canadian hospitals as part of the CANWARD study, 2011-2012. *Antimicrobial Agents and Chemotherapy, 58*(5), 2554–2563. doi:10.1128/AAC.02744-13 PMID:24550325

Wang, G., Li, X., & Wang, Z. (2016). APD3: The antimicrobial peptide database as a tool for research and education. *Nucleic Acids Research, 44*(D1), D1087–D1093. doi:10.1093/nar/gkv1278 PMID:26602694

Wei, M.-Y., Li, D., Shao, C.-L., Deng, D.-S., & Wang, C.-Y. (2013). (+/-)-Pestalachloride D, an antibacterial racemate of chlorinated benzophenone derivative from a soft coral-derived fungus Pestalotiopsis sp. *Marine Drugs, 11*(4), 1050–1060. doi:10.3390/md11041050 PMID:23538869

Wright, A. J. (1999). The penicillins. *Mayo Clinic Proceedings, 74*(3), 290–307. doi:10.4065/74.3.290 PMID:10090000

Wright, J. L. C. (1983). *The lincomycin-celesticetin-anthramycin group*. Biotechnology Series.

Yates, J. D., & Schaible, P. J. (1962). Virginiamycin as an antibiotic for poultry feeds. *Nature, 194*(4824), 183–184. doi:10.1038/194183b0 PMID:14009058

Zhanel, G. G., Calic, D., Schweizer, F., Zelenitsky, S., Adam, H., Lagac-Wiens, P. R. S., Rubinstein, E., Gin, A. S., Hoban, D. J., & Karlowsky, J. A. (2010). New lipoglycopeptides: A comparative review of dalbavancin, oritavancin and telavancin. *Drugs, 70*(7), 859–886. doi:10.2165/11534440-000000000-00000 PMID:20426497

Zhanel, G. G., Cheung, D., Adam, H., Zelenitsky, S., Golden, A., Schweizer, F., Gorityala, B., Lagacé-Wiens, P. R. S., Walkty, A., Gin, A. S., Hoban, D. J., & Karlowsky, J. A. (2016). Review of Eravacycline, a Novel Fluorocycline Antibacterial Agent. *Drugs, 76*(5), 567–588. doi:10.100740265-016-0545-8 PMID:26863149

## KEY TERMS AND DEFINITIONS

**Adverse Effects:** When medical drugs, like antibiotics, have harmful or distracting effects on person being used or applied.

**Antibacterial:** Class of substances that can kill or inhibit the growth of bacteria.

**Antibiotic:** Class of substances that can kill or inhibit the growth of some groups of microorganisms. Despite being a wide term, herein the word antibiotic resembles a drug that kills or stops the growth of bacteria particularly. Examples include penicillin and streptomycin.

**Antifungal:** Class of substances that can kill or inhibit the growth of fungi.

**Antimicrobial:** Any substance (including an antibiotic) used to kill or inhibit the growth of bacteria, viruses, fungi, or parasites. This term is a non-specific one which is a wider term than antibiotic.

**Antimicrobial Resistance:** The ability of a microbe to survive and continue to multiply during and after the encounter with a certain antimicrobial agent. Antimicrobial resistance includes antibiotic, antibacterial, antifungal, and antiviral resistance.

**Infection:** An invasion of an organism by a pathogen such as bacteria or viruses, often causing an immune response from the host. Some infections lead to disease.

**Natural Product:** Substances, compounds or molecules isolated from natural sources.

**Pathogens:** Bacteria, viruses, parasites, or fungi that can cause disease.

# Chapter 8
# Alternative Therapies:
## Toolbox to Combat Antibiotic-Resistant Bugs

**Vijay Singh Gondil**
*Panjab University, India*

**Sanjay Chhibber**
*Panjab University, India*

## ABSTRACT

*Antibiotic resistance is one of the leading public health concerns across the globe. Antibiotics are losing their effectiveness, leading to uncertainty in available treatment options to clinicians. Resistance to antibiotics is at an all-time high, and there is a pressing demand to look for alternative antimicrobial candidates other than antibiotics. Alternative therapies include use of bacteriophages, lytic proteins, nanoparticles, phytochemicals, quorum quenchers, and other antibacterial or antivirulent agents that can eradicate bacterial infection alone or in conjunction with antibiotics. Alternative therapies can replace or lower the effective antibiotic dose, which can help to tackle antibiotic resistance as well as counter its side effects. For sustainable development of antimicrobials against drug resistant bugs, novel alternative strategies need to be explored in the near future. Alternative therapies can help researchers to construct a toolbox containing a variety of antimicrobial agents, which can be used alone, in combination with other agents, or in rotation.*

DOI: 10.4018/978-1-7998-0307-2.ch008

## ANTIMICROBIAL RESISTANCE

Antibiotic resistance is a state when microorganisms show resistance to an antibiotic upon exposure. Antibiotic resistance is a global problem which can cross international boundaries and spread between continents in recent times. Various forms of resistance spread with astonishing speed which could not be paralleled by the development of novel antimicrobial products. Researchers have described antibiotic resistant microorganisms as “nightmare bacteria” that can “pose a catastrophic threat” to global population. The problem is so serious that majority of pathogenic bacteria such as *Streptococcus pneumoniae, Mycobacterium tuberculosis, Acinetobacter baumannii, Neisseria gonorrhoeae, Klebsiella pneumoniae, Enterococcus faecium, Staphylococcus aureus, Pseudomonas aeruginosa*, and species of *Enterobacter, Shigella* and *Salmonella* are now resistant to most of the available antibiotics (Nathan, 2014). Perhaps the single biggest public health threat today is antibiotic resistance (Laxminaraya et al, 2013). For instance gonorrhea, was treatable with penicillin in the 1970s but now has become resistant even to third generation oral cephalosporins (Fernandes et al, 2017). Drug resistance causes 25 000 deaths per year (“ECDC/ EMEA Joint Technical Report”, 2009). Similarly, in same epoch in the USA and China, resistance caused 100 000 and 80 000 deaths, respectively (Prestinaci et al, 2015). Antibiotic resistance can lead to escalating costs and deterioration of health care systems. Patients suffering from drug resistant nosocomial (hospital acquired) bloodstream infections or infection following consumption of food contaminated with antibiotic resistant pathogens experience longer recovery and a higher occurrence of septicemia and mortality (Angulo et al,2004).

There are several reasons that contribute towards antimicrobial resistance crisis. These includes overuse of antibiotics, inappropriate prescription, extensive non-clinical use, and slow development of newer antibiotics. The overuse and misuse of antibiotics in clinical practice clearly leads the evolution of antibiotic resistance. A number of epidemiological studies have established a direct correlation between antibiotic utilization and emergence of drug resistant bacterial strains. In bacteria, genes which confer antibiotic resistance can be inherited from closely related or from nonrelated on the mobile genetic elements which primarily includes plasmids (Reed et al, 2014). Antibiotics are poorly regulated and easily available over the counter without a proper prescription (Michael et al, 2014). Poor regulatory system results in antibiotic misuse due to their easy availability and low cost. Faultily prescribed antibiotics have dubious therapeutic effects and may expose patients to the possible complications of antibiotic treatment (Murthy et al, 2015). Subinhibitory antibiotic concentrations can also promote the development of antibiotic resistance by supporting genetic alterations (Viswanathan et al, 2014). The antibiotics used in farm animals are ingested by humans when they consume such foods (Golkar et al, 2014). Even environmental use of antibiotics also affects the microbiome (Bartlett et al, 2013). Antibacterial products used frequently for hygienic and cleaning purposes, also contribute to the current problem, as they may lead to development of low level of immunity to these environmental pathogens. The advancement in developing new antibiotics, an approach that had been successful at tackling bacterial pathogens in the past decades, had essentially slowed down due to economic and regulatory barriers (Bartlett et al, 2013). The emergence of antibiotic resistance has become a critical problem in modern medicine and because of this mankind is reentering the "pre-antibiotics" era.

## Alternative Therapies, a New Tool from the Old Books

Increasing antibiotic resistance has prompted clinicians to prescribe fewer antibiotics, which, in turn has led to search for alternatives to antibiotics. In post antibiotic era, treatment of drug resistant bacterial pathogens is cumbersome and there is a pressing need to consider alternative options. Alternative therapies has gained attention in antibiotic resistance era especially lytic bacteriophages, endolysins, antibacterial nanoparticles, plant extracts, quorum quenching and other non-antibiotic antimicrobial products. These have been acquired the highest priority and possible options in the modern health care system. Alternative therapies contain the potential to reduce the antibiotic usage, tend to be safer upon usage and increases the immune potential of the host to counter pathogen.

*Figure 1. Image showing potential alternative therapeutic agents to antibiotics*

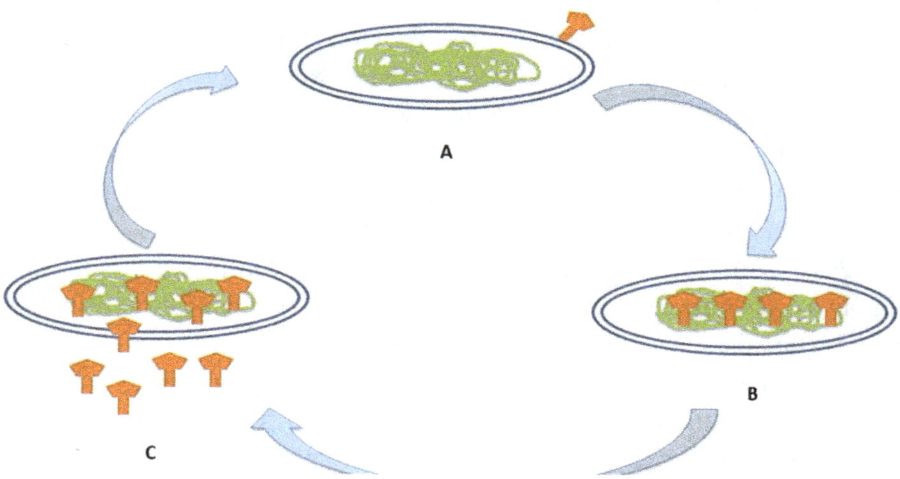

## Lytic Bacteriophages

Fredrick Twort and Felix D'Herelle proposed the use of bacteriophages/phages (viruses that infect bacteria) to treat bacterial infections. The first reported bacteriophage based clinical trial was performed by Richard Bruynoghe and Joseph Maisin in 1921 ("BACTERIOPHAGE THERAPY", n.d.). In their clinical trails, they used phages to treat Staphylococcal skin disease by injecting the phages into/around surgically opened lesions. However, following the discovery of antibiotics, attention on bacteriophages as potential antimicrobial tools for controlling bacterial infections decreased swiftly (Sulakvelidze et al, 2001). With the emergence of antibiotic resistance, application of bacteriophages or their products in eradicating infections caused by *E. coli, P. aeruginosa, K. pneumoniae, S. aureus* and proteus strains in human and animal models has increased (Barrow et al, 1997). Phage therapy is beneficial over antibiotic therapy because of the high specificity of phages whereby they have an effect on the target bacteria only and are non-toxic to normal microflora. Phages are self-replicating, self-limiting and are successful against multidrug resistant pathogens. Phage production is uncomplicated, fast and relatively economical (Azeredo et al, 2008). Phages can be used in concoction with other antibacterial agents including other phages (phage cocktails), the lytic range of phage products can be much wider than the range of activity of monophage (Liu et al, 2020; Yang et al, 2020). Phage cocktail loaded liposomes also lead to fast resolution of the infection process as compared to non-liposomal free phage cocktail. Phages showed antibody mediated inactivation when used without delivery system, but liposomal delivery system provided 100% immune protection from anti-phage antibodies as constant phage titre was seen and can be used further to evade antibody mediated phage inactivation (Singla et al, 2016). Use of model organisms to treat cumbersome pathogens such as *Mycobacterium tuberculosis* using phage therapy has also been suggested in literature (Gondil et al, 2008). A recent clinical trial of intravesical application of bacteriophages for treatment of urinary tract infections showed non-inferiority to standard antibiotic treatment but opened a new prospective of well-designed clinical trials for therapeutic application of phages (Leitner et al, 2020).

## Endolysins

Endolysins are phage encoded peptidoglycans hydrolases (PGHs) that are employed by most lytic phages to degrade the peptidoglycan layer of the bacterial host at the end of their lytic-multiplication cycle (Young et al, 1992). Peptidoglycans, as we know, is a most important structural component of bacterial cells supporting the internal structures inside the cell. This membrane also helps in maintaining the turgor

*Figure 2. Phage mediated bacterial lysis (a) attachment (b) multiplication (c) lysis (d) image showing clear plaques after phage infection*

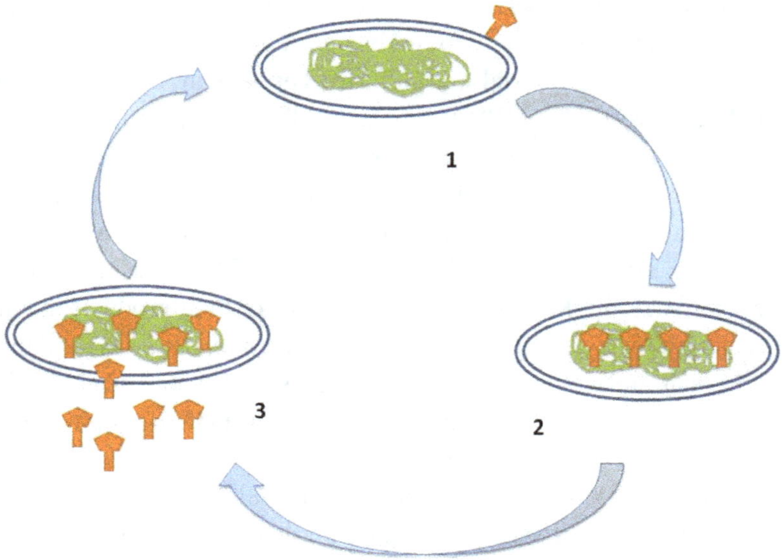


pressure of 20-50 atmosphere in case of gram-positive organisms (Whatmore et al, 1990). Breaching of this peptidoglycan layer leads to changes in turgor pressure and lysis of bacterial cell that results in liberation of virion progeny along with other bacterial cell components in the outer environment. This disruption of bacterial host cells is time dependent as the liberation of phage progeny occurs with the help of another protein called 'Holin' which is produced during the release of double-stranded bacteriophage exactly at the end of lytic cycle (Wang et al, 2000). Holin forms pores in host's cell membrane that helps endolysins to reach and degrade peptidoglycans. Cpl-1, a lytic enzyme produced by phage Cp-1, exclusively and rapidly lyses several serotypes of *S. pneumoniae*, including antibiotic susceptible and antibiotic resistant strains, and exerts synergistic antibacterial activity with well-known antibiotics (for example penicillin and gentamicin) *in vitro* (Loeffler et al, 2003). Cpl-1 was shown to be beneficial in animal models of pneumococcal diseases such as sepsis, endocarditis and meningitis (Entenza et al, 2005). Other than Cpl-1 several potent anti-streptococcal phage lysins have also been reported in literature which mainly includes PlyC, ClyV, ClyR, Cpl-7, ClyJ, and Cpl-711 (40-45). Recently a chitosan based Cpl-1 lysin delivery system has been reported to increase its *in-vivo* antibacterial potential (Gondil et al, 2021). Anti-staphylococcal phage lysins have been also extensively studied in literature as various endolysins (lysK, ClyF, $CHAP_{K,}$ Ply187, LysGH15 and MR10) has been reported so far (Gondil et al, 2020). *S. aureus* specific lysin MV-L derived from phage φMR11 rescued mice from MRSA fatal infection. MV-L also showed the activity against vancomycin-resistant strains when used in combination with vancomycin resulting in the increase of efficacy of a non-effective antibiotic (Rashel et al, 2007). ClyS, an endolysin bioengineered construct from Twort phage lysin and phiNM3, has also showed synergy with oxacillin to treat murine oxacillin resistant MRSA infections. It opened an avenue for use of discontinued antibiotics due to resistance issues (Daniel et al, 2010). Chopra *et al* also showed efficacy of MR-10 endolysin alone and along with antibiotic as a therapeutic agent to control MRSA induced burn wounds (Chopra et al, 2016). The exogenous use of endolysins in case of gram-negative pathogens is limited because of the outer membrane which acts as a permeability barrier for entry of endolysins into the peptidoglycan layer. However, some endolysins are known to permeate the outer membrane as LysAB2 derived from phage φAB2 can disrupt *A. baumannii* by facilitating the formation of transmembrane pores by its amphipathic α-helix (Lai et al, 2011). LysPA26, an endolysin against *P. aeruginosa* showed its antibacterial activity without any pretreatment for destabilizing the outer cell membrane. LysPA26 showed its potent antibacterial activity as it can reduce 4 log units of *P. aeruginosa* in 30 minutes time interval (Guo et al, 2017). In a recent study, LysAB54 also exhibited high degree of antibacterial activity against number of gram-negative pathogens (Khan VSG, et al, 2021). Bioengineered Artilysins such as Art-175 and Art-240 have also shown their potential in controlling gram-negative pathogens (Briers et al, 2015) . The spectrum of gram negative endolysins is expanding markedly with advancement of molecular engineering techniques.

## Antibacterial-Nanoparticles

Silver in it salt form has been used for treatment of burn wounds and chronic wounds. Carl S.F Creed's in 1881 treated ophthalmia neonatrum, and cured it as the solution used for eye drop was silver nitrate. Recently, due to emergence of drug resistant strains, antibiotic usage is limited (Chopra et al, 2007). These factors renewed the interest of scientific community in silver salts and led to exploitation of silver salts to give better and effective alternative therapy. Nanoparticles exhibit a size range of 1-1000 nm. The use of metallic nanoparticles is increasing as potent antibacterial agent because of their large surface area to volume ratio (56). Nanotechnology is emerging as a rapidly growing field along with its applications. Many types of metallic antibacterial nanoparticles like zinc, silver, copper, titanium, magnesium, and gold have been reported but silver nanoparticles are known to have significant antimicrobial activity (Schabes-Retchkiman et al, 2006). In recent studies, Histidine capped silver nanoparticles, sodium acetate capped silver nanoparticles, glycolic acid functionalized silver nanoparticles, sodium benzoate capped silver nanoparticles and Seabuckthorn capped silver nanoparticles showed significant activity against drug resistant and food-borne pathogens (Kumar et al, 2017). Silver nanoparticles can also be used as disinfectant, however, at high concentrations it may be toxic to mammalian tissues. The recent studies support that silver nanoparticles can be exploited in medicine for burn, dental treatments and coating material used in hospital so that biofilm formation on instruments can be reduced (Abbasi et al, 2016). Moreover, silver nanoparticles can be used to make water potable for drinking. It can also be used as a sunscreen agent in various cosmetic preparations. Wound dressings containing silver as an antimicrobial agent are very popular (Castellano et al, 2007). Silver nanoparticle diffusion is limited within biofilms matrix because of the presence of meshes of exopolysaccharides (EPS) network and environmental DNA (eDNA). Peulen *et al* showed the diffusion coefficient of 2 nm silver nanoparticles in a biofilm, that remained at 86% of its actual value (Peulen et al, 2011). In nanometric form, antibacterial activity of silver is accentuated. As it can easily enter the cells, binds to various metabolic enzymes, inhibit respiratory chain, alter DNA synthesis and induce oxidative stress to bacteria resulting in degradation of bacterium.

Copper in metallic micrometric size do not exhibit any antibacterial activity as compared to copper nanoparticles. As copper nanoparticles can oxidize easily and soluble ions can confer toxicity to bacterial cells. Copper nanoparticles in polylactic matrix have been shown to inhibit the growth of *Pseudomonas* spp (Longano et al, 2012). Ananth *et al* showed the antibacterial activity of copper nanoparticles against a range of gram positive bacteria *Streptococcus parauberis, Streptococcus iniae* and gram negative bacteria *E. coli and Vibrio anguillarum* (Ananth et al, 2015). Titanium is always a metal of choice for making implants during past few decades (Chouirfa et al, 2019). Nowadays titanium nanoparticles are being investigated for their antibacterial efficacy. Titanium also showed potent antibacterial activity against gram positive strains such as *S. aureus* and *S. epidermidis*. Antibacterial activity of titanium resulted from photocatalytic reaction under solar light or exposure to light, producing OH that can distrupt bacterial cell wall (Vimbela et al, 2017). Zinc has also been used in biomedical applications due to its abundance and low toxicity (Jiang et al, 2018). Zinc oxide nanoparticles confer antibacterial effects which possibly result from reactive oxygen species generation and $Zn^{2+}$ release. Oxidative stress resulting from superoxide anions, hydrogen peroxide and hydroxide, may damage lipids and proteins, whereas zinc ions can possibly disrupt important metabolic machinery and pathways. Zinc oxide nanoparticles showed potent antibacterial activity against *E. coli*, *Salmonella*, *L. monocytogenes*, and *S. aureus* and minimal cytotoxicity on mammalian cells (Jones et al, 2018). Magnesium is biocompatible and biodegradable metal which can inhibit bacterial growth by generating oxidative stress. Magnesium oxide nanoparticles also exhibit antibacterial, anti-inflammatory, antioxidant as well as bone regeneration capability which makes them suitable candidate for use in bone and connective tissue infections (Sushma et al, 2016).

## Plant Extracts

Plant extracts or phytochemicals constitute a range of chemical compounds of natural origin that confers color, flavor, aroma and texture. These compounds are widely distributed in fruits, vegetables, legumes, seeds and herbs. Phytochemicals are classified as alkaloids, sulphur containing groups, terpenoids, carotenoids and flavanoids according to the biosynthetic grouping. Alkaloids are organic nitrogenous bases with their variable chemical structure, produced by a variety of organisms such as plants, bacteria, fungi. These have been used in medicine since ancient times. Various alkaloids possess antibacterial and antiviral activity. Magesh showed that berberine, an alkaloid inhibits the biofilm formation in *K. pneumoniae* (Magesh et al, 2013) Another alkaloid reserpine, isolated from Rauwolfia (Apocynaceae) showed inhibitory activity against *K. pnuemoniae* biofilm. Sulfur-containing compounds such as allicin, ajoene and isothiocyanates have shown promising antibacterial activity against both gram-positive and gram-negative bacteria by interfering with biofilm formation and quorum sensing process. Allicin found in garlic inhibited *P. aeroginosa* PAO1 biofilm adhesion and also reduced the expression of quorum sensing (QS) regulated virulence factors (Militz et al, 2014). Terpenoids represent the large and most diverse class of chemicals derived from plants. The major function of these is in growth and development whereas some have a specialized function like protection from biotic and abiotic factors (82). Different types of terpenes like monoterpenes, limonoids, and triterpenes have been evaluated for their anti-biofilm and anti-QS activities. Cunha evaluated the antibacterial activity of oleanolic acid and ursolic acid (triterpenes) extracted from *Miconia ligustroides* against *K. pneumoniae* (Cunha et al, 2010). Carotenoids (tetraterpenoids) are well known to have nutritional properties and health benefits. Several studies have evaluated their biological and antibacterial activity against gram-positive and gram-negative bacteria (Honda et al, 2019). Polyphenols constitutes a diverse range of phytochemicals found abundantly in vegetables, fruits, and its products. In recent years, polyphenols have been labeled as good therapeutic and chemopreventive agents because of their direct antimicrobial and antibiotic modulation activity (Stermitz et al, 2009). Cheng and Huang studied 4 different flavonoids (galangin, kaempferol, quercetin and myricetin) and these inhibited the *K. pneumoniae* growth by inhibiting DNA B helicase with dNTPs at 10μM concentration (Chen et al, 2011). Quercetin mainly found in the diet (onion and propolis) is most extensively studied flavonoid due to its anti-cancer, anti-inflammatory, and antibacterial activities (Harborne et al,2000). Epigallocatechin (EGCG) (flavonoids with catechin) shows stronger effects on gram positive bacteria than against gram negative bacteria as latter posses lipopolysaccharide layer. Cho showed weak and moderate activity of ECGC against imipenem-resistant *K. pneumoniae* (89). Flavonoids also show great antibacterial activity in synergy with antibiotics. Gopu et al has shown the synergistic effect of phytochemicals present in *Syzygium cumini*, an Indian blackberry with antibiotics against *K. pneumoniae* (Gopu et al, 2015).

*Figure 3. Image showing representative chemical structures of each class of phytochemicals A:Resperine (Alkaloid), B:Allicin (sulphur containing compound), C:oleanolic acid (Terpenoid), D:β-carotene (carotenoid) and E:Quercetin (flavonoid)*

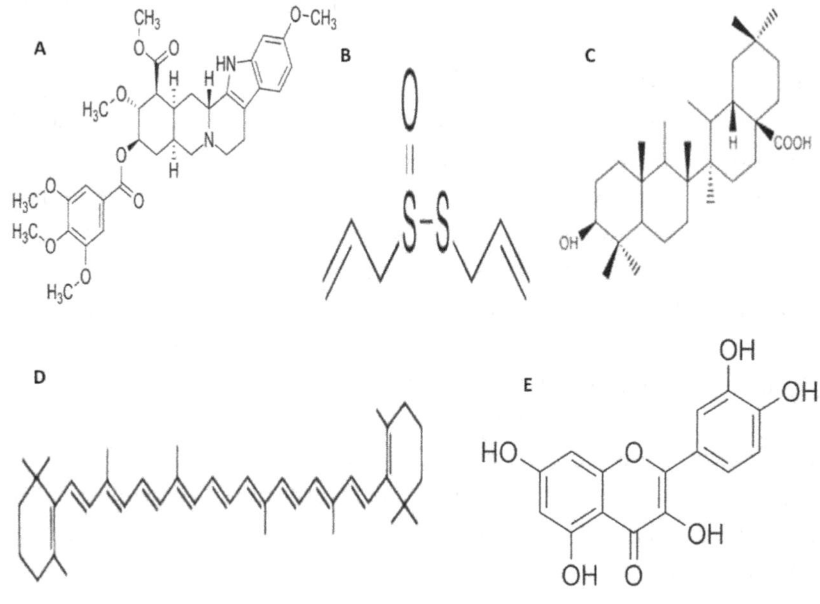

## Quorum Quenching

Majority of the bacterial population is capable of monitoring their population density and gene expression through a specific controlling mechanism known as quorum sensing (QS), which helps to keep a check on the multicellular activities (growth and development of biofilm), horizontal gene transfer, host-microbe (symbiosis and pathogenesis), microbe-microbe interactions and switching 'on-off' of a gene that has direct effect on virulence factors of the microorganisms (Fuqua WC et al, 1994). This control is performed by synthesis, exchange, and perception of bacterial compounds, known as autoinducers (e.g. N-acyl homoserine lactone). N-acyl homoserine lactone

(AHL) molecules are highly conserved in different organisms as they contain the basic homoserine lactone ring, but differ in the length and substitution of acyl chain. Inhibition of bacterial quorum sensing system by disrupting or blocking quorum sensing molecules can be an attractive way of combating bacterial infection because it can think to exert a reduced pressure to select resistant bacterial strains from the population. This is known as quorum quenching and is mediated either through structural mimics of quorum sensing molecules or through inhibition of enzymes involved in the synthesis of quorum sensing molecules. Enzymes like AHL-lactonase and AHL-acylase can degrade quorum sensing molecules by hydrolyzing lactone ring or by liberating free homoserine lactone along with fatty acid (Czajkowski et al, 2009). The major advantage of using quorum sensing inhibitors is that as it does not enforce selection pressure, it restricts the evolution of drug-resistant strain. Both virulence and biofilm formation require cell to cell communication via quorum sensing mechanism. Inhibition of the same enhances efficacy of antibiotics that can be considered as an extra advantage (Kalia et al, 2014). Gupta *et al* showed that a combination of lactonase and antibiotic can potentially attenuate the virulence of *P. aeruginosa* in murine burn wound model (Gupta et al, 2016).

*Figure 4. Image showing activity of AHL-lactonase and AHL-acylase.*

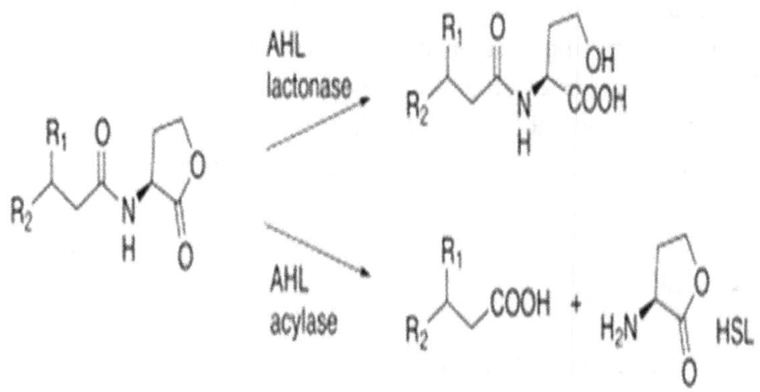


## Other Antimicrobial Products

Non-antibiotic antimicrobial products which include killing factors, ant microbial peptides, capsular depolymerases and pigments can be used to treat antibiotic resistant infections in near future. Killing factors is the term used to define the factors released by bacterial cells to kill the sibling cells during starvation. The nutrients released from the lysed cells are utilized by killer cells for their survival and spore formation.

Antimicrobial peptides are short antibacterial positively charged ribosomal or non-ribosomal peptides such as esculentin and bacteriocins. Esculentin-1a showed potent anti-bacterial activity against pseudomonas as well as pro wound healing activity (Grazia et al, 2015). Bacteriocin is generally defined as small ribosomally synthesized peptides of 12-70 amino acid long that are secreted by bacteria and inhibit the growth of closely related species and decrease the competition in the surroundings. Lactic acid bacteria are mostly considered as bacteriocin producers, which can infiltrate the outer membrane of gram-negative bacteria and control their growth. Bacteriocins can limit a growth of number of food spoilers/pathogens such as *S. aureus, E. coli, S. enterica, B. cereus* (de la Fuente-Salcido N, et al, 2008). Nisin, commercially used bacteriocin is currently used in more than 50 countries as an antibacterial agent (Liu et al, 2005). Pompilio showed reduction in biofilm formation of multidrug-resistant (MDR) *P. aeroginosa* strains isolated from a cystic fibrosis patient treated with bacteriocin (Pompilio et al, 2011). In a recent study, bacteriocins also showed potent antibiofilm activity against *P. aeruginosa* biofilms (Sharma et al, 2018). Capsular depolymerases disrupt the integrity of capsular matrix of bacterium and increases the susceptibility of pathogen to antibiotics. Treatment with capsular depolymerase increased the susceptibility to gentamicin and other antibiotics to *K. pneumoniae in-vivo and in-vitro* (Bansal et al, 2015). Microbial pigments also showed significant anticancer and antibacterial activites. Lee *et al*, Priya *et al* and Gondil *et al* showed an antibacterial activity of prodigiosin, a pigment from Serratia spp against several potent pathogenic microorganisms (Leet et al, 2014). The antibacterial activity of pigments is attributed to their ability to pass the bacterial cell membrane cause inhibition of cellular enzymes, induces ROS generation and DNA fragmentation (Darshan et al, 2016).

## CONCLUSION

Alternative therapies present themselves as an attracting option to combat drug resistant superbugs, which are exceedingly difficult to eradicate with established drug therapies. Alternative therapies have gained success in their initial attempts such as *in-vitro* models, *ex-vivo* as well as *in-vivo* animal models but the studies are extremely limited in terms of human healthcare system. Some therapies such as phytochemicals, silver nanoparticles lack exact elucidation of their mechanism which makes treatment process trickier. On other hand personalized therapy with bacteriophages have gained attention of scientists in developed countries. Alternative therapies can help in combating antibiotic resistance and assist sustainable use of accessible antibiotics but their validation in human healthcare system should be encouraged and supported.

## REFERENCES


Abbasi, E., Milani, M., Fekri Aval, S., Kouhi, M., Akbarzadeh, A., Tayefi Nasrabadi, H., & ... . (2016). Silver nanoparticles: Synthesis methods, bio-applications and properties. *Critical Reviews in Microbiology*, *42*(2), 173–180. PMID:24937409

Ananth, A., Dharaneedharan, S., Heo, M.-S., & Mok, Y. S. (2015). Copper oxide nanomaterials: Synthesis, characterization and structure-specific antibacterial performance. *Chemical Engineering Journal*, *262*, 179–188. doi:10.1016/j.cej.2014.09.083

Angulo, F. J., Nunnery, J. A., Bair, H. D., & Wint, W. (2004). Antimicrobial resistance in zoonotic enteric pathogens. *Revue Scientifique et Technique (International Office of Epizootics)*, *23*(2), 485–496. doi:10.20506/rst.23.2.1499 PMID:15702715

*Antibiotic Resistance Threats in the United States 2019*. (2019). Academic Press.

Arnoldi, M., Fritz, M., Bäuerlein, E., Radmacher, M., Sackmann, E., & Boulbitch, A. (2000). Bacterial turgor pressure can be measured by atomic force microscopy. *Physical Review E: Statistical Physics, Plasmas, Fluids, and Related Interdisciplinary Topics*, *62*(1, 1 Pt B), 1034–1044. doi:10.1103/PhysRevE.62.1034 PMID:11088560

Azeredo, J., & Sutherland, I. W. (2008). The use of phages for the removal of infectious biofilms. *Current Pharmaceutical Biotechnology*, *9*(4), 261–266. doi:10.2174/138920108785161604 PMID:18691087

Bacteriophage Therapy. (n.d.). Available from: http://www.bterfoundation.org/bacteriophage-therapy/

Bansal, S., Harjai, K., & Chhibber, S. (2015). Aeromonas punctata derived depolymerase improves susceptibility of Klebsiella pneumoniae biofilm to gentamicin. *BMC Microbiology*, *15*(1), 119. doi:10.118612866-015-0455-z PMID:26063052

Barrow, P. A., & Soothill, J. S. (1997). Bacteriophage therapy and prophylaxis: Rediscovery and renewed assessment of potential. *Trends in Microbiology*, *5*(7), 268–271. doi:10.1016/S0966-842X(97)01054-8 PMID:9234508

Bartlett, J. G., Gilbert, D. N., & Spellberg, B. (2013). Seven ways to preserve the miracle of antibiotics. *Clinical Infectious Diseases*, *56*(10), 1445–1450. doi:10.1093/cid/cit070 PMID:23403172

Briers, Y., & Lavigne, R. (2015). Breaking barriers: Expansion of the use of endolysins as novel antibacterials against Gram-negative bacteria. *Future Microbiology*, *10*(3), 377–390. doi:10.2217/fmb.15.8 PMID:25812461

Brussow, H. (2005). Phage therapy: The Escherichia coli experience. *Microbiology (Reading), 151*(Pt 7), 2133–2140. doi:10.1099/mic.0.27849-0 PMID:16000704

Bustamante, N., Campillo, N. E., Garcia, E., Gallego, C., Pera, B., Diakun, G. P., Sáiz, J. L., García, P., Díaz, J. F., & Menéndez, M. (2010). Cpl-7, a lysozyme encoded by a pneumococcal bacteriophage with a novel cell wall-binding motif. *The Journal of Biological Chemistry, 285*(43), 33184–33196. doi:10.1074/jbc.M110.154559 PMID:20720016

Castellano, J. J., Shafii, S. M., Ko, F., Donate, G., Wright, T. E., Mannari, R. J., Payne, W. G., Smith, D. J., & Robson, M. C. (2007). Comparative evaluation of silver-containing antimicrobial dressings and drugs. *International Wound Journal, 4*(2), 114–122. doi:10.1111/j.1742-481X.2007.00316.x PMID:17651227

Chen, C. C., & Huang, C. Y. (2011). Inhibition of Klebsiella pneumoniae DnaB helicase by the flavonol galangin. *The Protein Journal, 30*(1), 59–65. doi:10.100710930-010-9302-0 PMID:21210194

Chhibber, S., Gondil, V. S., & Kaur, J. (2018). Isolation, characterization, statistical optimization, and application of a novel broad-spectrum capsular depolymerase against Klebsiella pneumoniae from Bacillus siamensis SCVJ30. *Biomedical and Biotechnology Research Journal, 2*(2), 125. doi:10.4103/bbrj.bbrj_40_18

Chhibber, S., Gondil, V. S., Sharma, S., Kumar, M., Wangoo, N., & Sharma, R. K. (2017). A novel approach for combating Klebsiella pneumoniae biofilm using histidine functionalized silver nanoparticles. *Frontiers in Microbiology, 8*, 1104. doi:10.3389/fmicb.2017.01104 PMID:28670301

Chhibber, S., Gondil, V. S., Singla, L., Kumar, M., Chhibber, T., Sharma, G., Sharma, R. K., Wangoo, N., & Katare, O. P. (2019). Effective Topical Delivery of H-AgNPs for Eradication of Klebsiella pneumoniae-Induced Burn Wound Infection. *AAPS PharmSciTech, 20*(5), 169. doi:10.120812249-019-1350-y PMID:31004249

Chopra, I. (2007). The increasing use of silver-based products as antimicrobial agents: A useful development or a cause for concern? *The Journal of Antimicrobial Chemotherapy, 59*(4), 587–590. doi:10.1093/jac/dkm006 PMID:17307768

Chopra, S., Harjai, K., & Chhibber, S. (2016). Potential of combination therapy of endolysin MR-10 and minocycline in treating MRSA induced systemic and localized burn wound infections in mice. *International Journal of Medical Microbiology, 306*(8), 707–716. doi:10.1016/j.ijmm.2016.08.003 PMID:27616281

Chouirfa, H., Bouloussa, H., Migonney, V., & Falentin-Daudré, C. (2019). Review of titanium surface modification techniques and coatings for antibacterial applications. *Acta Biomaterialia*, *83*, 37–54. doi:10.1016/j.actbio.2018.10.036 PMID:30541702

Cunha, W. R., de Matos, G. X., Souza, M. G., Tozatti, M. G., Andrade e Silva, M. L., Martins, C. H., Silva, R., & Da Silva Filho, A. A. (2010). Evaluation of the antibacterial activity of the methylene chloride extract of Miconia ligustroides, isolated triterpene acids, and ursolic acid derivatives. *Pharmaceutical Biology*, *48*(2), 166–169. doi:10.3109/13880200903062648 PMID:20645834

Czajkowski, R., & Jafra, S. (2009). Quenching of acyl-homoserine lactone-dependent quorum sensing by enzymatic disruption of signal molecules. *Acta Biochimica Polonica*, *56*(1), 1–16. doi:10.18388/abp.2009_2512 PMID:19287806

Daniel, A., Euler, C., Collin, M., Chahales, P., Gorelick, K. J., & Fischetti, V. A. (2010). Synergism between a novel chimeric lysin and oxacillin protects against infection by methicillin-resistant Staphylococcus aureus. *Antimicrobial Agents and Chemotherapy*, *54*(4), 1603–1612. doi:10.1128/AAC.01625-09 PMID:20086153

Darshan, N., & Manonmani, H. K. (2016). Prodigiosin inhibits motility and activates bacterial cell death revealing molecular biomarkers of programmed cell death. *AMB Express*, *6*(1), 50. doi:10.118613568-016-0222-z PMID:27460563

de la Fuente-Salcido, N., Guadalupe Alanís-Guzmán, M., Bideshi, D. K., Salcedo-Hernández, R., Bautista-Justo, M., & Barboza-Corona, J. E. (2008). Enhanced synthesis and antimicrobial activities of bacteriocins produced by Mexican strains of Bacillus thuringiensis. *Archives of Microbiology*, *190*(6), 633–640. doi:10.100700203-008-0414-2 PMID:18654760

Di Grazia, A., Cappiello, F., Imanishi, A., Mastrofrancesco, A., Picardo, M., Paus, R., & Mangoni, M. L. (2015). The Frog Skin-Derived Antimicrobial Peptide Esculentin-1a(1-21)NH2 Promotes the Migration of Human HaCaT Keratinocytes in an EGF Receptor-Dependent Manner: A Novel Promoter of Human Skin Wound Healing? *PLoS One*, *10*(6), e0128663. doi:10.1371/journal.pone.0128663 PMID:26068861

Diez-Martinez, R., De Paz, H. D., Garcia-Fernandez, E., Bustamante, N., Euler, C. W., Fischetti, V. A., Menendez, M., & Garcia, P. (2015). A novel chimeric phage lysin with high in vitro and in vivo bactericidal activity against Streptococcus pneumoniae. *The Journal of Antimicrobial Chemotherapy*, *70*(6), 1763–1773. doi:10.1093/jac/dkv038 PMID:25733585

Djurkovic, S., Loeffler, J. M., & Fischetti, V. A. (2005). Synergistic killing of Streptococcus pneumoniae with the bacteriophage lytic enzyme Cpl-1 and penicillin or gentamicin depends on the level of penicillin resistance. *Antimicrobial Agents and Chemotherapy*, *49*(3), 1225–1228. doi:10.1128/AAC.49.3.1225-1228.2005 PMID:15728935

Doehn, J. M., Fischer, K., Reppe, K., Gutbier, B., Tschernig, T., Hocke, A. C., Fischetti, V. A., Loffler, J., Suttorp, N., Hippenstiel, S., & Witzenrath, M. (2013). Delivery of the endolysin Cpl-1 by inhalation rescues mice with fatal pneumococcal pneumonia. *The Journal of Antimicrobial Chemotherapy*, *68*(9), 2111–2117. doi:10.1093/jac/dkt131 PMID:23633685

*ECDC/EMEA Joint Technical Report: The bacterial challenge: time to react.* (2009). Academic Press.

Entenza, J. M., Loeffler, J. M., Grandgirard, D., Fischetti, V. A., & Moreillon, P. (2005). Therapeutic effects of bacteriophage Cpl-1 lysin against Streptococcus pneumoniae endocarditis in rats. *Antimicrobial Agents and Chemotherapy*, *49*(11), 4789–4792. doi:10.1128/AAC.49.11.4789-4792.2005 PMID:16251333

Fazal Mehmood Khanm Li, Jiang, Li, Yu, Wei, & Yang. (n.d.). A novel Acinetobacter baumannii bacteriophage endolysin LysAB54 with high antibacterial activity against multiple Gram-negative microbes. *Front Cell Infect Microbiol.*

Fernandes, P., & Martens, E. (2017). Antibiotics in late clinical development. *Biochemical Pharmacology*, *133*, 152–163. doi:10.1016/j.bcp.2016.09.025 PMID:27687641

Fuqua, W. C., Winans, S. C., & Greenberg, E. P. (1994). Quorum sensing in bacteria: The LuxR-LuxI family of cell density-responsive transcriptional regulators. *Journal of Bacteriology*, *176*(2), 269–275. doi:10.1128/jb.176.2.269-275.1994 PMID:8288518

Gemmell, C. G., Edwards, D. I., Fraise, A. P., Gould, F. K., Ridgway, G. L., & Warren, R. E. (2006). Guidelines for the prophylaxis and treatment of methicillin-resistant Staphylococcus aureus (MRSA) infections in the UK. *The Journal of Antimicrobial Chemotherapy*, *57*(4), 589–608. doi:10.1093/jac/dkl017 PMID:16507559

Golkar, Z., Bagasra, O., & Pace, D. G. (2014). Bacteriophage therapy: A potential solution for the antibiotic resistance crisis. *Journal of Infection in Developing Countries*, *8*(2), 129–136. doi:10.3855/jidc.3573 PMID:24518621

Gondil, Asif, & Bhalla. (2017). Optimization of physicochemical parameters influencing the production of prodigiosin from Serratia nematodiphila RL2 and exploring its antibacterial activity. *3 Biotech, 7*(5), 338.

Gondil, Kalaiyarasan, Bharti, & Chhibber. (2019). Antibiofilm potential of Seabuckthorn silver nanoparticles (SBT@AgNPs) against Pseudomonas aeruginosa. *3 Biotech, 9*(11), 402.

Gondil, V. S., & Chhibber, S. (2017). Evading Antibody Mediated Inactivation of Bacteriophages Using Delivery Systems. *Juniper Online Journal of Immuno Virology., 1*(5), 555–574.

Gondil, V. S., & Chhibber, S. (2018). Exploring potential of phage therapy for tuberculosis using model organism. *Biomedical and Biotechnology Research Journal, 2*(1), 9. doi:10.4103/bbrj.bbrj_93_17

Gondil, V. S., Dube, T., Panda, J. J., Yennamalli, R. M., Harjai, K., & Chhibber, S. (2020). Comprehensive evaluation of chitosan nanoparticle based phage lysin delivery system; a novel approach to counter S. pneumoniae infections. *International Journal of Pharmaceutics, 573*, 118850. doi:10.1016/j.ijpharm.2019.118850 PMID:31759993

Gondil, V. S., Harjai, K., & Chhibber, S. (2020). Endolysins as emerging alternative therapeutic agents to counter drug-resistant infections. *International Journal of Antimicrobial Agents, 55*(2), 105844. doi:10.1016/j.ijantimicag.2019.11.001 PMID:31715257

Gondil, V. S., Harjai, K., & Chhibber, S. (2021). Investigating the potential of endolysin loaded chitosan nanoparticles in the treatment of pneumococcal pneumonia. *Journal of Drug Delivery Science and Technology, 61*, 102142. doi:10.1016/j.jddst.2020.102142

Gong, P., Li, H., He, X., Wang, K., Hu, J., Tan, W., Zhang, S., & Yang, X. (2007). Preparation and antibacterial activity of Fe3O4@Ag nanoparticles. *Nanotechnology, 18*(28), 285604. doi:10.1088/0957-4484/18/28/285604

Gopu, V., Kothandapani, S., & Shetty, P. H. (2015). Quorum quenching activity of Syzygium cumini (L.) Skeels and its anthocyanin malvidin against Klebsiella pneumoniae. *Microbial Pathogenesis, 79*, 61–69. doi:10.1016/j.micpath.2015.01.010 PMID:25637095

Gorski, A., & Weber-Dabrowska, B. (2005). The potential role of endogenous bacteriophages in controlling invading pathogens. *Cellular and Molecular Life Sciences, 62*(5), 511–519. doi:10.100700018-004-4403-6 PMID:15747058

Gu, H., Ho, P. L., Tong, E., Wang, L., & Xu, B. (2003). Presenting Vancomycin on Nanoparticles to Enhance Antimicrobial Activities. *Nano Letters, 3*(9), 1261–1263. doi:10.1021/nl034396z

Guo, M., Feng, C., Ren, J., Zhuang, X., Zhang, Y., Zhu, Y., Dong, K., He, P., Guo, X., & Qin, J. (2017). A Novel Antimicrobial Endolysin, LysPA26, against Pseudomonas aeruginosa. *Frontiers in Microbiology, 8*, 293. doi:10.3389/fmicb.2017.00293 PMID:28289407

Gupta, P., Chhibber, S., & Harjai, K. (2016). Subinhibitory concentration of ciprofloxacin targets quorum sensing system of Pseudomonas aeruginosa causing inhibition of biofilm formation & reduction of virulence. *The Indian Journal of Medical Research, 143*(5), 643–651. doi:10.4103/0971-5916.187114 PMID:27488009

Harborne, J. B., & Williams, C. A. (2000). Advances in flavonoid research since 1992. *Phytochemistry, 55*(6), 481–504. doi:10.1016/S0031-9422(00)00235-1 PMID:11130659

Honda, M., Kageyama, H., Hibino, T., Zhang, Y., Diono, W., Kanda, H., Yamaguchi, R., Takemura, R., Fukaya, T., & Goto, M. (2019). Improved Carotenoid Processing with Sustainable Solvents Utilizing Z-Isomerization-Induced Alteration in Physicochemical Properties: A Review and Future Directions. *Molecules (Basel, Switzerland), 24*(11), 2149. doi:10.3390/molecules24112149 PMID:31181605

Huang, L., Luo, D. H., Gondil, V. S., Gong, Y. J., Jia, M. H., Yan, D. Z., He, J., Hu, S., Yang, H., & Wei, H. (2020). Construction and characterization of a chimeric lysin ClyV with improved bactericidal activity against Streptococcus agalactiae in vitro and in vivo. *Applied Microbiology and Biotechnology, 104*(4), 1609–1619. doi:10.100700253-019-10325-z PMID:31900556

Jiang, J., Pi, J., & Cai, J. (2018). The Advancing of Zinc Oxide Nanoparticles for Biomedical Applications. *Bioinorganic Chemistry and Applications, 2018*, 1062562. doi:10.1155/2018/1062562 PMID:30073019

John Sushma, N., Prathyusha, D., Swathi, G., Madhavi, T., Deva Prasad Raju, B., Mallikarjuna, K., & Kim, H.-S. (2016). Facile approach to synthesize magnesium oxide nanoparticles by using Clitoria ternatea—Characterization and in vitro antioxidant studies. *Applied Nanoscience, 6*(3), 437–444. doi:10.100713204-015-0455-1

Jones, N., Ray, B., Ranjit, K. T., & Manna, A. C. (2008). Antibacterial activity of ZnO nanoparticle suspensions on a broad spectrum of microorganisms. *FEMS Microbiology Letters, 279*(1), 71–76. doi:10.1111/j.1574-6968.2007.01012.x PMID:18081843

Kalia, V. C., Wood, T. K., & Kumar, P. (2014). Evolution of resistance to quorum-sensing inhibitors. *Microbial Ecology, 68*(1), 13–23. doi:10.100700248-013-0316-y PMID:24194099

Kaur, P., Gondil, V. S., & Chhibber, S. (2019). A novel wound dressing consisting of PVA-SA hybrid hydrogel membrane for topical delivery of bacteriophages and antibiotics. *International Journal of Pharmaceutics*, *572*, 118779. doi:10.1016/j.ijpharm.2019.118779 PMID:31740093

Kumar, M., Bala, R., Gondil, V. S., Jain, D., Chhibber, S., Sharma, R. K., & Wangoo, N. (2017). Efficient, green and one pot synthesis of sodium acetate functionalized silver nanoparticles and their potential application as food preservative. *BioNanoScience*, *7*(3), 521–529. doi:10.100712668-017-0417-z

Kumar, M., Bala, R., Gondil, V. S., Pandey, S. K., Chhibber, S., Jain, D., Sharma, R. K., & Wangoo, N. (2017). Combating food pathogens using sodium benzoate functionalized silver nanoparticles: Synthesis, characterization and antimicrobial evaluation. *Journal of Materials Science*, *52*(14), 8568–8575. doi:10.100710853-017-1072-z

Kumar, M., Bansal, K., Gondil, V. S., Sharma, S., Jain, D., Chhibber, S., Sharma, R. K., & Wangoo, N. (2018). Synthesis, characterization, mechanistic studies and antimicrobial efficacy of biomolecule capped and pH modulated silver nanoparticles. *Journal of Molecular Liquids*, *249*, 1145–1150. doi:10.1016/j.molliq.2017.11.143

Kumar, M., Wangoo, N., Gondil, V. S., Pandey, S. K., Lalhall, A., Sharma, R. K., & … . (2020). *Glycolic acid functionalized silver nanoparticles: A novel approach towards generation of effective antibacterial agent against skin infections*. J Drug Deliv Sci Tec.

Kumari, S., Harjai, K., & Chhibber, S. (2009). Efficacy of bacteriophage treatment in murine burn wound infection induced by klebsiella pneumoniae. *Journal of Microbiology and Biotechnology*, *19*(6), 622–628. PMID:19597322

Lai, M. J., Lin, N. T., Hu, A., Soo, P. C., Chen, L. K., Chen, L. H., & Chang, K.-C. (2011). Antibacterial activity of Acinetobacter baumannii phage ϕAB2 endolysin (LysAB2) against both gram-positive and gram-negative bacteria. *Applied Microbiology and Biotechnology*, *90*(2), 529–539. doi:10.100700253-011-3104-y PMID:21264466

Laxminarayan, R., & Chaudhury, R. R. (2016). Antibiotic Resistance in India: Drivers and Opportunities for Action. *PLoS Medicine*, *13*(3), e1001974. doi:10.1371/journal.pmed.1001974 PMID:26934098

Lee, J. S., Kim, Y. S., Park, S., Kim, J., Kang, S. J., Lee, M. H., Ryu, S., Choi, J. M., Oh, T.-K., & Yoon, J.-H. (2011). Exceptional production of both prodigiosin and cycloprodigiosin as major metabolic constituents by a novel marine bacterium, Zooshikella rubidus S1-1. *Applied and Environmental Microbiology*, *77*(14), 4967–4973. doi:10.1128/AEM.01986-10 PMID:21642414

Leitner, L., Ujmajuridze, A., Chanishvili, N., Goderdzishvili, M., Chkonia, I., Rigvava, S., & … . (2020). Intravesical bacteriophages for treating urinary tract infections in patients undergoing transurethral resection of the prostate: A randomised, placebo-controlled, double-blind clinical trial. *The Lancet. Infectious Diseases*. PMID:32949500

Lihua, L., Jianhuit, W., Jialini, Y., Yayin, L., & Guanxin, L. (2013). Effects of allicin on the formation of Pseudomonas aeruginosa biofinm and the production of quorum-sensing controlled virulence factors. *Polish Journal of Microbiology*, *62*(3), 243–251. doi:10.33073/pjm-2013-032 PMID:24459829

Liu, C., Liu, Y., & Chen, S. (2005). Effects of nutrient supplements on simultaneous fermentation of nisin and lactic acid from cull potatoes. *Applied Biochemistry and Biotechnology*, *121-124*(1-3), 475–483. doi:10.1385/ABAB:122:1-3:0475 PMID:15920257

Liu, N., Lewis, C., Zheng, W., & Fu, Z. Q. (2020). Phage Cocktail Therapy: Multiple Ways to Suppress Pathogenicity. *Trends in Plant Science*, *25*(4), 315–317. doi:10.1016/j.tplants.2020.01.013 PMID:32191865

Loeffler, J. M., Djurkovic, S., & Fischetti, V. A. (2003). Phage lytic enzyme Cpl-1 as a novel antimicrobial for pneumococcal bacteremia. *Infection and Immunity*, *71*(11), 6199–6204. doi:10.1128/IAI.71.11.6199-6204.2003 PMID:14573637

Longano, D., Ditaranto, N., Cioffi, N., Di Niso, F., Sibillano, T., Ancona, A., Conte, A., Del Nobile, M. A., Sabbatini, L., & Torsi, L. (2012). Analytical characterization of laser-generated copper nanoparticles for antibacterial composite food packaging. *Analytical and Bioanalytical Chemistry*, *403*(4), 1179–1186. doi:10.100700216-011-5689-5 PMID:22262051

Luo, D. H., Huang, L., Gondil, V. S., Zhou, W. L., Yang, W., Jia, M. H., Hu, S., He, J., Yang, H., & Wei, H. (2020). A Choline-Recognizing Monomeric Lysin, ClyJ-3m, Shows Elevated Activity against Streptococcus pneumoniae. *Antimicrobial Agents and Chemotherapy*, *64*(12). Advance online publication. doi:10.1128/AAC.00311-20 PMID:32958710

Magesh, Kumar, & Alam, Priyam, Sekar, & Sumantran. (2013). Identification of natural compounds which inhibit biofilm formation in clinical isolates of Klebsiella pneumoniae. *Indian Journal of Experimental Biology*, *51*(9), 764–772. PMID:24377137

Michael, C. A., Dominey-Howes, D., & Labbate, M. (2014). The antimicrobial resistance crisis: Causes, consequences, and management. *Frontiers in Public Health*, *2*, 145. doi:10.3389/fpubh.2014.00145 PMID:25279369

Middleton, E. Jr, Kandaswami, C., & Theoharides, T. C. (2000). The effects of plant flavonoids on mammalian cells: Implications for inflammation, heart disease, and cancer. *Pharmacological Reviews*, *52*(4), 673–751. PMID:11121513

Militz, T. A., Southgate, P. C., Carton, A. G., & Hutson, K. S. (2014). Efficacy of garlic (Allium sativum) extract applied as a therapeutic immersion treatment for Neobenedenia sp. management in aquaculture. *Journal of Fish Diseases*, *37*(5), 451–461. doi:10.1111/jfd.12129 PMID:23952605

Murthy, V. H. (2015). Surgeon General's Perspectives. *Public Health Reports (Washington, D.C.)*, *130*(4), 299. doi:10.1177/003335491513000403 PMID:26346094

Nathan, C., & Cars, O. (2014). Antibiotic resistance—Problems, progress, and prospects. *The New England Journal of Medicine*, *371*(19), 1761–1763. doi:10.1056/NEJMp1408040 PMID:25271470

Nelson, D., Schuch, R., Chahales, P., Zhu, S., & Fischetti, V. A. (2006). PlyC: A multimeric bacteriophage lysin. *Proceedings of the National Academy of Sciences of the United States of America*, *103*(28), 10765–10770. doi:10.1073/pnas.0604521103 PMID:16818874

Pacheco, S. D. G., Gasparin, A. T., Jesus, C. H. A., Sotomaior, B. B., Ventura, A., Redivo, D. D. B., Cabrini, D. A., Gaspari Dias, J. F., Miguel, M. D., Miguel, O. G., & da Cunha, J. M. (2019). Antinociceptive and Anti-Inflammatory Effects of Bixin, a Carotenoid Extracted from the Seeds of Bixa orellana. *Planta Medica*, *85*(16), 1216–1224. doi:10.1055/a-1008-1238 PMID:31546267

Pérez-Giraldo, C., Cruz-Villalón, G., Sánchez-Silos, R., Martínez-Rubio, R., Blanco, M. T., & Gómez-García, A. C. (2003). In vitro activity of allicin against Staphylococcus epidermidis and influence of subinhibitory concentrations on biofilm formation. *Journal of Applied Microbiology*, *95*(4), 709–711. doi:10.1046/j.1365-2672.2003.02030.x PMID:12969283

Peulen, T. O., & Wilkinson, K. J. (2011). Diffusion of nanoparticles in a biofilm. *Environmental Science & Technology*, *45*(8), 3367–3373. doi:10.1021/es103450g PMID:21434601

Pompilio, A., Pomponio, S., Crocetta, V., Gherardi, G., Verginelli, F., Fiscarelli, E., Dicuonzo, G., Savini, V., D'Antonio, D., & Di Bonaventura, G. (2011). Phenotypic and genotypic characterization of Stenotrophomonas maltophilia isolates from patients with cystic fibrosis: Genome diversity, biofilm formation, and virulence. *BMC Microbiology*, *11*(1), 159. doi:10.1186/1471-2180-11-159 PMID:21729271

Prestinaci, F., Pezzotti, P., & Pantosti, A. (2015). Antimicrobial resistance: A global multifaceted phenomenon. *Pathogens and Global Health*, *109*(7), 309–318. doi:10.1179/2047773215Y.0000000030 PMID:26343252

Priya, K. A., Satheesh, S., Ashokkumar, B., Varalakshmi, P., Selvakumar, G., & Sivakumar, N. (2013). Antifouling Activity of Prodigiosin from Estuarine Isolate of Serratia marcescens CMST 07. In R. K. Velu (Ed.), *Microbiological Research In Agroecosystem Management* (pp. 11–21). Springer India. doi:10.1007/978-81-322-1087-0_2

Rashel, M., Uchiyama, J., Ujihara, T., Uehara, Y., Kuramoto, S., Sugihara, S., Yagyu, K.-I., Muraoka, A., Sugai, M., Hiramatsu, K., Honke, K., & Matsuzaki, S. (2007). Efficient elimination of multidrug-resistant Staphylococcus aureus by cloned lysin derived from bacteriophage phi MR11. *The Journal of Infectious Diseases*, *196*(8), 1237–1247. doi:10.1086/521305 PMID:17955443

Read, A. F., & Woods, R. J. (2014). Antibiotic resistance management. *Evolution, Medicine, and Public Health*, *2014*(1), 147. doi:10.1093/emph/eou024 PMID:25355275

Reddy, K. M., Feris, K., Bell, J., Wingett, D. G., Hanley, C., & Punnoose, A. (2007). Selective toxicity of zinc oxide nanoparticles to prokaryotic and eukaryotic systems. *Applied Physics Letters*, *90*(213902), 2139021–2139023. doi:10.1063/1.2742324 PMID:18160973

Rodríguez-Rubio, L., Chang, W. L., Gutiérrez, D., Lavigne, R., Martínez, B., Rodríguez, A., Govers, S. K., Aertsen, A., Hirl, C., Biebl, M., Briers, Y., & García, P. (2016). 'Artilysation' of endolysin λSa2lys strongly improves its enzymatic and antibacterial activity against streptococci. *Scientific Reports*, *6*(1), 35382. doi:10.1038rep35382 PMID:27775093

Roh, H. S., Lee, C. M., Hwang, Y. H., Kook, M. S., Yang, S. W., Lee, D., & Kim, B.-H. (2017). Addition of MgO nanoparticles and plasma surface treatment of three-dimensional printed polycaprolactone/hydroxyapatite scaffolds for improving bone regeneration. *Materials Science and Engineering C*, *74*, 525–535. doi:10.1016/j.msec.2016.12.054 PMID:28254327

Schabes-Retchkiman, P. S., Canizal, G., Herrera-Becerra, R., Zorrilla, C., Liu, H. B., & Ascencio, J. A. (2006). Biosynthesis and characterization of Ti/Ni bimetallic nanoparticles. *Optical Materials*, *29*(1), 95–99. doi:10.1016/j.optmat.2006.03.014

Sharma, V., Harjai, K., & Shukla, G. (2018). Effect of bacteriocin and exopolysaccharides isolated from probiotic on P. aeruginosa PAO1 biofilm. *Folia Microbiologica*, *63*(2), 181–190. doi:10.100712223-017-0545-4 PMID:28905285

Singla, S., Harjai, K., Katare, O. P., & Chhibber, S. (2016). Encapsulation of Bacteriophage in Liposome Accentuates Its Entry in to Macrophage and Shields It from Neutralizing Antibodies. *PLoS One*, *11*(4), e0153777. doi:10.1371/journal.pone.0153777 PMID:27115154

Stermitz, F. R., Lorenz, P., Tawara, J. N., Zenewicz, L. A., & Lewis, K. (2000). Synergy in a medicinal plant: Antimicrobial action of berberine potentiated by 5′-methoxyhydnocarpin, a multidrug pump inhibitor. *Proceedings of the National Academy of Sciences of the United States of America*, *97*(4), 1433–1437. doi:10.1073/pnas.030540597 PMID:10677479

Sulakvelidze, A., & Morris, J. G. Jr. (2001). Bacteriophages as therapeutic agents. *Annals of Medicine*, *33*(8), 507–509. doi:10.3109/07853890108995959 PMID:11730156

Tholl, D. (2015). Biosynthesis and biological functions of terpenoids in plants. *Advances in Biochemical Engineering/Biotechnology*, *148*, 63–106. doi:10.1007/10_2014_295 PMID:25583224

Verma, V., Harjai, K., & Chhibber, S. (2009). Characterization of a T7-like lytic bacteriophage of Klebsiella pneumoniae B5055: A potential therapeutic agent. *Current Microbiology*, *59*(3), 274–281. doi:10.100700284-009-9430-y PMID:19484297

Vimbela, G. V., Ngo, S. M., Fraze, C., Yang, L., & Stout, D. A. (2017). Antibacterial properties and toxicity from metallic nanomaterials. *International Journal of Nanomedicine*, *12*, 3941–3965. doi:10.2147/IJN.S134526 PMID:28579779

Viswanathan, V. K. (2014). Off-label abuse of antibiotics by bacteria. *Gut Microbes*, *5*(1), 3–4. doi:10.4161/gmic.28027 PMID:24637595

Wang, I. N., Smith, D. L., & Young, R. (2000). Holins: The protein clocks of bacteriophage infections. *Annual Review of Microbiology*, *54*(1), 799–825. doi:10.1146/annurev.micro.54.1.799 PMID:11018145

Whatmore, A. M., & Reed, R. H. (1990). Determination of turgor pressure in Bacillus subtilis: A possible role for K+ in turgor regulation. *Journal of General Microbiology*, *136*(12), 2521–2526. doi:10.1099/00221287-136-12-2521 PMID:2127801

Yang, A. Y. (2020). Comparison of long-term efficacy and renal safety of febuxostat and allopurinol in patients with chronic kidney diseases. *International Journal of Clinical Pharmacology and Therapeutics*, *58*(1), 21–28. doi:10.5414/CP203466 PMID:31657714

Yang, H., Linden, S. B., Wang, J., Yu, J., Nelson, D. C., & Wei, H. (2015). A chimeolysin with extended-spectrum streptococcal host range found by an induced lysis-based rapid screening method. *Scientific Reports*, *5*(1), 17257. doi:10.1038rep17257 PMID:26607832

Young, R. (1992). Bacteriophage lysis: Mechanism and regulation. *Microbiological Reviews*, *56*(3), 430–481. doi:10.1128/mr.56.3.430-481.1992 PMID:1406491

# Chapter 9
# Ancient Pediocin to Innovative Antimicrobial

**Tejinder Kaur**
*Punjabi University, India*

**Praveen P. Balgir**
*Punjabi University, India*

## ABSTRACT

*Multi-drug resistance among patients suffering from infectious diseases has reached such proportions as to render them ineffective. WHO has to put out advisories time and again as to regulate their use. The presently available antibiotics are targeted at inhibiting vital biochemical pathways of pathogens, like nucleotide, protein, or cell wall synthesis in a very specific manner. Antibiotics have been rendered ineffective due to chemical modification, gene mutation, or transport mechanisms employed by pathogens. The novel approach to this problem can be naturally occurring antimicrobial peptides like bacteriocins produced by food grade bacteria. Pediocins produced by pediococcal strains have been found to inhibit a broad spectrum of pathogens by mechanisms that are robust enough to withstand development of resistance. Thus, these pediocins are attractive molecular precursors to develop novel antimicrobials. However, their application as such poses challenges that can be overcome with developing innovative technologies of chemical modifications and delivery strategies.*

DOI: 10.4018/978-1-7998-0307-2.ch009

## INTRODUCTION

The emergence of Multi Drug Resistant pathogens in last few decades is posing a real threat to the world community. With emerging new pathogenic diseases, the repertoire of drugs suddenly seems to be woefully inadequate. This scenario becomes scarier when we factor in lack of reported novel antimicrobials in the pharmaceutical research pipelines. A cause of concern is also an increasing population living in closely packed clusters in cities and increased mobility, creating conditions conducive to quick spread of infectious agents as well as their antimicrobial resistance.

Such is the extent of the problem that in 2017, World Health Organization (WHO) published its first ever list of antibiotic-resistant "priority pathogens" –basically cataloguing 12 families of bacteria which were identified as greatest threat to human health. So much so that WHO further prioritized them into critical, high and medium based on the urgency to develop novel antibiotics specially for pathogens spreading in hospitals, threatening the medical fraternity itself (Table 1). ESKAPE is the acronym of six such antibiotic resistant species of pathogenic bacteria, the six species are ***E**nterococcus faecium*, ***S**taphylococcus aureus*, ***K**lebsiella pneumoniae*, ***A**cinetobacter baumannii*, ***P**seudomonas aeruginosa* and ***E**nterobacter.*

*Table 1. WHO prioritization of antibiotic resistant pathogens (WHO, 2017)*

| Priority 1: CRITICAL | Priority 2: HIGH | Priority 3: MEDIUM |
|---|---|---|
| *Acinetobacter baumannii*, carbapenem-resistant<br>*Pseudomonas aeruginosa*, carbapenem-resistant<br>*Enterobacteriaceae*, carbapenem-resistant, extended spectrum β-lactamase (ESBL) producing | *Enterococcus faecium*, vancomycin-resistant<br>*Staphylococcus aureus*, methicillin-resistant, vancomycin-intermediate and resistant<br>*Helicobacter pylori*, clarithromycin-resistant<br>*Campylobacter* spp., fluoroquinolone-resistant<br>*Salmonellae*, fluoroquinolone-resistant<br>*Neisseria gonorrhoeae*, cephalosporin-resistant, fluoroquinolone-resistant | *Streptococcus pneumoniae*, penicillin-non-susceptible<br>*Haemophilus influenzae*, ampicillin-resistant<br>*Shigella* spp., fluoroquinolone-resistant |

## BACKGROUND

In order to curb this spread, a few restrictions were imposed on widespread use of antibiotics and only the permitted antibiotics and other chemicals to be used in feed supplements were allowed. However, the non-compliance with these regulations at country and regional level is being encountered all over the world, thus, compromising the health and overall wellbeing of mankind. To cite an example the recently released WHO (2018) guidelines for treatment of MDR/RR-TB, recommended with drawing kanamycin and capreomycin; from long term treatment regime, keeping in view the spread of resistance to the same. Thus both consumer and manufacturer are searching and looking forward to newer and safer alternatives to antibiotics.

The mechanisms of action of presently available antimicrobials involve inhibiting a pathway essential for pathogen survival like synthesis of nucleotides, cell wall and proteins. Most of these interventions are inhibitors of proteins involved in the above mentioned pathways as depicted in Figure 1.

A functional mutation in such proteins as mentioned above, that influences their activity or expression, can result in the survival and proliferation of the microorganism in presence of such antimicrobials resulting in propagation of antibiotic resistance phenomenon. The functional mutation may influence either the activity or the expression level of the target protein; thus the occurrence and evolution of antibiotic resistance basically involves mutagenesis of particular target protein. This then acts as the main source of multi drug resistance (MDR) which spreads with the exchange of genetic material in nature via spread of extrachromosomal plasmids, transposons and acquired genetic elements among microbial populations. Apart from these mechanisms the overexpression of efflux pumps capable of pumping out variety of antibiotics from the bacterial cells, also result in MDR ; which once established provides the pathogen a selective advantage and helps in its spread.

Various mechanisms have been discovered that underlie antibiotic resistance. Inactivation of the drug, its modification or alteration of its binding site, change in bacterial cell's membrane permeability and biofilm formation are some such mechanisms that render the pathogen resistant to the drug. Bacterial production of enzymes that can modify the antibiotic molecule and lead to its inactivation has been observed to be the mechanism underlying carbapenem, aminoglycosidic antibiotic and chloramphenicol resistance by producing $\beta$-lactamases, aminoglycoside-modifying enzymes, and chloramphenicol acetyltransferases respectively.

Another mechanism involves mutation in the gene encoding for drug binding protein like the penicillin-binding proteins (PBPs), leading to penicillin resistance. Bacterial membrane permeability determines the level of antibiotic within a bacterial cell. The membrane permeability is dependent on the presence of porins in the membranes of Gram-negative bacteria, which form channels for passage of hydrophilic antibiotics. A reduction in porin protein OprD in *P. aeroginosa* results in less drug entering bacterial cell rendering it to be imipenem resistance (Papp-Wallace et al., 2011). Similar multidrug resistance mechanisms underlie loss of outer membrane proteins known as OmpK35 and OmpK36 along with production of $\beta$-lactamase enzymes in K. *pneumonia* (Sugawara et al., 2016). The intracellular level of antibiotic is also dependent on the efflux pumps present in bacterial cell membranes. An increase in such pumps in some bacteria render them to be multi drug resistant. Efficient secretion of biofilm forming exopolysaccharides also render many bacterial strains resistant to antibiotics.

*Figure 1. Popularly used antibiotics and their target protein pathways*

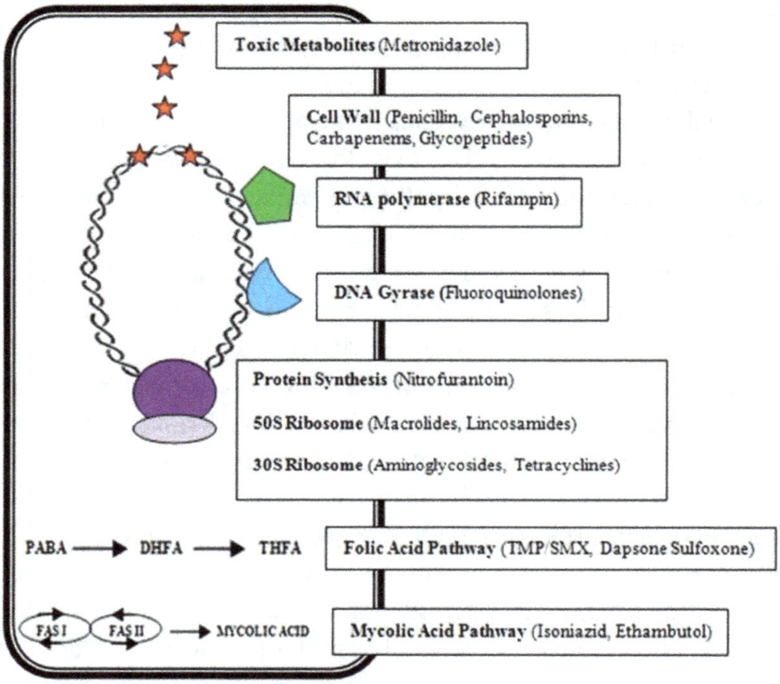


Hence for development of novel antibiotics, such moieties need to be identified as are based on newer mechanisms of action other than protein inhibition. Presently scientists are investigating bacteriophages, lytic enzymes of phages, probiotics, bacteriocins and antimicrobial peptides from various organisms; as potential antibiotics. The main premise underlying these biologics is the robustness of these systems to evolution of resistance. Biologists are also looking at other common traits of pathogens like quorum sensing and siderophores, though such approaches are very limited in their inhibitory potential presently and cannot be used solely as standalone antibiotics (Allen et al., 2014; Ross-Gillespie et al., 2014; Czaplewski et al., 2016). An example of bacteriophage use for prophylaxis of respiratory tract infections like tonsillitis among Russian military personnel was reported recently (Akimkin et al.,

2016). Since the mechanism underlying phage lysis is known even lytic enzymes of phages when tested, gave positive result in a mouse model in gastrointestinal and oral infections (Lood et al., 2014). More recently, bacteriophage derived lysins (endolysins and holins) have been investigated for their antimicrobial properties. These lyase-based antibacterial drugs kill the bacteria by destruction of the bacterial cell wall structure. The endolysins act on the amide bond between peptidoglycans, and holin being a transmembrane protein acts by creating holes in the bacterial cell membrane (Gondil et al., 2020). Although the phage based lysins are mainly known for their activity against Gram-positive bacteria. A recent study has reported the use of engineered lysin complexes called "artilysins" that are capable of penetrating the lipopolysaccharide-rich cell wall of Gram-negative pathogenic bacteria such as *Helicobacter pylori* with bacteriolytic effect against the pathogen (Xu et al., 2021).

Antimicrobial peptides (AMPs) are another class of naturally occurring molecules found in organisms ranging from prokaryotes to humans. These peptides are active against a broad spectrum of bacteria, yeasts, fungi, viruses and cancerous cells. These properties form the basis for development and use of AMPs as antibiotics to overcome the challenge of drug resistance (Zasloff, 2002; Haney and Hancock, 2013; Zhang & Gallo, 2016; Boparai & Sharma, 2020). AMPs expressed in insects and plants protect them against pathogenic microorganisms and microbial cells produce these peptides to safeguard their ecological niche from other fellow microorganisms (Hoffmann & Hetru, 1992; Boman, 2003; Lemaitre et al., 1996). AMPs are known to play a wide array of roles including immunomodulatory properties, immune homeostasis. More than 60 peptide drugs have already reached the market and several hundreds of novel therapeutic peptides are in preclinical and clinical development. Rational designing can be used further to modify the chemical and physical properties of existing peptides. The Antimicrobial Peptide Database (http://aps.unmc.edu/AP/main.php) contains more than 2,500 entries for AMPs at present (Wang, 2020). There are open access resources available with comprehensive research data and machine learning prediction tools. Some examples include collection of antimicrobial peptides or CAMP (www.camp.bicnirrh.res.in), CAMPSign, and ClassAMP that have been developed to enhance research on AMPs (Waghu et al., 2016). Table 2 provides a comprehensive list of naturally occurring AMPs along with their antimicrobial activity.

Probiotic Lactic Acid Bacteria (LAB) produce potent antimicrobial metabolites and peptides that are considered to be suitable for exploration as antimicrobials since their mode of action is different from the antibiotics mentioned above. Probiotics are commonly defined as viable microorganisms (bacteria or yeasts) capable of exerting a beneficial effect on the health of the host upon ingestion (FAO/WHO,

2002; ICMR, 2011). These probiotics produce antimicrobial like bacteriocins, organic acids (lactic and acetic acid), hydrogen peroxide, and antifungal peptides making them efficient at eliminating related and not so related strains of bacteria and fungi from their surroundings. An extensive research on exploring this antibiotic potential of different probiotic LAB strains is being carried out worldwide.

*Table 2. Comprehensive list of naturally occurring antimicrobial peptides*

| Source | AMPs | Antimicrobial activity | Reference |
|---|---|---|---|
| Insects | Acaloleptin | $G^+$, $G^-$ | Vogel et al., 2014; Mylonakis et al., 2016 |
| | Andropin | $G^+$ | McCaskey et al., 2016; Abry et al., 2017 |
| | Apidaecin IA | $G^-$ | Mylonakis et al., 2016; Farouk et al., 2017 |
| | Cecropin | $G^-$ | Lee & Lee, 2015; Mylonakis et al., 2016 |
| | Defensin- α | $G^+$, $G^-$ | Price et al., 2015; Mylonakis et al., 2016 |
| | Drosomycin | F | Mylonakis et al., 2016; Allocca et al., 2018 |
| | Holotricin | $G^+$, $G^-$ | Mylonakis et al., 2016; Thiyonila et al., 2018 |
| | Sapecin- α | $G^+$, $G^-$ | Lee et al., 2014; Manabe & Kawasaki,017 |
| | Tenicin 1 | $G^+$, $G^-$ | Yang et al., 2017 |
| | Thanatin | $G^+$, $G^-$ | Duwadi et al., 2018; Ma et al., 2019 |
| Humans | Cathelicidins | F, $G^-$, $G^+$ | Sheehan et al., 2018 |
| | A Defensins | F, $G^-$, $G^+$ | Candille et al., 2007; Schaal et al., 2018 |
| | Human Histatin 8 | F, $G^-$, $G^+$ | Khurshid et al., 2017; Sun et al., 2020 |
| | LL37 | F, $G^-$, $G^+$ | Singh et al., 2013; Baxter et al., 2017 |
| Animals | Androctonin | F, $G^-$, $G^+$ | Hetru et al., 2000; Panteleev et al., 2017 |
| | Bactenecin | $G^-$, $G^+$ | Young-Speirs et al., 2018 |
| | Brevinin | $G^-$, $G^+$ | Savelyeva et al., 2014; Timmons et al., 2019 |
| | Buforin II | F, $G^-$, $G^+$ | Lee et al., 2008; Sun et al., 2015 |
| | Cupiennin | $G^-$, $G^+$ | Upadhyay, 2018 |
| | Dermaseptin S1 | $G^-$, $G^+$ | Belmadani et al., 2018 |
| | Lycotoxin | $G^-$, $G^+$ | Tahir et al., 2018 |
| | Tachyplesins | $G^-$ | Kuzmin et al., 2017 |
| Plants | Hevein | F | Rojas et al., 2001; Coulen et al., 2017 |
| | Purothionins | $G^+$, $G^-$ | Thao et al., 2017 |

*continues on following page*

*Table 2. Continued*

| Source | AMPs | Antimicrobial activity | Reference |
|---|---|---|---|
| Microorganisms | Nisin | $G^+$ | Gharsallaoui et al., 2016; Mills et al., 2017 |
| | Alamethicin | $G^+$ | Su et al., 2018 |
| | Enterocin | $G^+$, $G^-$ | Braiek et al., 2018 |
| | Hominicin | $G^+$, $G^-$ | Kim et al., 2010; Ebrahimipour et al., 2014 |
| | Ericin S | $G^+$ | Sharma et al., 2018 |
| | Plantaricin A | $G^+$, $G^-$ | Jiang et al., 2018 |
| | Carnobacteriocin B2 | $G^+$, $G^-$ | Hammi et al., 2016 |
| | Leucocin A, C | $G^+$, $G^-$ | Chen et al., 2018; Li et al., 2021 |
| | Subtilin | $G^+$ | Singh et al., 2017 |
| | Pyrularia thionin | $G^+$, $G^-$ | Guzman-Rodriguez et al., 2015 |
| | Microcin J25 | $G^-$ | Zhao et al., 2016 |
| | Gramicidin A | $G^+$, $G^-$ | Muhammad et al., 2016 |
| | Pediocin PA-1/ AcH | $G^+$ | Araujo et al., 2016 |
| | Mesentericin | $G^+$ | Arakawa et al., 2016 |
| | Carnobacteriocin BM1, B2 | $G^+$, $G^-$ | McCormick et al., 1996; Tulini et al., 2014 |
| | Streptin 1 | $G^+$ | Bosma et al., 2011 |
| | Planosporicin | $G^+$, $G^-$ | Gajalakshmi, 2017 |
| | Gassericin A | $G^+$, $G^-$ | Pandey et al., 2013; Maldonadi-Barragan et al., 2016 |
| | Circularin A | $G^+$, $G^-$ | Perez et al., 2016 |
| | Divercin V41 | $G^+$ | Brillet-Viel et al., 2016 |
| | Listeriocin 743A | $G^+$ | Kalmokoff et al., 2001; Wan et al., 2013 |
| | Plantaricin C19 | $G^+$ | Wang et al., 2018 |
| | Enterocin P | $G^+$ | Le et al., 2014; Ben Braïek et al., 2017 |
| | Subtilosin A | $G^+$, $G^-$ | Thennarasu et al., 2005; Venturina et al., 2016 |
| | Plantaricin ASM1 | $G^+$ | Bhat et al., 2018 |
| | Lichenin | $G^+$, $G^-$ | Hollmann et al., 2018 |

## ANTIMICROBIALS FROM PROBIOTIC LAB (MAIN FOCUS OF THE CHAPTER)

Lactic acid bacteria have been extensively exploited in Food industry to produce various food products, pointing to their nutritional value and safety status on consumption. Further, these are being prescribed along with antibiotic therapy as additives, to overcome the mass extermination of microbes in human system caused by antibiotic therapy (FAO/WHO, 2002). These bacteria include a diverse group of Gram-positive, heterotrophic and relatively fastidious bacteria, which produce lactic acid as an end product of carbohydrate fermentation. Genus *Lactococcus, Lactobacillus, Leuconostoc, Pediococcus*, and *Streptococcus* form the microbial core are of the family Lactobacillaceae. The ubiquitous occurrence of LAB in natural food products, along with their Generally Recognized as Safe (GRAS) status and ability to exert beneficial health effects beyond basic nutrition have led to their exploration for other applications. Especially in therapeutics as a recent study pointed to use of probiotic strains of Lactobacillus showing antimicrobial effect against MDR strains isolated from urinary tract infections (Naderi et al., 2014). LAB produce various antimicrobial compounds, such as lactic acid, hydrogen peroxide, carbon dioxide, diacetyl and uncharacterized compounds, which are of low molecular mass and other high molecular mass compounds like bacteriocins (Ammor et al., 2006; Mobolaji & Wuraola, 2011; Sanlibaba & Gucer, 2015).

Some of these attributes are discussed in detail below:

### 1. Lactic Acid

The primary antimicrobial effect exerted by LAB species is attributed to the production of organic acid such as lactic acid, as a result of fermentation (Mobolaji & Wuraola, 2011). These bacteria produce lactic acid in L- or D-isomeric forms. L-lactic acid is desirable for food and pharmaceutical applications and is also used as starting material for biopolymer production. D-lactic acid is not associated with any applications due to its toxicity in human hosts. Due to this reason, metabolic engineering studies have been aimed at the production of pure L-lactic acid by homofermentative LAB. Furthermore, both the stereoisomeric forms of lactic acid display variations towards their antimicrobial activity, with L-lactic acid being more potent inhibitor than the D-form (Papagianni, 2012). The mechanism of antimicrobial action exerted by lactic acid is through interference with the maintenance of cell membrane potential. The inhibition of active transport across cell membrane leads to a reduction in intracellular pH and inhibition of metabolic functions (Rattanachaikunsopon & Phumkhachorn, 2010). Lactic acid production and lowering of pH has been found to be affected by type of bacterial strain or species, culture composition and growth conditions (Olaoye

& Onilude, 2011). The acid production inhibits the growth of Gram-positive and Gram-negative bacteria, yeast and moulds (Rattanachaikunsopon & Phumkhachorn, 2010). Most of the lactic acid is in undissociated form at low pH, with toxicity towards several bacterial, fungal and yeast strains. Due to differential sensitivity of different microorganisms towards lactic acid, it was found to be toxic in case of spore-forming bacteria whereas non-toxic against yeasts and moulds (Yang, 2000).

## 2. Diacetyl

Diacetyl is a flavor compound in dairy products produced as a result of co-fermentation of citrate and lactose. Strains belonging to genera *Lactococcus lactis*, *Leuconostoc* and *Weissella* are commonly used as diacetyl producers (Kleerebezem et al., 2000). Citrate-utilizing LAB produce diacetyl during fermentation of milk for the production of butter, buttermilk and various types of cheese. Diacetyl is responsible for the typical buttery aroma in these dairy products. These properties have lead to development of metabolic engineering strategies for efficient production of diacetyl from lactose instead of citrate. Gram-negative bacteria are much more sensitive to diacetyl as compared to Gram-positive bacteria. Diacetyl affects the arginine utilization pathway by interaction with arginine-binding protein leading to growth inhibition of Gram-negative bacteria. Strains of *Listeria, Salmonella, Yersinia, Esherichia coli*, and *Aeromonas* are sensitive to diacetyl at a concentration of 350μg/ml (Yang, 2000; Ammor et al., 2006; Papagianni, 2012).

## 3. Hydrogen Peroxide

Hydrogen peroxide ($H_2O_2$) has been widely used in food and pharmaceutical products, dental products, textiles, environmental protection. It is also involved in advanced oxidation and biochemical processes (Abbas et al., 2010). $H_2O_2$ produced by LAB strains under aerobic conditions, leads to denaturation of several enzymes by oxidation of their sulfhydryl groups. Further, it also has a strong oxidizing effect on membrane lipids and cellular proteins. It increases membrane permeability by peroxidation of membrane lipids, and generates bactericidal free radicals such as superoxide ($O^{-2}$) and hydroxyl ($OH^-$) radicals which can damage DNA (Yang, 2000; Ammor et al., 2006, Sunil & Narayana, 2008; Rattanachaikunsopon & Phumkhachorn, 2010). It can inhibit the growth of pathogenic microorganisms such as *Aeromonas hydrophila, Listeria monocytogenes, Yersinia enterocolitica,* and *Clostridium botulinum* type E (Yang, 2000; Ito et al., 2003; Zalan et al., 2005; Abbas et al., 2010).

## 4. Reuterin

Reuterin is a glycerol-derived antimicrobial compound produced under anaerobic conditions by a number of lactobacilli. Its production has been found to enhance in the presence of glycerol. It is capable of inhibiting the growth of fungal strains of *Aspergillus* and *Fusarium*, and plays an important role in neutralizing mycotoxins. It has potential antimicrobial activity against Gram-positive and Gram-negative bacteria, including enteropathogens, yeast, fungi, protozoa, as well as viruses (Nes et al., 2012). Many pathogenic species of *Candida, Clostridium, Listeria, Salmonella, Shigella, Staphylococcus,* and *Trypanosoma* are sensitive to reuterin (Yang, 2000).

## 5. Enzymes

Probiotics also inhibit pathogens by producing enzymes that either block toxin mediated effects or modify the toxin receptors, for example, degradation of *C. difficile* toxin receptors by *S .boulardii* enzymes in rabbit ileum (Pouthoulakis et al., 1993). Probiotics have also been reported to improve immunity of intestinal mucosa as shown by Kaila et al. (1995). Majamaa et al. (1995) pointed to adjuvant like effects on intestinal and systemic immune response by oral intake of probiotics and a special stimulation of Immunoglobulin A was observed in response to viruses. Probiotics are observed to enhance phagocytosis of intracellular pathogens.

## 6. Bacteriocins

The above discussed antimicrobial compounds are important when the live probiotics produce them and are used as whole cells, whereas bacteriocins have recently attracted attention as antimicrobial peptides as antimicrobial drug precursors for exploration as alternative therapy. Bacteriocins are ribosomally synthesized antimicrobial peptides with activity against other bacteria of the same or different genus and species (Soomro et al., 2002; Yang et al., 2012). Some LAB species also produce antimicrobial bacteriocins and bacteriocin-like compounds (Mobolaji & Wuraola, 2011). These are small thermostable or large thermolabile proteins or protein complexes with antimicrobial effect against other Gram-positive bacteria. The bacteriocin-producers are immune against their own bacteriocins and antimicrobial action is often displayed against closely related members (Zacharof & Lovitt, 2012). Bacteriocin production is affected by pH, nutrient sources and incubation temperature (Todorov and Dicks, 2005c; de Vuyst & Leroy, 2007). On the basis of biochemical and genetic characterization, LAB bacteriocins have been classified into four distinct classes, namely Class I lantibiotics, Class 2 comprises of small, heat-stable nonlanthionine peptides, Class 3 includes large heat-labile proteins and Class 4 with complex bacteriocins having lipid and carbohydrate chemical moieties (Hernandez et al., 2005).

Bacteriocins harbor several attractive features which render them suitable for use as therapeutics. Their proteinaceous nature proves detrimental to microbes but harmless to humans. The non-toxicity of bacteriocins has already been proven in case of laboratory animals. These are generally non-immunogenic and thermo resistant. Some bacteriocins have broad bactericidal activity affecting majority of Gram-positive bacteria, while others affect pathogenic Gram-negative bacteria with compromised outer membrane integrity after osmotic shock or low pH or detergent treatment. The genetic determinants of bacteriocins are generally located on plasmids. These attributes facilitate genetic manipulation in order to increase the variety and efficacy of natural peptide analogues with desirable traits (Juodeikiene et al., 2012).

Nisin is the most thoroughly studied bacteriocin till date with commercial applications in food as biopreservation. It is produced by *Lactococcus lactis* and is the only FDA approved bacteriocin to date with activity against a wide range of Gram-positive and Gram-negative pathogenic microorganisms. It is most effective when applied at high concentration or on target cells pre-treated with EDTA (Todorov & Dicks, 2005c). The solubility and stability of nisin increases substantially with increasing levels of acidity. Nisin is stable at a low pH of 2 and can be autoclaved at 121°C. Under alkaline conditions, the antimicrobial activity is reduced substantially, with complete inactivation at 63°C and pH 11 after 30 min (Jozala et al., 2005). Studies have reported the production of various bacteriocins like acidophilin, alctocidin, lactolin, acidolin, lactobrevin and lactobacillin by strains of lactobacilli (Zacharof & Lovitt, 2012; Mokoena, 2017) *Lactobacillus rhamnosus* produces an antimicrobial compound inhibitory to the growth of *Bacteroides fragilis, Clostridium difficile, Escherichia coli, Streptococci* sp., and *Salmonella* sp. (Banna et al., 2017). Several studies have advocated the therapeutic utility of bacteriocins (Evangelin et al., 2015; Mokoena, 2017).

## ANTIMICROBIAL PEDIOCINS FROM PEDIOCOCCAL SP.

One genera belonging to Lactic acid bacteria is Pediococcus, recognized as a probiotic (FAO/WHO, 2002; ICMR, 2011) These are Gram-positive, highly fastidious, non-motile, non-sporulating, catalase-negative, homofermentative facultative anaerobes of family lactobacillaceae (Kawai et al., 2004). They produce lactic acid as a result of sugar fermentation and can survive at very low pH and high temperature as well (Gonzalez & Kunka, 1987). These were first isolated and characterized from plants by

Mundt et al. (1969). Apart from plants, these are commonly found in dairy products, and in alcoholic beverages (Galvez et al., 2007) and are used in the food industry and as silage additives (Nes et al., 1996; Kumar et al., 2012). They have also been found in human saliva and faeces (Jimenez–Diaz et al., 1993; Jong et al., 2006). The genus Pediococcus is comprised of *P. acidilactici, P. cellicola, P. claussenii, P. damnosus, P. dextrinicus, P. ethanolidurans, P. halophilus, P. inopinatus, P. parvulus, P. pentosaceus,* and *P. stilesii* representative species. Among genus *Pediococcus*, two species namely *P. acidilactici* and *P.pentosaceus* (Golledge et al., 1990; Mastro et al., 1990; Sire et al., 1992; Facklam & Elliot, 1995) are mostly studied and are commonly used in the fermentation of vegetables (Pederson, 1949) and meats (Smith & Palumbo, 1983; Kumar et al., 2012). *P. acidilactici* include food grade strains which secrete bacteriocin, pediocin PA-1.

Pediococci are capable of producing antimicrobial bacteriocins termed as 'Pediocins' which form an integral component of natural bacterial defense against other microorganisms. A number of pediocins have been identified and characterized from these prokaryotic organisms. These peptides are recently gaining attention due to their remarkable heat stability, activity and stability over a wide pH range, broad antimicrobial spectrum; high specificity and efficacy even in very low concentrations (Jack et al., 1995; Lavermicocca et al., 2000; Galvez et al., 2007; de Vuyst & Leroy, 2007; Kumar et al., 2012; Anu & Singh, 2018; Balandin et al., 2019).

Pediococcal strains have also been shown to fulfil the mandatory criterion for use as a probiotic. These acid and bile tolerant bacteria exert beneficial effects on host by efficiently adhering to intestinal epithelial cells (Kumar et al., 2012; Balgir et al., 2013). A study by Balgir et al. (2014) reported improved iron bioavailability and haemoglobin status of anemic subjects after oral consumption of a strain of *Pediococcus acidilactici* MTCC 5101 mainly due to production of acid and pediocin leading to killing of iron-leaching pathogens. Like other LAB, the antimicrobial activity in members of Genus Pediococcus is partly due to production of organic acids such as lactic acid and $H_2O_2$ and majorly attributable tothe presence of gene cluster for production and expression of pediocins (Ray & Daeschel, 1994; Jack et al., 1995; Galvez et al., 2007; de Vuyst & Leroy, 2007; Kaur et al., 2019). A large number of pediocins have been isolated and characterized from different Pediococcal sp. (Table 3).

*Table 3. Pediocin producers and antimicrobial range of pediocins produced*

| Bacteriocin | Producer strain | Antibacterial range | Reference |
|---|---|---|---|
| PediocinAcH | *P. acidilactici* H, E, F, M | *Aeromonashydrophila, Bacillus cereus, Brochothrix, Clostridium perfringens, C. botulinum, Enterococcus faecalis, E. faecium, E. hirae, Escherichia, Listeria monocytogenes, L. innocua, L. seeligeri, Lactococcuslactis, Leuconostocmesenteroides, Micrococcus sedentarius, Pediococcusacidilactici, P. pentosaceus, Pseudomonas putida, Salmonella, Staphylococcus aureus, S. xylosus, Yersinia* | Bhunia et al., 1987, 1988, 1990; Ennahar et al., 1996 |
| Pediocin PA-1 | *P. acidilactici* PAC1.0 NRRL-5627 | *B. cereus, L. bifermentans, L. brevis, L. plantarum, L. lactis, L. dextranicum, L. mesenteroides, L. monocytogenes, P. acidilactici, P. pentosaceus* | Pucci et al., 1988; Marugg et al., 1992 |
| Pediocin PO2 | *P. acidilactici* PO2 | *B. coagulans, E. faecalis, L. curvatus, L. monocytogenes, L. mesenteroides, S. aureus, Streptococcus faecalis* | Hoover *et al.*, 1988; Liao *et al.*, 1993; |
| Pediocin JD | *P. acidilactici* JD-123 | *L. monocytogenes* | Berry *et al.*, 1990; Berry *et al.*, 1991 |
| Pediocin PC | *P. acidilactici* PC | *C. perfringens, Listeria, Leuconostoc, Pediococcus* | Jager & Harlander, 1992 |
| Pediocin SJ-1 | *P. acidilactici* SJ-1 | *C. perfringens, L. brevis, L. plantarum, L. leichmanni, L. monocy-togenes* | Schved et al., 1993 |
| Pediocin L50 | *P. acidilactici* L50 | *B. cereus, C. botulinum, C. perfringens, E. faecalis, L. monocytogenes, S. aureus, L. brevis, L. plantarum, L. sake* 148, *L. innocua, L. lactis, L. mesenteroides, P. acidilactici, P. pentosaceus* | Cintas et al., 1995 |
| PediocinAcM | *P. acidilactici* M | *A. hydrophila, B. coagulans, B. cereus, C. perfringens, L. monocytogenes, S. aureus* | Elegado et al., 1997 |
| Pediocin F | *P. acidilactici* F | - | Osmanagaoglou et al., 1998; Osmanagaoglou et al., 2000 |
| Pediocin CP2 | *P. acidilactici* MTCC 5101 | *Aspergillus flavus, C. sporogenes, E. faecalis, L. brevis, L. bulgaricus, L. mesenteroides, L. monocytogenes, Micrococcus flavus, Neisseria mucosa, P. acidilactici, P. pentosaceus, Pseudomonas putida, P. aeruginosa, Staphylococcus albus, S. aureus, Streptococcus mutans, S. pyogenes* | Kaur & Balgir, 2004; Kaur & Balgir, 2006; Kaur et al., 2009 |
| Pediocin SA-1 | *P. acidilactici* NRRL B5627 | *B. cereus, C. sporogenes, C. thiaminolyticum, E. faecalis, L. brevis, L. bulgaricus, L. casei, L. curvatus, L. jensenii, L. plantarum, L. sakei, L. lactis, L. monocytogenes, L. innocua, L. mesenteroides, M. flavus, M. luteus, P. acidilactici, P. pentosaceus, S. carnosus* | Anastasiadou et al., 2008 |
| Pediocin A | *P. pentosaceus*ATCC 43200, ATCC 43201 | *B. cereus, C. botulinum, C. perfringens, C. sporogenes, C. tyrobutyricum, E. faecalis, E. coli, L. monocytogenes, L. innocua, L. sake, L. brevis, L. plantarum, L. lactis, L. mesenteroides, P. acidilactici,*<br>*P. pentosaceus, Salmonella typhimurium, S. aureus* | Ray & Daeschel (1994), Daeschel & Klaenhammer (1985), Grahm & McKay (1985) |

*continues on following page*

*Table 3. Continued*

| Bacteriocin | Producer strain | Antibacterial range | Reference |
|---|---|---|---|
| Pediocin N5p | *P. pentosaceus* | *Lactobacillus hilgardii, Leuconostocoenos, P. pentosaceus*E5p | Strasser de Saad & Manca de Nadra (1993), Strasser de Saad & Manca de Nadra (1995) |
| Pediocin PD-1 | *P. damnosus* NCFB-1832 | *Oenococcusoeni*, several food spoilage and pathogenic bacteria | Green et al., 1997 |
| Pediocin ISK-1 (nukacin ISK-1) | Pediococcus sp. ISK-1 | *Bacillus subtilis, L. casei ssp. casei, L. lactis, M. luteus, P. acidilactici* | Kimura et al., 1997, Sashihara et al., 2000 |
| Pediocin K1 | Pediococcus sp. KCA1303-10 | *E. faecalis, E. faecium, L. monocytogenes* | Kim et al., 2000 |
| Pentocin L | *P. pentosaceus* L | Broad inhibition spectrum, *B. subtilis, B. cereus* | Yin et al., 2003 |
| Pentocin S | *P. pentosaceus* S | Broad inhibition spectrum, *B. subtilis, B. cereus* | Yin et al., 2003 |
| Pediocin ACCEL | *P. pentosaceus* ACCEL | *B. subtilis, B. cereus, C. perfringens, L. helveticus, L. plantarum, L. monocytogenes L. lactis, P. pentosaceus,* *S. faecalis, S. epidermidis* | Wu et al., 2004 |
| Pediocin ST18 | *P. pentosaceus* ST18 | *L. innocua, L. plantarum, Pediococcus* spp. | Todorov & Dicks, 2005a |
| Pediocin SM-1 | *P. pentosaceus* SM-1 | *C. thiaminolyticum, C. sporogenes,* *L. monocytogenes, L. innocua,* *Pediococcus* spp., several LAB species | Anastasiadou et al., 2008 |
| Pediocin pK23-2 | *P. pentosaceus* K23-2 | Gram-positive bacteria, especially *L. monocytogenes* | Shin et al., 2008 |
| Pediocin 05-10 | *P. pentosaceus* 05-10 | *Enterococcus, Lactobacillus, Leuconostoc,* *Listeria, Pediococcus, Streptococcus* | Huang et al., 2009 |
| Bacteriocin ST44AM | *P. pentosaceus* ST44AM | *E. coli, Klebsiella pneumoniae, L. monocytogenes, L. innocua, L. ivanovii subsp. ivanovii, P. aeruginosa,* other LAB | Todorov & Dicks, 2005b |
| pediocin 2292 | *Pediococcus acidilactici NCIM 2292* | *st Listeria monocytogenes, Staphylococcus aureus and Bacillus subtilis.* | Mandal et al., 2014 |

Pediocins are attractive antimicrobial peptides for developing alternative antibiotic therapies as: (i) they are safe for consumption, as they are completely digested in the gastrointestinal tract of humans and animals, (ii) being $10^3$ to $10^6$ times more potent than conventional antibiotics, (iii) being resistant to common thermal treatments for pasteurization or even sterilization (Abriouel et al., 2010).They display a broad antimicrobial range and have been found specially inhibitory to *Listeria monocytogenes*, a pathogen that causes listeriosis, fatal in as many as 20–30% of cases, and has been involved in neonatal deaths, miscarriages, severe meningitis and sepsis (Jackson et al., 2010). Another commercially produced bacteriocin is pediocin PA-1 produced by *Pediococcus acidilactici* and marketed as ATTATM 2431 (Yang et al., 2012; Sanlibaba & Gucer, 2015). Khalaf et al. (2016) reported the antimicrobial activity of bacteriocins against periodontal pathogens. A recent study has compared the antimicrobial activity of pediocins for root canal disinfection with conventional agents such as chlorhexidine (CHX) and calcium hydroxide (Ca(OH)2) and found it active against *Enterococcus faecalis* and *Staphylococcus epidermidis* biofilms (Ooi et al., 2019).Pediocin is an attractive antimicrobial agent against many pathogenic bacteria. Certain strains of pediococci also act as potent probiotics in modulating gut microbiota and aids in lowering cholesterol along with antidiabetic and antihypertensive properties. All these properties of bacteriocin-producing pediococci highlight their use as antimicrobials in food and health sector for various applications. Further research investigations need to be carried out in future in order to provide valuable insights into mechanism and mode of action behind probiotic and antimicrobial properties displayed by members of this not-so-famous LAB genera.

## 1. Genetic Organization and Structure of Pediocin Operon

The genetic organization of plasmid-linked pediocin operon in pediococcal strains has been studied by site-specific mutation and deletion analysis. The results indicate the presence of a cluster of four genes with common promoter and terminator sequences (Marugg et al., 1992; Bukhtiyarova et al., 1994; Motlagh et al., 1994; Kaur & Balgir, 2006). Gene *pedA* encodes a 62 amino acid long precursor prepediocin, from which 44 amino acid long mature pediocin is formed by cleavage of 18 residue long leader sequence at N-terminal of pre-pediocin during processing and translocation of pediocin through cell membrane (Henderson et al., 1992; Miller et al., 1998). Gene *pedB* following *pedA* is responsible for providing immunity to producer organism by production of a 112 amino acid protein PedB. Gene *pedC* encodes a 174 amino acid long amphiphilic protein PedC, which regulates trans-membrane export of prepediocin along with PedD protein. Gene *pedD* specifies a 724 amino acid residues polypeptide, showing homology to ATP dependent transport proteins and is essential for secretion of pediocin (Marugg et al., 1992). *P. acidilactici* PAC1.0

contains *pedABCD* operon which expresses Pediocin PA-1 (Venema et al., 1995) and *P. acidilactici* H produces Pediocin AcH linked to *papABCD* operon (Bukhtiyarova et al., 1994). The sizes and organization of the various pediocin-encoding plasmids from different pediococcal strains are similar (Ray et al., 1992; Bhunia et al., 1994; Rodriguez et al., 1997; Cui et al., 2012). Structure, immunity and secretion system genes are linked together in the operons, with similar promoter sequences. Figure 2 depicts genetic organization of pediocin operon.

*Figure 2. Genetic organization of the pediocin gene cluster found in different plasmids of Pediococcal sp.*

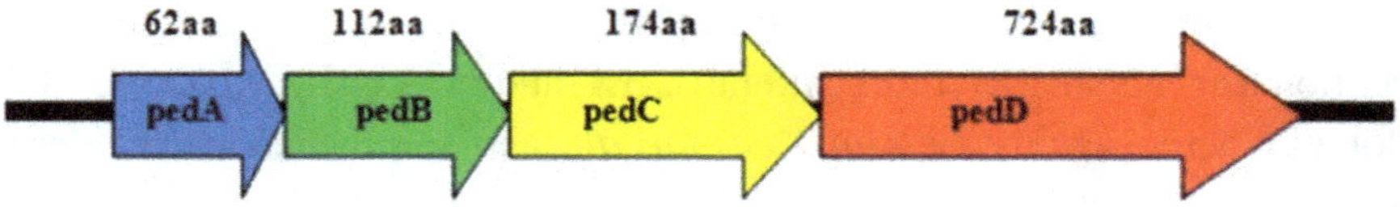


Pediocin PA-1 is a class IIa bacteriocin produced and secreted by *P. acidilactici* PA-1. It is a 62aa protein encoded by gene *pedA* of pediocin operon. The protein is synthesized as a propeptide which is 18 amino acid in length (position 1 – 18) and a 44 amino acid residue long peptide chain (position 19 – 62). Pediocin PA-1 peptide contains two disulfide bonds between Cys9–Cys14 and Cys24–Cys44 (Figure 3). The Cys9–Cys14 bond is conserved among the class IIa bacteriocins, as is the N-terminal YGNGV sequence, both structures being signatures of this family (Bedard et al., 2018).

*Figure 3. Secondary structure of Pediocin PA-1 peptide predicted using Swiss Model*

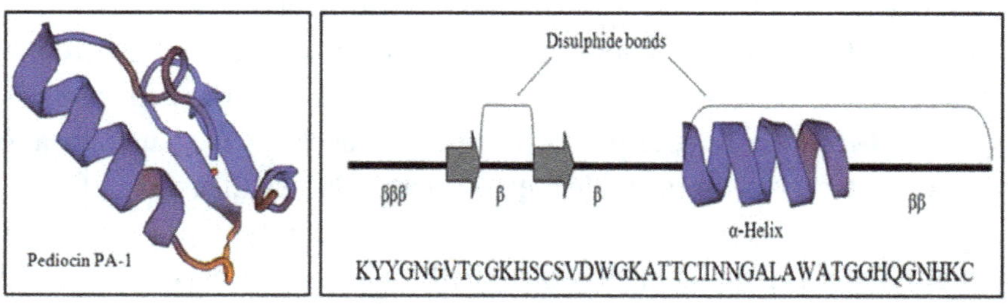


The sequence of hydrophobic peptide chain is conserved among pediocins derived from Genus Pediococcus (Figure 4).

*Figure 4. Multiple sequence alignment of PA-1 with other pediocins using Clustal omega*

```
BA        -----------MANIIGGKYYGNGVTCGKHSCSVDWGKATTCIINNGAMAWATGGHQGNH  49
PA-1      MKKIEKLTEKEMANIIGGKYYGNGVTCGKHSCSVDWGKATTCIINNGAMAWATGGHQGNH  60
AcH       MKKIEKLTEKEMANIIGGKYYGNGVTCGKHSCSVDWGKATTCIINNGAMAWATGGHQGND  60
                     ************************************************.

BA        KCEQFIKLGLLRGEFLF      66
PA-1      KC---------------      62
AcH       KC---------------      62
          **
```

## 2. Mechanism of Pediocin Action

Antimicrobial peptides are rich in cationic amino acid residues which interact strongly with anionic bacterial cell membranes. The sensitive bacteria are killed due to creation of pores in their cell membranes, which disrupts the trans-membrane potential leading to imbalance between microorganisms and their environment (Chikindas et al., 1993). Liposome-mediated delivery of pediocin PA-1 into membrane vesicles in both sensitive and immune cells initiates efflux of small ions in a concentration dependent manner (Chikindas et al., 1993). Pediocin is capable of effectively releasing high molecular weight compounds at a high concentration. These peptides are amphipathic in nature with properly segregated polar and non-polar residues, which aids in more peptide internalization and membrane perturbation. Trans-membrane potential in bacteria, acts as a potential driving force for insertion and internalization of these peptides promoting peptide interaction (Manuel et al., 2009). Pediocin monomers accumulate on the bacterial surface forming circular patterns, and adopt an orientation parallel to the lipid bilayer. Then they reorient and penetrate into the membrane with their hydrophobic surfaces touching the lipid and hydrophilic moieties oriented towards the center forming a barrel shaped pore (Figure 5).

Pediocin PA-1 is known to exert bactericidal or bacteriolytic effect depending on the type and sensitivity of species (Bhunia et al., 1991). Pediocin ST18 and Pediocin CP2 act in a bacteriostatic manner inhibiting the proliferation of sensitive cells with antifungal and spore-inhibitory acitivity against *A. niger*i solates (Kumar et al., 2012). There are fundamental differences in structure of membrane of bacterial and mammalian cells which makes them resistant to antimicrobial pediocin. The interaction of pediocin with bacterial cells is electrostatic in nature whereas in case of mammalian cells, a relatively weak hydrophobic interaction has been reported. This implies that any therapeutics based on pediocins will be safe for animal and human consumption.

*Figure 5. Mechanism of pediocin action*

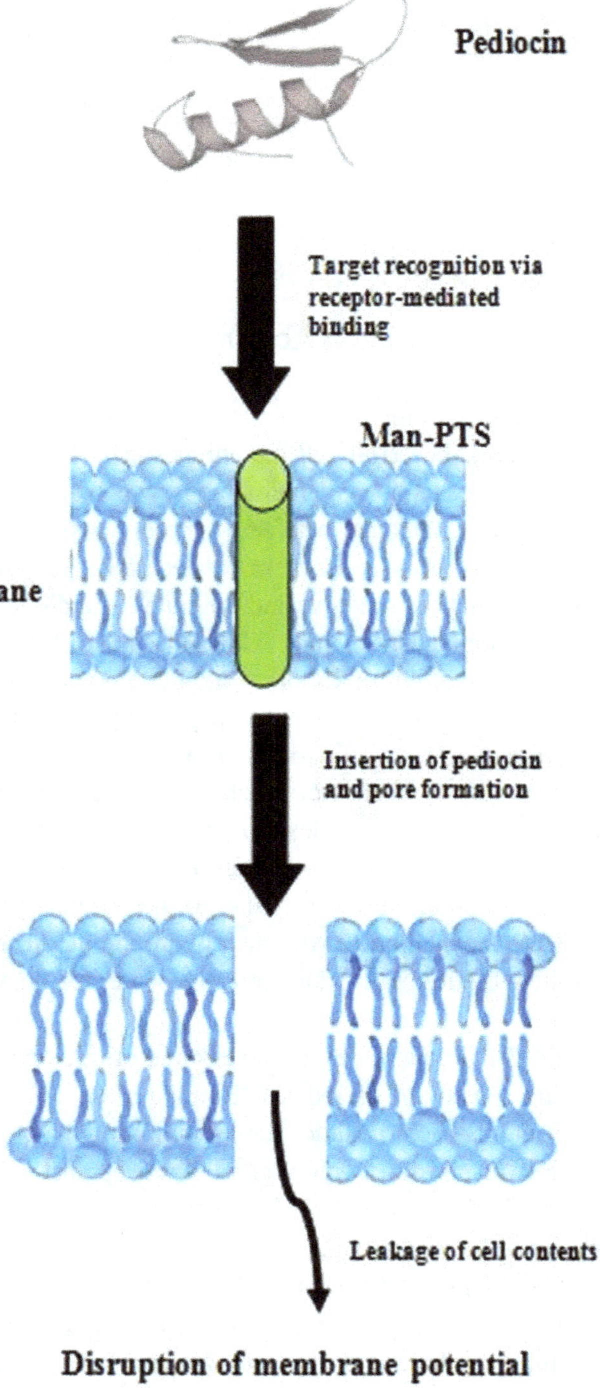

## PRESENT SCENARIO (SOLUTIONS AND RECOMMENDATIONS)

Current research is focused on using antimicrobial peptides as an alternative line of treatment for curing bacterial infections. A study has reported the effectiveness and efficiency of using bacteriocin 'nisin' as a therapeutic for the treatment of Staphylococcal mastitis (Falagas et al., 2006; McFarland, 2007; Fernandez et al., 2008; Kaur et al., 2010; Bastos et al., 2010). *In vitro* and *in vivo* studies performed during another study reported the successful application of a class III bacteriocin 'lysostaphin' alone or in combination with other antibacterial agents in prevention or treatment of staphylococcal infections (Bastos et al., 2010). Further, purified bacteriocins such as pyocin, colicin, pediocin, and microcin are available now-a-days which are known to possess anti-neoplastic activity (Zacharof & Lovitt, 2012; Mokoena, 2017). The effectiveness of modified bacteriocins has also been proven in a glioblastoma xenograft mouse model (Gilbert et al., 2008).

Pediocin like other bacteriocins have stability issues as far as its systemic application is concerned, since it is prone to protease activity. Studies suggest introduction of D-amino acids which will not only resist action of proteases but also prohibit aggregation of bacteriocin monomers and reduce their toxicity. Bacteriocins are also reported to be protease resistant when modified at their terminal regions by amidation, acetylation, or other hydrophobic tags. Substituting Tryptophan and pegylation are other ways to improve stability of bacteriocins in presence of proteases (Carmona-Ribeiro & de Melo Carrasco, 2014; Kumar et al., 2018).

Naturally occurring antimicrobial peptides like pediocin to be used as antimicrobial therapeutics pose a number of challenges (Jenssen et al., 2006; Vaara, 2009):

- The susceptibility of degradation by proteases in the digestive tract if given orally or in the blood if i.v. or intra muscular route is chosen.
- Rapid clearance by kidneys
- Toxicity observed with systemic application

To overcome these challenges chemically modifying these molecules to improve their activity as well as biocompatibility is an option selected in number of studies (Gentilucci et al., 2010; Berthold et al., 2013). Various strategies to protect peptides against protease activity include incorporation of D-amino acids, cyclization of the peptide, and addition of acetyl moiety at N-terminal as well as designing peptidomimetic backbone in such molecules; have been investigated. Though non-natural D-amino acids protect the molecules from protease activity, presently it is a very costly option, thus alternatives are needed, keeping in view the economics of the application.

Apart from chemical modifications half life of peptides can also be improved by the delivery systems used. These systems may include covalent linkage to delivery molecule or non-covalent encapsulation in various materials (Reinhardt & Neundorf, 2016; Nordstrom & Malmsten, 2017). These authors have listed silica, metal nanoparticles, including titanium, gold and silver, quantum dots, graphene and carbon nanotubes for the purpose. Graphene oxide nanotubes with covalently attach nisin were found to enhance its antimicrobial activity against methicillin resistant *Staphylococcus aureus* (MRSA) (Kanchanapally et al., 2015). Micelles, liposomes and micro-emulsions have been employed to deliver antimicrobials as in case of nisin against *S. aureus* (Sadiq et al., 2016). Polymer based gels, fibres, multilayers hydrogels and conjugates are other options available for loading the antimicrobials. Polyethylene glycol conjugation or PEGylation is known to protect proteins from degradation by proteases. Other polymers used include chitosan, hyaluronic acid, hyperbranched polyglycerol, all increased the therapeutic index of drugs loaded onto them (Kumar et al., 2017).

Lam et al. (2016) have reported developing star shaped structurally nano-engineered antimicrobial peptide polymers (SNAPPs) S16 and S32 as antimicrobials effective in killing of Gram-negative bacteria, such as clinical MDR isolates of *Acinetobacter baumannii* and *Pseudomonas aeruginosa.* A peptide synthesis approach termed ring-opening polymerization to α-amino acid N-carboxyanhydrides was applied. The antimicrobial activity was found to involve disruption of *E. coli* cell membrane as well as trigger an apoptotsis like process that killed the gram negative cells. Further no drug resistance was detected even after 600 generations of *A. baumannii* growth in the presence of low S16 concentrations. The SNAPPs have been reported to display low toxicity to human RBCs.

The therapeutic application of Pediocin has faced the challenge of low level production both from native strains as well as recombinant, which recently has been tackled by chemical synthesis using a solid and solution based technology, which also provides with the handle of making changes to its sequence and structure (Bedard et al., 2018). They produced a number of Pediocin PA1 linear analogues eliminating disulfide bonds, which displayed enhanced antimicrobial activity against listeria as well as *Clostridium perfringens.*

## CONCLUSION (FUTURE RESEARCH DIRECTIONS)

Bacteriocin 'Pediocin' from one of LAB Genus Pediococcus is an attractive antimicrobial agent against many pathogenic bacteria. Certain strains of pediococci also act as potent probiotics in modulating gut microbiota and aid in lowering cholesterol along with antidiabetic and antihypertensive properties. All these properties of bacteriocin-producing pediococci highlight their use as aids for health improvement. Pediocins as antimicrobials for application in health sector for various applications are attractive potential drug molecules. Further investigations need to be carried out in future in order to provide valuable insights into mechanism and mode of action behind probiotic and antimicrobial properties displayed by members of this LAB genera in order to widen the horizons of strain specific applications of probiotics in general and Pediococcus in particular.

Since FDA is yet to approve of a bacteriocin for antibiotic use, at current level of research a combination therapy of bacteriocin along with antibiotic can be useful in preventing the development of antibiotic resistance as well as reducing the amount of antibiotic drug used. The strategy of using bacteriocin or bacteriocin producer probiotic synergistically with the antibiotic certainly helps in dose reduction as it facilitates the entry of antibiotic into the pathogen. Such dose reduction will reduce the side effects associated with antibiotics affecting the whole microbiome instead of killing the pathogen alone. For future, a number of paths for improving the druggability of the bacteriocins are being researched, ranging from synthetic peptides with substituted D-amino acids, modified non natural amino acids, end modified and encapsulated or immobilized antimicrobial bacteriocin molecules. Using a range of molecular engineering strategies, novel and more potent antimicrobial peptides can be produced. The engineered antimicrobial peptides are increasingly gaining attention as therapeutic agent due to robust action mechanisms and low production costs.

## CONFLICT OF INTERESTS

The authors declare that they have no conflict of interests.

## ACKNOWLEDGMENT

The authors would like to acknowledge Indian Council of Medical Research, New Delhi for providing financial assistance to Tejinder Kaur (Senior Research Fellowship vide letter no. 3/1/1&2/36/2014-Nut. dated 24/05/2016).

## REFERENCES

Abbas, M. E., Luo, W., Zhu, L., Zou, J., & Tang, H. (2010). Fluorometric determination of hydrogen peroxide in milk by using a Fenton reaction system. *Food Chemistry*, *120*(1), 327–331. doi:10.1016/j.foodchem.2009.10.024

Abriouel, H., Lucas, R., Omar, N. B., Valdivia, E., & Galvez, A. (2010). Potential applications of the cyclic peptide enterocin AS-48 in the preservation of vegetable foods and beverages. *Probiotics and Antimicrobial Proteins*, *2*(2), 77–89. doi:10.100712602-009-9030-y PMID:26781116

Abry, M. F., Kimenyi, K. M., Masiga, D., & Kulohoma, B. W. (2017). Comparative genomics identifies male accessory gland proteins in five Glossina species. *Wellcome Open Research*, *2*, 73. doi:10.12688/wellcomeopenres.12445.2 PMID:29260004

Akimkin, V., Kalmykov, A., Aminev, R., Polyakov, V., & Artebyakin, S. (2016). Experience of using bacteriophages and bitsillin-5 to reduce the incidence of respiratory diseases of bacterial etiology in military personnel. *Voenno-Meditsinskii Zhurnal*, *337*, 36–40. PMID:27263210

Allen, R. C., Popat, R., Diggle, S. P., & Brown, S. P. (2014). Targeting virulence: Can we make evolution-proof drugs? *Nature Reviews. Microbiology*, *12*(4), 300–308. doi:10.1038/nrmicro3232 PMID:24625893

Allocca, M., Zola, S., & Bellosta, P. (2018) The Fruit Fly, Drosophila melanogaster: Modeling of Human Diseases (Part II). In Drosophila melanogaster-Model for Recent Advances in Genetics and Therapeutics. InTech Open.

Ammor, S., Tauveron, G., Dufour, E., & Chevallier, I. (2006). Antibacterial activity of lactic acid bacteria against spoilage and pathogenic bacteria isolated from the same meat small-scale facility 1- Screening and caharacterization of the antimicrobial compounds. *Food Control*, *17*(6), 454–461. doi:10.1016/j.foodcont.2005.02.006

Anastasiadou, S., Papagianni, M., Filiousis, G., Ambrosiadis, I., & Koidis, P. (2008). Pediocin SA–1, an antimicrobial peptide from *Pediococcus acidilactici* NRRL B5627: Production conditions, purification and characterization. *Bioresource Technology*, *99*(13), 5384–5390. doi:10.1016/j.biortech.2007.11.015 PMID:18093831

Anu, & Singh, H. (2018) Applications of the Bacteriocin, Pediocin. *International Journal of Recent Scientific Research, 9*(3), 24962-24965.

Arakawa, K., Yoshida, S., Aikawa, H., Hano, C., Bolormaa, T., Burenjargal, S., & Miyamoto, T. (2016). Production of a bacteriocin-like inhibitory substance by *Leuconostoc mesenteroides* subsp. *dextranicum* 213M0 isolated from Mongolian fermented mare milk, Airag. *Animal Science Journal*, *87*(3), 449–456. doi:10.1111/asj.12445 PMID:26388181

Araújo, C., Muñoz-Atienza, E., Poeta, P., Igrejas, G., Hernández, P. E., Herranz, C., & Cintas, L. M. (2016). Characterization of *Pediococcus acidilactici* strains isolated from rainbow trout (*Oncorhynchus mykiss*) feed and larvae: Safety, DNA fingerprinting, and bacteriocinogenicity. *Diseases of Aquatic Organisms*, *119*(2), 129–143. doi:10.3354/dao02992 PMID:27137071

Balandin, S. V., Sheremeteva, E. V., & Ovchinnikova, T. V. (2019). Pediocin-Like Antimicrobial Peptides of Bacteria. *Biochemistry (Moscow)*, *84*(5), 464–478. doi:10.1134/S000629791905002X PMID:31234762

Balgir, P. P., Kaur, B., & Kaur, T. (2014). A Preliminary Clinical Evaluation of Probiotics *Pediococcus acidilactici* MTCC5101 and *Bacillus coagulans* MTCC492 on Young Anemic Women. *International Journal of Fermented Foods*, *3*(1), 45–59. doi:10.5958/2321-712X.2014.01308.8

Balgir, P. P., Kaur, B., Kaur, T., Daroch, N., & Kaur, G. (2013). *In Vitro* and *In Vivo* Survival and Colonic Adhesion of *Pediococcus acidilactici* MTCC5101 in Human Gut. *BioMed Research International*, *2013*, 1–9. doi:10.1155/2013/583850 PMID:24175293

Banna, G. L., Torino, F., Marletta, F., Santagati, M., Salemi, R., Cannarozzo, E., Falzone, L., Ferraù, F., & Libra, M. (2017). *Lactobacillus rhamnosus* GG: An Overview to Explore the Rationale of Its Use in Cancer. *Frontiers in Pharmacology*, *8*, 603. doi:10.3389/fphar.2017.00603 PMID:28919861

Bastos, M. C. F., Coutinho, B. G., & Coelho, M. L. V. (2010). Lysostaphin: A Staphylococcal Bacteriolysin with Potential Clinical Applications. *Pharmaceuticals*, *3*(4), 1139–1161. doi:10.3390/ph3041139 PMID:27713293

Baxter, A. A., Lay, F. T., Poon, I. K. H., Kvansakul, M., & Hulett, M. D. (2017). Tumor cell membrane-targeting cationic antimicrobial peptides: Novel insights into mechanisms of action and therapeutic prospects. *Cellular and Molecular Life Sciences*, *74*(20), 3809–3825. doi:10.100700018-017-2604-z PMID:28770291

Bedard, F., Hammami, R., Zirah, S., Rebuffat, S., Fliss, I., & Biron, E. (2018). Synthesis, antimicrobial activity and conformational analysis of the class IIa bacteriocin pediocin PA-1 and analogs thereof. *Scientific Reports*, *8*(1), 9029. doi:10.103841598-018-27225-3 PMID:29899567

Belmadani, A., Semlali, A., & Rouabhia, M. (2018). Dermaseptin-S1 decreases *Candida albicans* growth, biofilm formation and the expression of hyphal wall protein 1 and aspartic protease genes. *Journal of Applied Microbiology, 125*(1), 72–83. doi:10.1111/jam.13745 PMID:29476689

Ben Braïek, O., Ghomrassi, H., Cremonesi, P., Morandi, S., Fleury, Y., Le Chevalier, P., Hani, K., Bel Hadj, O., & Ghrairi, T. (2017). Isolation and characterisation of an enterocin P-producing *Enterococcus lactis* strain from a fresh shrimp (*Penaeus vannamei*). *Antonie van Leeuwenhoek, 110*(6), 771–786. doi:10.100710482-017-0847-1 PMID:28265787

Berry, E. D., Hutkins, R. W., & Mandigo, R. W. (1991). The use of bacteriocin-producing *Pediococcus acidilactici* to control post processing *Listeria monocytogenes* contamination of frankfurters. *Journal of Food Protection, 54*(9), 681–686. doi:10.4315/0362-028X-54.9.681 PMID:31051573

Berry, E. D., Liewen, M. B., Mandigo, R. W., & Hutkins, R. W. (1990). Inhibition of *Listeria monocytogenes* by bacteriocin producing *Pediococcus* during the manufacture of fermented semidry sausage. *Journal of Food Protection, 53*(3), 194–197. doi:10.4315/0362-028X-53.3.194 PMID:31018396

Berthold, N., Czihal, P., Fritsche, S., Sauer, U., Schiffer, G., Knappe, D., Alber, G., & Hoffmann, R. (2013). Novel Apidaecin 1b analogs with superior serum stabilities for treatment of infections by Gram-negative pathogens. *Antimicrobial Agents and Chemotherapy, 57*(1), 402–409. doi:10.1128/AAC.01923-12 PMID:23114765

Bhat, S. G. (2018). Modelling and computational sequence analysis of a bacteriocin Isolated from *Bacillus licheniformis* strain BTHT. *International Journal of Computational Biology, 7*(1), 29–34. doi:10.34040/IJCB.7.1.2018.93

Bhunia, A. K., Bhowmik, T. K., & Johnson, M. G. (1994). Determination of bacteriocin-encoding plasmids of *Pediococcus acidilactici* strains by southern hybridization. *Letters in Applied Microbiology, 18*(3), 168–170. doi:10.1111/j.1472-765X.1994.tb00836.x

Bhunia, A. K., Johnson, M. C., & Ray, B. (1987). Direct detection of an antimicrobial peptide of *Pediococcus acidilactici*in sodium dodecyl sulphate-polyacrilamide gel electrophoresis. *Journal of Industrial Microbiology, 2*(5), 319–322. doi:10.1007/BF01569434

Bhunia, A. K., Johnson, M. C., & Ray, B. (1988). Purification, characterization and antimicrobial spectrum of a bacteriocin produced by *Pediococcus acidilactici*. *The Journal of Applied Bacteriology, 65*(4), 261–268. doi:10.1111/j.1365-2672.1988.tb01893.x PMID:2906056

Bhunia, A. K., Johnson, M. C., Ray, B., & Belden, E. L. (1990). Antigenic property of pediocin AcH produced by *Pediococcus acidilactici* H. *The Journal of Applied Bacteriology*, *69*(2), 211–215. doi:10.1111/j.1365-2672.1990.tb01511.x PMID:2272942

Bhunia, A. K., Johnson, M. C., Ray, B., & Kalchayanand, N. (1991). Mode of action of pediocin AcH from *Pediococcus acidilactici* H on sensitive bacterial strains. *The Journal of Applied Bacteriology*, *70*(1), 25–33. doi:10.1111/j.1365-2672.1991.tb03782.x

Boman, H. G. (2003). Antibacterial peptides: Basic facts and emerging concepts. *Journal of Internal Medicine*, *254*(3), 197–215. doi:10.1046/j.1365-2796.2003.01228.x PMID:12930229

Boparai, J. K., & Sharma, P. K. (2020). Mini Review on Antimicrobial Peptides, Sources, Mechanism and Recent Applications. *Protein and Peptide Letters*, *27*(1), 4–16. doi:10.2174/0929866526666190822165812 PMID:31438824

Bosma, T., Kuipers, A., Bulten, E., de Vries, L., Rink, R., & Moll, G. N. (2011). Bacterial Display and Screening of Posttranslationally Thioether-Stabilized Peptides. *Applied and Environmental Microbiology*, *77*(19), 6794–6801. doi:10.1128/AEM.05550-11 PMID:21821759

Braïek, O. B., Morandi, S., Cremonesi, P., Smaoui, S., Hani, K., & Ghrairi, T. (2018). Biotechnological potential, probiotic and safety properties of newly isolated enterocin-producing *Enterococcus lactis* strains. *Lebensmittel-Wissenschaft + Technologie*, *92*, 361–370. doi:10.1016/j.lwt.2018.02.045

Brillet-Viel, A., Pilet, M. F., Courcoux, P., Prévost, H., & Leroi, F. (2016). Optimization of growth and bacteriocin activity of the food bioprotective *Carnobacterium divergens* V41 in an animal origin protein free medium. *Frontiers in Marine Science*, *3*, 128. doi:10.3389/fmars.2016.00128

Bukhtiyarova, M., Yang, R., & Ray, B. (1994). Analysis of the pediocin AcH gene cluster from plasmid pSMB74 and its expression in a pediocin-negative *Pediococcus acidilactici* strain. *Applied and Environmental Microbiology*, *60*(9), 3405–3408. doi:10.1128/aem.60.9.3405-3408.1994 PMID:7944372

Candille, S. I., Kaelin, C. B., Cattanach, B. M., Yu, B., Thompson, D. A., Nix, M. A., Kerns, J. A., Schmutz, S. M., Millhauser, G. L., & Barsh, G. S. (2007). A -defensin mutation causes black coat color in domestic dogs. *Science*, *318*(5855), 1418–1423. doi:10.1126cience.1147880 PMID:17947548

Carmona-Ribeiro, A. M., & de Melo Carrasco, L. D. (2014). Novel formulations for antimicrobial peptides. *International Journal of Molecular Sciences*, *15*(10), 18040–18083. doi:10.3390/ijms151018040 PMID:25302615

Chen, Y. S., Wu, H. C., Kuo, C. Y., Chen, Y. W., Ho, S., & Yanagida, F. (2018). Leucocin C-607, a novel bacteriocin from the multiplebacteriocin-producing *Leuconostoc pseudomesenteroides* 607 isolated from Persimmon. *Probiotics and Antimicrobial Proteins*, *10*(2), 148–156. doi:10.100712602-017-9359-6 PMID:29177756

Chikindas, M. L., Garcia–Garcera, M. J., Driessen, A. J. M., Ledeboer, A. M., Nissen–Meyer, J., Nes, I. F., Abee, T., Konings, W. N., & Venema, G. (1993). Pediocin PA–1, a bacteriocin from *Pediococcus acidilactici* PAC1.0, forms hydrophilic pores in the cytoplasmic membrane of target cells. *Applied and Environmental Microbiology*, *59*(11), 3577–3584. doi:10.1128/aem.59.11.3577-3584.1993 PMID:8285666

Cintas, L. M., Rodriguez, J. M., Fernandez, M. F., Sletten, K., Nes, I. F., Hernandez, P. E., & Holo, H. (1995). Isolation and characterization of pediocin L50, a new bacteriocin from *Pediococcus acidilactici* with a broad inhibitory range. *Applied and Environmental Microbiology*, *61*(7), 2643–2648. doi:10.1128/aem.61.7.2643-2648.1995 PMID:7618877

Coulen, S. C., Sanders, J. P., & Bruins, M. E. (2017). Valorisation of proteins from rubber tree. *Waste and Biomass Valorization*, *8*(4), 1027–1041. doi:10.100712649-016-9688-9

Cui, Y. H., Zhang, C., Wang, Y. F., Shi, J., Zhang, L. W., Ding, Z. Q., Qu, X. J., & Cui, H. Y. (2012). Class IIa bacteriocins, diversity and new developments. *International Journal of Molecular Sciences*, *13*(12), 16668–16707. doi:10.3390/ijms131216668 PMID:23222636

Czaplewski, L., Bax, R., Clokie, M., Dawson, M., Fairhead, H., Fischetti, V. A., Foster, S., Gilmore, B. F., Hancock, R. E. W., Harper, D., Henderson, I. R., Hilpert, K., Jones, B. V., Kadioglu, A., Knowles, D., Ólafsdóttir, S., Payne, D., Projan, S., Shaunak, S., ... Rex, J. H. (2016). Alternatives to antibiotics-a pipeline portfolio review. *The Lancet. Infectious Diseases*, *16*(2), 239–251. doi:10.1016/S1473-3099(15)00466-1 PMID:26795692

Daeschel, M. A., & Klaenhammer, T. R. (1985). Association of a 13.6 M Dal plasmid in *Pediococcus pentosaceus* with bacteriocin activity. *Applied and Environmental Microbiology*, *50*(6), 1538–1541. doi:10.1128/aem.50.6.1538-1541.1985 PMID:16346954 de Vuyst, L., & Leroy, F. (2007). Bacteriocins from lactic acid bacteria: Production, purification, and food applications. *Journal of Molecular Microbiology and Biotechnology*, *13*(4), 194–199. doi:10.1159/000104752 PMID:17827969

Duwadi, D., Shrestha, A., Yilma, B., Kozlovski, I., Sa-Eed, M., Dahal, N., & Jukosky, J. (2018). Identification and screening of potent antimicrobial peptides in arthropod genomes. *Peptides*, *103*, 26–30. doi:10.1016/j.peptides.2018.01.017 PMID:29501691

Ebrahimipour, G. H., Khosravibabadi, Z., Sadeghi, H., & Aliahmadi, A. (2014). Isolation, partial purification and characterization of an antimicrobial compound, produced by *Bacillus atrophaeus*. *Jundishapur Journal of Microbiology*, *7*(9), e11802. doi:10.5812/jjm.11802 PMID:25485057

Elegado, F. B., Kim, W. J., & Kwon, D. Y. (1997). Rapid purification, partial characterization, and antimicrobial spectrum of the bacteriocin, Pediocin AcM, from *Pediococcus acidilactici* M. *International Journal of Food Microbiology*, *37*(1), 1–11. doi:10.1016/S0168-1605(97)00037-8 PMID:9237116

Ennahar, S., Aoude-Werner, D., Sorokine, O., van Dorsselaer, A., Bringel, F., Hubert, J. C., & Hasselmann, C. (1996). Production of pediocin AcH by *Lactobacillus plantarum* WHE92 isolated from cheese. *Applied and Environmental Microbiology*, *62*(12), 4381–4387. doi:10.1128/aem.62.12.4381-4387.1996 PMID:8953710

Evangelin, Y., Venkateswarulu, T. C., John Babu, D., & Kasturi, K. (2015). Bacteriocins from lactic acid bacteria and its potential applications. *International Journal of Pharmaceutical Sciences Review and Research*, *32*(1), 306–309.

Facklam, R., & Elliott, J. A. (1995). Identification, classification, and clinical relevance of catalase-negative, Gram-positive cocci, excluding Streptococci and Enterococci. *Clinical Microbiology Reviews*, *8*(4), 479–495. doi:10.1128/CMR.8.4.479 PMID:8665466

Falagas, M. E., Betsi, G. I., Tokas, T., & Athanasiou, S. (2006). Probiotics for prevention of recurrent urinary tract infections in women: A review of the evidence from microbiological and clinical studies. *Drugs*, *66*(9), 1253–1261. doi:10.2165/00003495-200666090-00007 PMID:16827601

FAO/WHO. (2002). Guidelines for the Evaluation of Probiotics in Food. Report of a Joint FAO/WHO working group of drafting guidelines for the Evaluation of Probiotics in Food, London, Ontario, Canada.

Farouk, A. E., Ahamed, N. T., AlZahrani, O., Alghamdi, A., & Bahobail, A. (2017). Inducible antimicrobial compounds (Halal) production in Honey Bee Larvae (*Apis mellifera*) from Rumaida, Taif by injecting of various dead microorganisms extracts. *Journal of Applied Biology and Biotechnology*, *5*(02), 23–29.

Fernandez, L., Delqado, S., Herrero, H., Maldonado, A., & Rodriquez, J. M. (2008). The bacteriocin nisin, an effective agent for the treatment of staphylococcal mastitis during lactation. *Journal of Human Lactation*, *24*(3), 311–316. doi:10.1177/0890334408317435 PMID:18689718

Gajalakshmi, P. (2017). Selective isolation and characterization of rare actinomycetes adopted in glacier soil of Manali ice point and its activity against *Mycobacterium* spp. *Journal of Microbiology and Biotechnology Research*, *7*(5), 1–10. doi:10.24896/jmbr.2017751

Galvez, A., Abriouel, H., Lopez, R. L., & Omar, N. B. (2007). Bacteriocin–based strategies for food biopreservation. *International Journal of Food Microbiology*, *120*(1-2), 51–70. doi:10.1016/j.ijfoodmicro.2007.06.001 PMID:17614151

Gentilucci, L., De Marco, R., & Cerisoli, L. (2010). Chemical modifications designed to improve peptide stability: Incorporation of non-natural amino acids, pseudo-peptide bonds, and cyclization. *Current Pharmaceutical Design*, *16*(28), 3185–3203. doi:10.2174/138161210793292555 PMID:20687878

Gharsallaoui, A., Oulahal, N., Joly, C., & Degraeve, P. (2016). Nisin as a Food Preservative: Part 1: Physicochemical Properties, Antimicrobial Activity, and Main Uses. *Critical Reviews in Food Science and Nutrition*, *56*(8), 1262–1274. doi:10.1080/10408398.2013.763765 PMID:25675115

Gilbert, C., Claude, F., & Denis, S. (2008). Antineoplastic Properties of Bacteriocins: Revisiting Potential Active Agents. *American Journal of Clinical Oncology*, *31*(4), 399–404. doi:10.1097/COC.0b013e31815e456d PMID:18846002

Golledge, C. L., Stingemore, N., Aravena, M., & Joske, D. (1990). Septicemia caused by vancomycin-resistant *Pediococcus acidilactici*. *Journal of Clinical Microbiology*, *28*(7), 1678–1679. doi:10.1128/jcm.28.7.1678-1679.1990 PMID:2199529

Gondil, V. S., Harjai, K., & Chhibber, S. (2020). Endolysins as emerging alternative therapeutic agent to counter drug resistant infections. *International Journal of Antimicrobial Agents*, *55*(2), 105844. doi:10.1016/j.ijantimicag.2019.11.001 PMID:31715257

Gonzalez, C. F., & Kunka, B. S. (1987). Plasmid– associated bacteriocin production and sucrose fermentation in *Pediococcus acidilactici. Applied and Environmental Microbiology*, *53*(10), 2534–2538. doi:10.1128/aem.53.10.2534-2538.1987 PMID:16347470

Grahm, D. C., & McKay, L. L. (1985). Plasmid DNA in strains of *Pediococcus cerevisiae* and *Pediococcus pentosaceus. Applied and Environmental Microbiology*, *50*(2), 532–534. doi:10.1128/aem.50.2.532-534.1985 PMID:16346868

Green, G., Dicks, L. M., Bruggeman, G., Vandamme, E. J., & Chikindas, M. L. (1997). Pediocin PD–1: A bactericidal antimicrobial peptide from *Pediococcus damnosus* NCFB1832. *The Journal of Applied Bacteriology*, *83*(1), 127–132. doi:10.1046/j.1365-2672.1997.00241.x PMID:9246779

Guzmán-Rodríguez, J. J., Ochoa-Zarzosa, A., López-Gómez, R., & López-Meza, J. E. (2015). Plant antimicrobial peptides as potential anticancer agents. *BioMed Research International*, *2015*, 735087. doi:10.1155/2015/735087 PMID:25815333

Hammi, I., Delalande, F., Belkhou, R., Marchioni, E., Cianferani, S., & Ennahar, S. (2016). Maltaricin CPN, a new class IIa bacteriocin produced by *Carnobacterium maltaromaticum* CPN isolated from mould-ripened cheese. *Journal of Applied Microbiology*, *121*(5), 1268–1274. doi:10.1111/jam.13248 PMID:27489131

Haney, E. F., & Hancock, R. E. W. (2013). Peptide design for antimicrobial and immunomodulatory applications. *Biopolymers*, *100*(6), 572–583. doi:10.1002/bip.22250 PMID:23553602

Henderson, J. T., Chopko, A. L., & Dyck van Wassenaar, P. D. (1992). Purification and primary structure of pediocin PA-1, produced by *Pediococcus acidilactici* PAC-1.0. *Archives of Biochemistry and Biophysics*, *295*(1), 5–12. doi:10.1016/0003-9861(92)90480-K PMID:1575516

Hernandez, D., Cardell, E., & Zarate, V. (2005). Antimicrobial activity of lactic acid bacteria isolated from Tenerife cheese: Initial characterization of plantaricin TF711, a bacteriocin like substance produced by *Lactobacillus plantarum* TF711. *Journal of Applied Microbiology*, *99*(1), 77–84. doi:10.1111/j.1365-2672.2005.02576.x PMID:15960667

Hetru, C., Letellier, L., Oren, Z., Hoffmann, J. A., & Shai, Y. (2000). Androctonin, a hydrophilic disulphide-bridged non-haemolytic anti-microbial peptide: A plausible mode of action. *The Biochemical Journal*, *345*(3), 653–664. doi:10.1042/bj3450653 PMID:10642525

Hoffmann, J. A., & Hetru, C. (1992). Insect defensins: Inducible antibacterial peptides. *Immunology Today*, *13*(10), 411–415. doi:10.1016/0167-5699(92)90092-L PMID:1418378

Hollmann, A., Martinez, M., Maturana, P., Semorile, L. C., & Maffia, P. C. (2018). Antimicrobial peptides: Interaction with model and biological membranes and synergism with chemical antibiotics. *Frontiers in Chemistry*, *6*, 204. doi:10.3389/fchem.2018.00204 PMID:29922648

Hoover, D. G., Walsh, P. M., Kolaetis, K. M., & Daly, M. M. (1988). A bacteriocin produced by *Pediococcus* species associated with a 5.5 kDa plasmid. *Journal of Food Protection*, *51*(1), 29–31. doi:10.4315/0362-028X-51.1.29 PMID:30978873

Huang, Y., Luo, Y., Zhai, Z., Zhang, H., Yang, C., Tian, H., Li, Z., Feng, J., Liu, H., & Hao, Y. (2009). Characterization and application of an anti-listeria bacteriocin produced by *Pediococcus pentosaceus* 05–10 isolated from Sichuan Pickle, a traditionally fermented vegetable product from China. *Food Control*, *20*(11), 1030–1035. doi:10.1016/j.foodcont.2008.12.008

ICMR. (2011) *Guidelines for evaluation of probiotics in food.* Retrieved from https://icmr.nic.in/

Ito, A., Sato, Y., Kudo, S., Sato, S., Nakajima, H., & Toba, T. (2003). The screening of hydrogen peroxide producing lactic acid bacteria and their application to inactivating psychotropic food-borne pathogens. *Current Microbiology*, *47*(3), 231–236. doi:10.100700284-002-3993-1 PMID:14570275

Jack, R. W., Tagg, J. R., & Ray, B. (1995). Bacteriocins of Gram positive bacteria. *Microbiological Reviews*, *59*(2), 171–200. doi:10.1128/mr.59.2.171-200.1995 PMID:7603408

Jackson, K. A., Iwamoto, M., & Swerdlow, D. (2010). Pregnancy-associated listeriosis. *Epidemiology and Infection*, *138*(10), 1503–1509. doi:10.1017/S0950268810000294 PMID:20158931

Jager, K., & Harlander, S. (1992). Characterization of a bacteriocin from *Pediococcus acidilactici* PC and comparison of bacteriocin-producing strains using molecular typing procedures. *Applied and Environmental Microbiology*, *37*, 631–637.

Jenssen, H., Hamill, P., & Hancock, R. E. W. (2006). Peptide Antimicrobial Agents. *Clinical Microbiology Reviews*, *19*(3), 491–511. doi:10.1128/CMR.00056-05 PMID:16847082

Jiang, H., Tang, X., Zhou, Q., Zou, J., Li, P., Breukink, E., & Gu, Q. (2018). Plantaricin NC8 from Lactobacillus plantarum causes cell membrane disruption to *Micrococcus luteus* without targeting lipid II. *Applied Microbiology and Biotechnology*, *102*(17), 7465–7473. doi:10.100700253-018-9182-3 PMID:29982926

Jimenez–Diaz, R., Rios–Sanchez, R. M., Desmazeaud, M., Ruiz–Barba, J. L., & Piard, J. C. (1993). Plantaricin S and T, two new bacteriocins produced by *Lactobacillus plantarum* LPCO10 isolated from a green olive fermentation. *Applied and Environmental Microbiology*, *59*(5), 1416–1424. doi:10.1128/aem.59.5.1416-1424.1993 PMID:16348933

Jong, A. D., van Hijum, S. A. F. T., Bijlsma, J. J. E., Kok, J., & Kuipers, O. P. (2006). BAGEL: A web–based bacteriocin genome mining tool. *Nucleic Acids Research*, *34*(2), 273–279. doi:10.1093/nar/gkl237 PMID:16845009

Jozala, A. F., Novaes, L. C. L., Cholewa, O., Moraes, D., & Penna, T. C. V. (2005). Increase of nisin production by *Lactococcus lactis* in different media. *African Journal of Biotechnology*, *4*(3), 262–265.

Juodeikiene, G., Bartkiene, E., Viskelis, P., Urbonaviciene, D., Eidukonyte, D., & Bobinas, C. (2012) Fermentation processes using lactic acid bacteria producing bacteriocins for preservation and improving functional properties of food products. In Advances in Applied Biotechnology. In Tech. doi:10.5772/30692

Kaila, M., Isolauri, E., Maija, S., Arvilommi, H., & Vesikari, T. (1995). Viable versus inactivated lactobacillus strain GG in acute rotavirus diarrhoea. *Archives of Disease in Childhood*, *72*(1), 51–53. doi:10.1136/adc.72.1.51 PMID:7717739

Kalmokoff, M. L., Banerjee, S. K., Cyr, T., Hefford, M. A., & Gleeson, T. (2001). Identification of a New Plasmid-Encoded *sec*-Dependent Bacteriocin Produced by *Listeria innocua* 743. *Applied and Environmental Microbiology*, *67*(9), 4041–4047. doi:10.1128/AEM.67.9.4041-4047.2001 PMID:11526003

Kanchanapally, R., Viraka Nellore, B. P., Sinha, S. S., Pedraza, F., Jones, S. J., Pramanik, A., Chavva, S. R., Tchounwou, C., Shi, Y., Vangara, A., Sardar, D., & Ray, P. C. (2015). Antimicrobial peptide-conjugated graphene oxide membrane for efficient removal and effective killing of multiple drug resistant bacteria. *RSC Advances*, *5*(24), 18881–18887. doi:10.1039/C5RA01321F PMID:26294958

Kaur, B., & Balgir, P. P. (2004). Purification, characterization and antimicrobial range of bacteriocin obtained from an isolate of *Pediococcus* spp. *Journal of Punjab Academy of Sciences*, *1*(2), 139–144.

Kaur, B., & Balgir, P. P. (2006). Pediocin CP2 Gene localisation to plasmid pCP289 of *Pediococcus acidilactici* MTCC 5101. *The Internet Journal of Microbiology*, 3.

Kaur, B., Balgir, P. P., Kumar, B., & Garg, N. (2010). *Helicobacter pylori* infection: Efficacy of probiotics and role of genome wide association studies. *Archives of Clinical Microbiology*, *1*(4), 3.

Kaur, B., Kumar, B., Balgir, P. P., & Bhatia, P. (2009). Comparative evaluation of media for pediocin production production by *Pediococcus acidilactici* CP2 isolate. *International Journal of Probiotics & Prebiotics*, *4*(4), 233–240.

Kaur, T., Balgir, P. P., & Kaur, B. (2019). Construction of a shuttle expression vector for lactic acid bacteria. *Journal of Genetic Engineering and Biotechnology*, *17*(1), 10. doi:10.118643141-019-0013-4 PMID:31736018

Kawai, Y., Ishii, Y., Arakawa, K., Uemura, K., Saitoh, B., Nishimura, J., Kitazawa, H., Yamazaki, Y., Tateno, Y., Itoh, T., & Saito, T. (2004). Structural and functional differences in two cyclic bacteriocins with the same sequences produced by *Lactobacilli. Applied and Environmental Microbiology*, *70*(5), 2906–2911. doi:10.1128/AEM.70.5.2906-2911.2004 PMID:15128550

Khalaf, H., Nakka, S. S., Sandén, C., Svard, A., Hultenby, K., Scherbak, N., Aili, D., & Bengtsson, T. (2016). Antibacterial effects of Lactobacillus and bacteriocin PLNC8 αβ on the periodontal pathogen *Porphyromonas gingivalis. BMC Microbiology*, *16*(1), 188. doi:10.118612866-016-0810-8 PMID:27538539

Khurshid, Z., Najeeb, S., Mali, M., Moin, S. F., Raza, S. Q., Zohaib, S., Sefat, F., & Zafar, M. S. (2017). Histatin peptides: Pharmacological functions and their applications in dentistry. *Saudi Pharmaceutical Journal*, *25*(1), 25–31. doi:10.1016/j.jsps.2016.04.027 PMID:28223859

Kim, C. H., Ji, G. E., & Ahn, C. (2000). Purification and molecular characterization of a bacteriocin from *Pediococcus* sp. KCA 1303–10 isolated from fermented flat fish. *Food Science and Biotechnology*, *9*(4), 270–276.

Kim, P. I., Sohng, J. K., Sung, C., Joo, H.-S., Kim, E.-M., Yamaguchi, T., Park, D., & Kim, B.-G. (2010). Characterization and structure identification of an antimicrobial peptide, hominicin, produced by *Staphylococcus hominis* MBBL 2–9. *Biochemical and Biophysical Research Communications*, *399*(2), 133–138. doi:10.1016/j.bbrc.2010.07.024 PMID:20654578

Kimura, H., Nagano, R., Matsusaki, H., Sonomoto, K., & Ishizaki, A. (1997). A bacteriocin of strain *Pediococcus* sp. ISK–I isolated from Nukadoko, bed of fermented rice bran. *Bioscience, Biotechnology, and Biochemistry*, *61*(6), 1049–1051. doi:10.1271/bbb.61.1049 PMID:9214771

Kleerebezem, M., Hols, P., & Hugenholtz, J. (2000). Lactic acid bacteria as a cell factory: Rerouting of carbon metabolism in *Lactococcus lactis* by metabolic engineering. *Enzyme and Microbial Technology*, *26*(9-10), 840–848. doi:10.1016/S0141-0229(00)00180-0 PMID:10862894

Kumar, B., Balgir, P. P., Kaur, B., Mittu, B., & Chauhan, A. (2012). *In vitro* cytotoxicity of native and rec-pediocin CP2 against cancer cell lines: A comparative study. *Pharmaceutica Analytica Acta*, *3*(8), 183. doi:10.4172/2153-2435.1000183

Kumar, P., Kizhakkedathu, J. N., & Straus, S. K. (2018). Antimicrobial peptides: Diversity, mechanism of action and strategies to improve the activity and biocompatibility *in vivo*. *Biomolecules*, *8*(1), 4. doi:10.3390/biom8010004 PMID:29351202

Kumar, P., Takayesu, A., Abbasi, U., Kalathottukaren, M. T., Abbina, S., Kizhakkedathu, J. N., & Straus, S. K. (2017). Antimicrobial peptide–polymer conjugates with high activity: Influence of polymer molecular weight and peptide sequence on antimicrobial activity, proteolysis, and biocompatibility. *ACS Applied Materials & Interfaces*, *9*(43), 37575–37586. doi:10.1021/acsami.7b09471 PMID:29019386

Kuzmin, D. V., Emelianova, A. A., Kalashnikova, M. B., Panteleev, P. V., & Ovchinnikova, T. V. (2017). Effect of N- and C-terminal modifications on cytotoxic properties of antimicrobial peptide tachyplesin I. *Bulletin of Experimental Biology and Medicine*, *162*(6), 754–757. doi:10.100710517-017-3705-2 PMID:28429216

Lam, S. J., O'Brien-Simpson, N. M., Pantarat, N., Sulistio, A., Wong, E. H., Chen, Y. Y., Lenzo, J. C., Holden, J. A., Blencowe, A., Reynolds, E. C., & Qiao, G. G. (2016). Combating multidrug-resistant Gram-negative bacteria with structurally nanoengineered antimicrobial peptide polymers. *Nature Microbiology*, *1*(11), 16162. doi:10.1038/nmicrobiol.2016.162 PMID:27617798

Lavermicocca, P., Valerio, F., Evidente, A., Lazzaroni, S., Corsetti, A., & Gobbetti, M. (2000). Purification and characterization of novel antifungal compounds from the sourdough *Lactobacillus plantarum* strain 21B. *Applied and Environmental Microbiology*, *66*(9), 4084–4090. doi:10.1128/AEM.66.9.4084-4090.2000 PMID:10966432

Le, T. N., Do, T. H., Nguyen, T. N., Tran, N. T., Enfors, S. O., & Truong, H. (2014). Expression and simple purification strategy for the generation of anti-microbial active Enterocin P from *Enterococcus faecium* expressed in *Escherichia coli* ER2566. *Iranian Journal of Biotechnology*, *12*(4), 17–25. doi:10.15171/ijb.1154

Lee, H. S., Park, C. B., Kim, J. M., Jang, S. A., Park, I. Y., Kim, M. S., Cho, J. H., & Kim, S. C. (2008). Mechanism of anticancer activity of buforin IIb, a histone H2A-derived peptide. *Cancer Letters*, *271*(1), 47–55. doi:10.1016/j.canlet.2008.05.041 PMID:18617323

Lee, J., Lee, D., Choi, H., Kim, H. H., Kim, H., Hwang, J. S., Lee, D. G., & Kim, J. I. (2014). Structure-activity relationships of the intramolecular disulfide bonds in coprisin, a defensin from the dung beetle. *BMB Reports*, *47*(11), 625–630. doi:10.5483/BMBRep.2014.47.11.262 PMID:24393527

Lee, J., & Lee, D. G. (2015). Antimicrobial peptides (AMPs) with dual mechanisms: Membrane disruption and apoptosis. *Journal of Microbiology and Biotechnology*, *25*(6), 759–764. doi:10.4014/jmb.1411.11058 PMID:25537721

Lemaitre, B., Nicolas, E., Michaut, L., Reichhart, J. M., & Hoffmann, J. A. (1996). The dorsoventral regulatory gene cassette spatzle/Toll/cactus controls the potent antifungal response in Drosophila adults. *Cell*, *86*(6), 973–983. doi:10.1016/S0092-8674(00)80172-5 PMID:8808632

Li, R., Wan, X., Takala, T. M., & Saris, P. E. J. (2021). Heterologous Expression of the *Leuconostoc* Bacteriocin Leucocin C in Probiotic Yeast *Saccharomyces boulardii*. *Probiotics and Antimicrobial Proteins*, *13*(1), 229–237. doi:10.100712602-020-09676-1 PMID:32567021

Liao, C. C., Yousef, A. E., Richter, R., & Chism, G. W. (1993). *Pediococcus acidilactici* PO2, bacteriocin production in whey permeate and inhibition of *Listeria monocytogenes* in foods. *Journal of Food Science*, *58*(2), 430–434. doi:10.1111/j.1365-2621.1993.tb04291.x

Lood, R., Raz, A., Molina, H., Euler, C. W., & Fischetti, V. A. (2014). A highly active and negatively charged *Streptococcus pyogenes* lysin with a rare d-Alanyl-l-Alanine endopeptidase activity protects mice against streptococcal bacteremia. *Antimicrobial Agents and Chemotherapy*, *58*(6), 3073–3084. doi:10.1128/AAC.00115-14 PMID:24637688

Ma, B., Fang, C., Lu, L., Wang, M., Xue, X., Zhou, Y., Li, M., Hu, Y., Luo, X., & Hou, Z. (2019). The antimicrobial peptide thanatin disrupts the bacterial outer membrane and inactivates the NDM-1 metallo-β-lactamase. *Nature Communications*, *10*(1), 3517. doi:10.103841467-019-11503-3 PMID:31388008

Majamaa, H., Isolauri, E., Saxelin, M., & Vesikari, T. (1995). Lactic acid bacteria in the treatment of acute rotavirus gastroenteritis. *Journal of Pediatric Gastroenterology and Nutrition*, *20*(3), 333–338. doi:10.1097/00005176-199504000-00012 PMID:7608829

Maldonado-Barragán, A., Caballero-Guerrero, B., Martín, V., Ruiz-Barba, J. L., & Rodríguez, J. M. (2016). Purification and genetic characterization of gassericin E, a novel co-culture inducible bacteriocin from *Lactobacillus gasseri* EV1461 isolated from the vagina of a healthy woman. *BMC Microbiology*, *16*(1), 37. doi:10.118612866-016-0663-1 PMID:26969428

Manabe, T., & Kawasaki, K. (2017). D-form KLKLLLLLKLK-NH2 peptide exerts higher antimicrobial properties than its L-form counterpart in an association with bacterial cell wall components. *Scientific Reports*, *7*(1), 43384. doi:10.1038rep43384 PMID:28262682

Mandal, B., Chowdhury, R., & Bhtattacharjee, C. (2014). Purification and Characterization of Pediocin produced by *Pediococcus acidilactici* NCIM 2292. *International Journal of Pharmacy and Pharmaceutical Sciences*, *6*(6), 357–361.

Manuel, N., Rafael, M., Miguel, A., & Castanho, R. B. (2009). Antimicrobial peptides: Linking partition, activity and high membrane–bound concentrations. *Nature Reviews. Microbiology*, *7*(3), 245–250. doi:10.1038/nrmicro2095 PMID:19219054

Marugg, J. D., Gonzalez, C. F., Kunka, B. S., Ledeboe, A. M., Pucci, M. J., Toonen, M. Y., ... Vandenbergh, P. A. (1992). Cloning, expression, and nucleotide sequence of genes involved in production of pediocin PA-1, and bacteriocin from *Pediococcus acidilactici* PAC1.0. *Applied and Environmental Microbiology*, *58*(8), 2360–2367. doi:10.1128/aem.58.8.2360-2367.1992 PMID:1514784

Mastro, T. D., Spika, J. S., Lozano, P., Appel, J., & Facklam, R. R. (1990). Vancomycin-resistant *Pediococcus acidilactici*: Nine cases of bacteremia. *The Journal of Infectious Diseases*, *161*(5), 956–960. doi:10.1093/infdis/161.5.956 PMID:2324545

McCaskey, M. K., & Masterson, L. R. (2016). Structural and Mechanistic Studies of Andropin, A Membrane-Selective Antimicrobial Peptide. *Biophysical Journal*, *110*(3), 414A. doi:10.1016/j.bpj.2015.11.2240

McCormick, J. K., Worobo, R. W., & Stiles, M. E. (1996). Expression of the antimicrobial peptide carnobacteriocin B2 by a signal peptide-dependent general secretory pathway. *Applied and Environmental Microbiology*, *62*(11), 4095–4099. doi:10.1128/aem.62.11.4095-4099.1996 PMID:8900000

McFarland, L. V. (2007). Meta–analysis of probiotics for the prevention of traveler's diarrhea. *Travel Medicine and Infectious Disease*, *5*(2), 97–105. doi:10.1016/j.tmaid.2005.10.003 PMID:17298915

Miller, K. W., Ray, P., Steinmetz, T., Hanekamp, T., & Ray, B. (2005). Gene organization and sequences of pediocin AcH/PA-1 production operons in *Pediococcus* and *Lactobacillus* plasmids. *Letters in Applied Microbiology*, *40*(1), 56–62. doi:10.1111/j.1472-765X.2004.01627.x PMID:15613003

Mills, S., Griffin, C., O'Connor, P. M., Serrano, L. M., Meijer, W. C., Hill, C., & Ross, R. P. (2017). A multibacteriocin cheese starter system, comprising nisin and lacticin 3147 in *Lactococcus lactis*, in combination with plantaricin from *Lactobacillus plantarum*. *Applied and Environmental Microbiology*, *83*(14), e00799–e17. doi:10.1128/AEM.00799-17 PMID:28476774

Mobolaji, O. A., & Wuraola, F. O. (2011). Assesment of the antimicrobial activity of lactic acid bacteria isolated from two fermented maize products-ogi and kunni-zaki. *Malaysian Journal of Microbiology*, *7*(3), 124–128.

Mokoena, M. P. (2017). Lactic acid bacteria and their bacteriocins: Classification, biosynthesis and applications against uropathogens: a mini-review. *Molecules (Basel, Switzerland)*, *22*(8), 1255. doi:10.3390/molecules22081255 PMID:28933759

Muhammad, S. A., Ali, A., Naz, A., Hassan, A., Riaz, N., Saeed-ul-Hassan, S., Andleeb, S., & Barh, D. (2016). A new broad-spectrum peptide antibiotic produced by *Bacillus brevis* strain MH9 isolated from Margalla Hills of Islamabad, Pakistan. *International Journal of Peptide Research and Therapeutics*, *22*(2), 271–279. doi:10.100710989-015-9508-2

Mundt, J. O., Bealtie, W. G., & Wieland, F. R. (1969). *Pediococci* residing on plants. *Journal of Bacteriology*, *98*(3), 938–942. doi:10.1128/jb.98.3.938-942.1969 PMID:5788719

Mylonakis, E., Podsiadlowski, L., Muhammed, M., & Vilcinskas, A. (2016). Diversity, evolution and medical applications of insect antimicrobial peptides. *Philosophical Transactions of the Royal Society of London. Series B, Biological Sciences*, *371*(1695), 20150290. doi:10.1098/rstb.2015.0290 PMID:27160593

Naderi, A., Kasra-Kermanshahi, R., Gharavi, S., Imani Fooladi, A. A., Abdollahpour Alitappeh, M., & Saffarian, P. (2014). Study of antagonistic effects of *Lactobacillus* strains as probiotics on multi drug resistant (MDR) bacteria isolated from urinary tract infections (UTIs). *Iranian Journal of Basic Medical Sciences.*, *17*(3), 201–208. PMID:24847423

Nes, I. F., Bao Diep, D., Havarstein, L. S., Brurberg, M. B., Eijsink, V., & Holo, H. (1996). Biosynthesis of bacteriocins of lactic acid bacteria. *Antonie van Leeuwenhoek, 70,* 113–128. doi:10.1007/BF00395929 PMID:8879403

Nes, I. F., Kjos, M., & Diep, D. B. (2012). Antimicrobial Components of lactic acid bacteria. In S. Lahtinen, A. C. Ouwehand, S. Salminen, & A. Wright (Eds.), *Lactic Acid Bacteria Microbiological and Functional Aspects* (pp. 285–311). CRC Press.

Nordstrom, R., & Malmsten, M. (2017). Delivery systems for antimicrobial peptides. *Advances in Colloid and Interface Science, 242,* 17–34. doi:10.1016/j.cis.2017.01.005 PMID:28159168

Olaoye, O. A., & Onilude, A. A. (2011). Quantitative estimation of antimicrobials produced by lactic acid bacteria isolated from Nigerian beef. *International Food Research Journal, 18*(3), 1155–1161.

Ooi, H. Y., Tee, W. Y., Davamani, F., & Nagendrababu, V. (2019). Comparing the antimicrobial efficacy of pediocin with chlorhexidine and calcium hydroxide as intracanal medicaments against persistent root canal infections. *Journal of Conservative Dentistry, 22*(3), 241–244. doi:10.4103/JCD.JCD_521_18 PMID:31367106

Osmanagaoglou, O., Gunduz, U., Beyatli, Y., & Cokmus, C. (1998). Purification and characterization of pediocin F, a bactericin produced by *Pediococcus acidilactici* F. *Turkish Journal of Biology, 22,* 217–228.

Osmanagaoglu, O., Beyatli, Y., & Gunduz, U. (2000). Cloning and expression of a plasmid–linked pediocin determinant trait of *Pediococcus acidilactici* F. *Journal of Basic Microbiology, 40*(1), 41–49. doi:10.1002/(SICI)1521-4028(200002)40:1<41::AID-JOBM41>3.0.CO;2-# PMID:10746198

Pandey, N., Malik, R. K., Kaushik, J. K., & Singroha, G. (2013). Gassericin A: A circular bacteriocin produced by Lactic acid bacteria *Lactobacillus gasseri. World Journal of Microbiology & Biotechnology, 29*(11), 1977–1987. doi:10.100711274-013-1368-3 PMID:23712477

Panteleev, P. V., Balandin, S. V., Ivanov, V. T., & Ovchinnikova, T. V. (2017). A therapeutic potential of animal β-hairpin antimicrobial peptides. *Current Medicinal Chemistry, 24*(17), 1724–1746. doi:10.2174/0929867324666170424124416 PMID:28440185

Papagianni, M. (2012). Metabolic engineering of lactic acid bacteria for the production of industrially important compounds. *Computational and Structural Biotechnology Journal, 3*(4), e201210003. doi:10.5936/csbj.201210003 PMID:24688663

Papp-Wallace, K. M., Endimiani, A., Taracila, M. A., & Bonomo, R. A. (2011). Carbapenems: Past, present, and future. *Antimicrobial Agents and Chemotherapy*, *55*(11), 4943–4960. doi:10.1128/AAC.00296-11 PMID:21859938

Pederson, C. S. (1949). The genus *Pediococcus. Bacteriological Reviews*, *13*(4), 225–232. doi:10.1128/br.13.4.225-232.1949 PMID:15407471

Perez, R. H., Ishibashi, N., Inoue, T., Himeno, K., Masuda, Y., Sawa, N., Zendo, T., Wilaipun, P., Leelawatcharamas, V., Nakayama, J., & Sonomoto, K. (2016). Functional analysis of genes involved in the biosynthesis of enterocin NKR-5-3B, a novel circular bacteriocin. *Journal of Bacteriology*, *198*(2), 291–300. doi:10.1128/JB.00692-15 PMID:26503847

Pothoulakis, C., Kelly, C. P., Joshi, M. A., Gao, N., O'Keane, C. J., Castagliuolo, I., & Lamont, J. T. (1993). *Saccharomyces boulardii* inhibits Clostridium difficile toxin A binding and enterotoxicity in rat ileum. *Gastroenterology*, *104*(4), 1108–1115. doi:10.1016/0016-5085(93)90280-P PMID:8462799

Price, D. P., Schilkey, F. D., Ulanov, A., & Hansen, I. A. (2015). Small mosquitoes, large implications: Crowding and starvation affects gene expression and nutrient accumulation in *Aedes aegypti. Parasites & Vectors*, *8*(1), 252. doi:10.118613071-015-0863-9 PMID:25924822

Pucci, M. J., Vedamuthu, E. R., Kunka, B. S., & Vandenbergh, P. A. (1988). Inhibition of Listeria monocytogenes by using bacteriocin PA-1 produced by *Pediococcus acidilactici* PAC 1.0. *Applied and Environmental Microbiology*, *54*(10), 2349–2353. doi:10.1128/aem.54.10.2349-2353.1988 PMID:3144243

Rattanachaikunsopon, P., & Phumkhachorn, P. (2010). Lactic acid bacteria: Their antimicrobial compounds and their uses in food production. *Annals of Biological Research*, *1*(4), 218–228.

Ray, B., & Daeschel, M. A. (1994). *Natural antimicrobial systems and food preservation*.

Ray, B., Motlagh, A. M., Johnson, M. C., & Bozoglu, F. (1992). Mapping of pSMB74, a plasmid encoding bacteriocin AcH production (Pap+) trait in *Pediococcus acidilactici* H. *Letters in Applied Microbiology*, *15*(2), 35–37. doi:10.1111/j.1472-765X.1992.tb00718.x

Reinhardt, A., & Neundorf, I. (2016). Design and application of antimicrobial peptide conjugates. *International Journal of Molecular Sciences*, *17*(5), 701. doi:10.3390/ijms17050701 PMID:27187357

Rodriguez, J. M., Cintas, L. M., Casaus, P., Martinez, M. I., Suarez, A., & Hernandez, P. E. (1997). Detection of pediocin PA-1 producing pediococci by rapid molecular producing by rapid molecular biology techniques. *Food Microbiology*, *14*, 363–371. doi:10.1006/fmic.1996.0084

Rojas, E., Llinas, P., Rodríguez-Romero, A., Hernández, C., Linares, M., Zenteno, E., & Lascurain, R. (2001). Hevein, an allergenic lectin from rubber latex, activates human neutrophils' oxidative burst. *Glycoconjugate Journal*, *18*(4), 339–345. doi:10.1023/A:1013621316647 PMID:11788802

Ross-Gillespie, A., Weigert, M., Brown, S. P., & Kümmerli, R. (2014). Gallium-mediated siderophore quenching as an evolutionarily robust antibacterial treatment. *Evolution, Medicine, and Public Health*, *2014*(1), 18–29. doi:10.1093/emph/eou003 PMID:24480613

Sadiq, S., Imran, M., Habib, H., Shabbir, S., Ihsan, A., Zafar, Y., & Hafeez, F. Y. (2016). Potential of monolaurin based food-grade nano-micelles loaded with nisin Z for synergistic antimicrobial action against *Staphylococcus aureus*. *Lebensmittel-Wissenschaft + Technologie*, *71*, 227–233. doi:10.1016/j.lwt.2016.03.045

Sanlıbaba, P., & Gucer, Y. (2015). Antimicrobial activity of Lactic Acid Bacteria. *Journal of International Scientific Publications*, *3*, 451–457.

Sashihara, T., Dan, M., Kimura, H., Matsusaki, H., Sonomoto, K., & Ishizaki, A. (2000). The effect of osmotic stress on the production of nukacin ISK–1 from *Staphylococcus warneri* ISK–1. *Applied Microbiology and Biotechnology*, *56*(3–4), 496–501. PMID:11549026

Savelyeva, A., Ghavami, S., Davoodpour, P., Asoodeh, A., & Łos, M. J. (2014). An overview of Brevinin superfamily: structure, function and clinical perspectives. In *Anticancer Genes* (pp. 197–212). Springer. doi:10.1007/978-1-4471-6458-6_10

Schaal, J. B., Maretzky, T., Tran, D. Q., Tran, P. A., Tongaonkar, P., Blobel, C. P., Ouellette, A. J., & Selsted, M. E. (2018). Macrocyclic θ-defensins suppress tumor necrosis factor-α (TNF-α) shedding by inhibition of TNF-α converting enzyme. *The Journal of Biological Chemistry*, *293*(8), 2725–2734. doi:10.1074/jbc.RA117.000793 PMID:29317500

Schved, F., Lalazar, A., Henis, Y., & Juven, B. J. (1993). Purification, partial characterization and plasmid linkage of pediocin SJ-1, a bacteriocin produced by *Pediococcus acidilactici*. *The Journal of Applied Bacteriology*, *74*(1), 67–77. doi:10.1111/j.1365-2672.1993.tb02998.x PMID:8420920

Sharma, G., Dang, S., Gupta, S., & Gabrani, R. (2018). Antibacterial activity, cytotoxicity, and the mechanism of action of bacteriocin from *Bacillus subtilis* GAS101. *Medical Principles and Practice*, *27*(2), 186–192. doi:10.1159/000487306 PMID:29402863

Sheehan, G., Bergsson, G., McElvaney, N. G., Reeves, E. P., & Kavanagh, K. (2018). The human cathelicidin antimicrobial peptide LL-37 promotes the growth of the pulmonary pathogen *Aspergillus fumigatus*. *Infection and Immunity*, *86*(7), e00097. doi:10.1128/IAI.00097-18 PMID:29712727

Shin, M. S., Han, S. K., Ryu, J. S., Kim, K. S., & Lee, W. K. (2008). Isolation and partial characterization of a bactericin produced by *Pediococcus pentosaceus* K23–2 isolated from Kimchi. *Journal of Applied Microbiology*, *105*(2), 331–339. doi:10.1111/j.1365-2672.2008.03770.x PMID:18540969

Singh, D., Qi, R., Jordan, J. L., San Mateo, L., & Kao, C. C. (2013). The human antimicrobial peptide LL-37, but not the mouse ortholog, mCRAMP, can stimulate signaling by poly(I:C) through a FPRL1-dependent pathway. *The Journal of Biological Chemistry*, *288*(12), 8258–8268. doi:10.1074/jbc.M112.440883 PMID:23386607

Singh, R., Miriyala, S. S., Giri, L., Mitra, K., & Kareenhalli, V. V. (2017). Identification of unstructured model for subtilin production through *Bacillus subtilis* using hybrid genetic algorithm. *Process Biochemistry*, *60*, 1–12. doi:10.1016/j.procbio.2017.06.005

Sire, J. M., Donnio, P. Y., Mensard, R., Pouedras, P., & Avril, J. L. (1992). Septicemia and hepatic abscess caused by *Pediococcus acidilactici*. *European Journal of Clinical Microbiology & Infectious Diseases*, *11*(7), 623–625. doi:10.1007/BF01961670 PMID:1396771

Smith, J. L., & Palumbo, S. A. (1983). Use of starter cultures in meats. *Journal of Food Protection*, *46*(11), 997–1006. doi:10.4315/0362-028X-46.11.997 PMID:30921859

Soomro, A. H., Masud, T., & Anwaar, K. (2002). Role of Lactic Acid Bacteria (LAB) in Food Preservation and Human Health – A Review. *Pakistan Journal of Nutrition*, *1*(1), 20–24.

Strasser de Saad, A. M., & Manca de Nadra, M. C. (1993). Characterization of bacteriocin produced by *Pediococcus pentosaceus* from wine. *The Journal of Applied Bacteriology*, *74*(4), 406–410. doi:10.1111/j.1365-2672.1993.tb05146.x PMID:8486547

Strasser de Saad, A. M., Pasteris, S. E., & Manca de Nadra, M. C. (1995). Production and stability of pediocin N5p in grape juice medium. *The Journal of Applied Bacteriology*, *78*(5), 473–476. doi:10.1111/j.1365-2672.1995.tb03087.x PMID:7759379

Su, Z., Leitch, J. J., Abbasi, F., Faragher, R. J., Schwan, A. L., & Lipkowski, J. (2018). EIS and PM-IRRAS studies of alamethicin ion channels in a tethered lipid bilayer. *Journal of Electroanalytical Chemistry (Lausanne, Switzerland)*, *812*, 213–220. doi:10.1016/j.jelechem.2017.12.039

Sugawara, E., Kojima, S., & Nikaido, H. (2016). *Klebsiella pneumoniae* major porins OmpK35 and OmpK36 allow more efficient diffusion of β-lactams than their *Escherichia coli* homologs OmpF and OmpC. *Journal of Bacteriology*, *198*(23), 3200–3208. doi:10.1128/JB.00590-16 PMID:27645385

Sun, P., Shi, A., Shen, C., Liu, Y., Wu, G., & Feng, J. (2020). Human salivary histatin-1 (Hst1) promotes bone morphogenetic protein 2 (BMP2)-induced osteogenesis and angiogenesis. *FEBS Open Bio*, *10*(8), 1503–1515. doi:10.1002/2211-5463.12906 PMID:32484586

Sun, T., Zhan, B., & Gao, Y. (2015). A novel cathelicidin from Bufo bufo gargarizans Cantor showed specific activity to its habitat bacteria. *Gene*, *571*(2), 172–177. doi:10.1016/j.gene.2015.06.034 PMID:26091834

Sunil, K., & Narayana, B. (2008). Spectrophotometric determination of hydrogen peroxide in water and cream samples. *Bulletin of Environmental Contamination and Toxicology*, *81*(4), 422–426. doi:10.100700128-008-9477-7 PMID:18594751

Tahir, H. M., Zaheer, A., Khan, A. A., & Abbas, M. (2018). Antibacterial potential of venom extracted from wolf spider, *Lycosa terrestris* (Araneae: Lycosiade). *The Indian Journal of Animal Sciences*, *52*(2), 286–290.

Thao, H. T., Lan, N. T. N., & Mau, C. H. (2017) Overexpression of VrPDF1 gene confers resistance to weevils in transgenic mung bean plant. *Peer J.* Preprints, 5:e3264v2.

Thennarasu, S., Lee, D. K., Poon, A., Kawulka, K. E., Vederas, J. C., & Ramamoorthy, A. (2005). Membrane permeabilization, orientation, and antimicrobial mechanism of subtilosin A. *Chemistry and Physics of Lipids*, *137*(1-2), 38–51. doi:10.1016/j.chemphyslip.2005.06.003 PMID:16095584

Thiyonila, B., Reneeta, N. P., Kannan, M., Shantkriti, S., & Krishnan, M. (2018). Dung beetle gut microbes: Diversity, metabolic and immunity related roles in host system. *International Journal of Scienctific Innovations*, *1*(2), 84–91.

Timmons, P. B., O'Flynn, D., Conlon, J. M., & Hewage, C. M. (2019). Structural and positional studies of the antimicrobial peptide brevinin-1BYa in membrane-mimetic environments. *Journal of Peptide Science*, *25*(11), e3208. doi:10.1002/psc.3208 PMID:31721374

Todorov, S. D., & Dicks, L. M. T. (2005a). Production of bacteriocin ST33LD, produced by *Leuconostoc mesenteroides* subsp. *mesenteroides*, as recorded in the presence of different medium components. *World Journal of Microbiology & Biotechnology*, *21*(8-9), 8–9. doi:10.100711274-005-8122-4

Todorov, S. D., & Dicks, L. M. T. (2005b). Pediocin ST18, an antilisterial bactericin produced by *Pedicoccus pentosacceus* ST18 isolated from boza, a traditional cereal beverage from Bulgaria. *Process Biochemistry*, *40*(1), 365–370. doi:10.1016/j.procbio.2004.01.011

Todorov, S. D., & Dicks, L. M. T. (2005c). *Lactobacillus plantarum* isolated from molasses produces bacteriocins active against Gram-negative bacteria. *Enzyme and Microbial Technology*, *36*(2-3), 318–326. doi:10.1016/j.enzmictec.2004.09.009

Tulini, F. L., Lohans, C. T., Bordon, K. C., Zheng, J., Arantes, E. C., Vederas, J. C., & De Martinis, E. C. (2014). Purification and characterization of antimicrobial peptides from fish isolate *Carnobacterium maltaromaticum* C2: Carnobacteriocin X and carnolysins A1 and A2. *International Journal of Food Microbiology*, *173*, 81–88. doi:10.1016/j.ijfoodmicro.2013.12.019 PMID:24412962

Upadhyay, R. K. (2018). Spider venom toxins, its purification, solubilization, and antimicrobial activity. *International Journal of Green Pharmacy*, *12*(1), S200–S2014.

Vaara, M. (2009). New approaches in peptide antibiotics. *Current Opinion in Pharmacology*, *9*(5), 571–576. doi:10.1016/j.coph.2009.08.002 PMID:19734091

Venema, K., Kok, J., Marugg, J. D., Toonen, M. Y., Ledeboer, A. M., Venema, G., & Chikindas, M. L. (1995). Functional analysis of the pediocin operon of *Pediococcus acidilactici* PAC 1.0: PedB is the immunity protein and PedD is the precursor processing enzyme. *Molecular Microbiology*, *17*(3), 515–522. doi:10.1111/j.1365-2958.1995.mmi_17030515.x PMID:8559070

Venturina, D. H., Villegas, L. C., Perez, M. T. M., & Elegado, F. B. (2016). Isolation and identification of subtilosin A-producing *Bacillus subtilis* from mongo sprouts, silage and soil samples in the Philippines. *Asia Life Sciences*, *25*(1), 123–136.

Vogel, H., Badapanda, C., Knorr, E., & Vilcinskas, A. (2014). RNA-sequencing analysis reveals abundant developmental stage-specific and immunity-related genes in the pollen beetle *Meligethes aeneus*. *Insect Molecular Biology*, *23*(1), 98–112. doi:10.1111/imb.12067 PMID:24252113

Waghu, F. H., Barai, R. S., Gurung, P., & Idicula-Thomas, S. (2016). CAMPR3: A database on sequences, structures and signatures of antimicrobial peptides. *Nucleic Acids Research*, *44*(D1), D1094–D1097. doi:10.1093/nar/gkv1051 PMID:26467475

Wan, X., Li, R., Saris, P. E., & Takala, T. M. (2013). Genetic characterisation and heterologous expression of leucocin C, a class IIa bacteriocin from *Leuconostoc carnosum* 4010. *Applied Microbiology and Biotechnology*, *97*(8), 3509–3518. doi:10.100700253-012-4406-4 PMID:23053070

Wang, G. (2020). The antimicrobial peptide database provides a platform for decoding the design principles of naturally occurring antimicrobial peptides. *Protein Science*, *29*(1), 8–18. doi:10.1002/pro.3702 PMID:31361941

Wang, Y., Shang, N., Qin, Y., Zhang, Y., Zhang, J., & Li, P. (2018). The complete genome sequence of *Lactobacillus plantarum* LPL-1, a novel antibacterial probiotic producing class IIa bacteriocin. *Journal of Biotechnology*, *266*, 84–88. doi:10.1016/j.jbiotec.2017.12.006 PMID:29229543

WHO. (2017). *WHO News Releases*. Retrieved from https://www.who.int/

WHO. (2018). *Rapid Communication: Key changes to treatment of multidrug- and rifampicin-resistant tuberculosis (MDR/RR-TB)*. Retrieved from https://www.who.int/

Wu, C. W., Yin, L. J., & Jiang, S. T. (2004). Purification and characterization of bacteriocin from *Pediococcus pentosaceus* ACCEL. *Journal of Agricultural and Food Chemistry*, *52*(5), 1146–1151. doi:10.1021/jf035100d PMID:14995112

Xu, D., Zhao, S., Dou, J., Xu, X., Zhi, Y., & Wen, L. (2021). Engineered endolysin-based "artilysins" for controlling the gram-negative pathogen *Helicobacter pylori*. *AMB Express*, *11*(1), 63. doi:10.118613568-021-01222-8 PMID:33913058

Yang, E., Fan, L., Jiang, Y., Doucette, C., & Fillmore, S. (2012). Antimicrobial activity of bacteriocin-producing lactic acid bacteria isolated from cheeses and yogurts. *AMB Express*, *2*(48), 1–12. doi:10.1186/2191-0855-2-48 PMID:22963659

Yang, Y. T., Lee, M. R., Lee, S. J., Kim, S., Nai, Y. S., & Kim, J. S. (2017). Tenebrio molitor Gram-negative-binding protein 3 (TmGNBP3) is essential for inducing downstream antifungal Tenecin 1 gene expression against infection with *Beauveria bassiana* JEF-007. *Insect Science*, *25*(6), 969–977. doi:10.1111/1744-7917.12482 PMID:28544681

Yang, Z. (2000) *Antimicrobial compounds and extracellular polysaaharides produced by lactic acid bacteria: Structures and properties* (Unpublished master's thesis). Faculty of Agriculture and Forestry, Department of Food Technology University, University of Helsinki.

Yin, L. J., Wu, C. W., & Jiang, S. T. (2003). Bacteriocins from *Pediococcus pentosaceus* L and S from pork meat. *Journal of Agricultural and Food Chemistry, 51*(4), 1071–1076. doi:10.1021/jf025838f PMID:12568574

Young-Speirs, M., Drouin, D., Cavalcante, P. A., Barkema, H. W., & Cobo, E. R. (2018). Host defense cathelicidins in cattle: Types, production, bioactive functions and potential therapeutic and diagnostic applications. *International Journal of Antimicrobial Agents, 51*(6), 813–821. doi:10.1016/j.ijantimicag.2018.02.006 PMID:29476808

Zacharof, M. P., & Lovitt, R. W. (2012). Bacteriocins produced by lactic acid bacteria. *APCBEE Procedia, 2*, 50–56. doi:10.1016/j.apcbee.2012.06.010

Zalan, Z., Nemeth, E., Barath, A., & Halasz, A. (2005). Influence of growth medium on hydrogen peroxide and bactteriocin production of *Lactobacillus* strains. *Food Technology and Biotechnology, 43*(3), 219–225.

Zasloff, M. (2002). Antimicrobial peptides of multicellular organisms. *Nature, 415*(6870), 389–395. doi:10.1038/415389a PMID:11807545

Zhang, L.-J., & Gallo, R. L. (2016). Antimicrobial peptides. *Current Biology, 26*(1), PR14–R19. doi:10.1016/j.cub.2015.11.017 PMID:26766224

Zhao, N., Pan, Y., Cheng, Z., & Liu, H. (2016). Lasso peptide, a highly stable structure and designable multifunctional backbone. *Amino Acids, 48*(6), 1347–1356. doi:10.100700726-016-2228-x PMID:27074719

# Chapter 10
# Nanomaterials to Overcome Emergence and Re-Emergence of Superbugs:
## Nanoarsenals for Superbugs

**Manju Nagpal**
*Chitkara University, India*

**Malkiet Kaur**
*Chitkara University, India*

**Manjinder Singh**
*Chitkara University, India*

**Geeta Aggarwal**
*Delhi Pharmaceutical Sciences and Research University (DPSRU), India*

## ABSTRACT

*Antimicrobial resistance remains a substantial global health concern, invigorating the critical need for alternate therapeutic options to combat chronic intracellular infections and biofilms so as to shorten the hospital stays, and hence mortality. Nanomaterials have been developed as delivery carriers for antibiotics to improve their penetration through these biofilms. Nanoformulations of existing antibiotics has led to enhanced bioavailability and site specificity. Moreover, diagnosis of infections using efficient nanosensors or probes may speed up the treatment process at earlier stages of infection.*

DOI: 10.4018/978-1-7998-0307-2.ch010

# 1. INTRODUCTION

## 1.1 Concept of Emergence and Re-Emergence of Superbugs

Approximately 70 years ago, antibiotics were introduced to cure infectious diseases. These are the drug molecules used to kill bacteria and they reduce the risk associated with various infections. Antibiotics exhibited their action by inhibiting the proliferation process of bacteria. Overtime, antibiotics started showing resistance to different bacteria's called superbugs (Drug-resistant microbes). Antibiotic resistance is the tolerance developed by bacteria to overcome the effects of antibiotics (https://www.davolterra.com/content/antibiotic-resistance-wonder-drugs-facing-rise-superbugs). Antibiotic resistance increases the duration of infection, cost of treatment and decreased success of treatment which finally led to economic loss (Zaman *et al.*, 2017). Drug resistance does not just prevail in healthcare environments but it is also increasing amongst community-acquired pathogens (Padhy *et al.*, 2016). For example, penicillin was the first antibiotic developed and was successfully used to control infections. Just after four years, it causes resistance to major microbe *i.e. Staphylococcus aureus* (https://www.statnews.com/2016/09/12/superbug-antibiotic-resistance-history/, Ventola, 2015). Also, *Klebsiella pneumonia* causes fatal and untreatable infections in a healthy population. It may cause a danger to life of individual (Gu *et al.*, 2018). Some other examples of antibacterial discovery with their resistance are given in Figure 1.

World Health Organization (WHO) has already warned about the emergence of infectious diseases at a faster rate which has never seen before (https://www.who.int/newsroom/factsheets/detail/antibiotic-resistance). Several human activities have led to the emergence and spread of new diseases such as moving into wildlife habitats, improvisation in agriculture techniques, modern transport and misuse of antibiotics (Bloom *et al.*, 2017). Re-emerging infectious diseases are those that once were foremost health problems across the globe and declined considerably, but they again became health problems for a significant population (https://www.sciencedaily.com/releases/2017/08/170831101508.html). Some of the re-emerging diseases with a causative agent are given in Table 1. The re-occurrence of old infections along with emergence of new infectious diseases and constant persistence of various intractable infections is a major challenge to the researchers.

One of the reasons for antibiotic resistance is a development of biofilms. Most of the bacterial strains occur in the form of a biofilm which are especially microbial aggregates that rely on a solid surface and extracellular products such as extracellular polymeric substances (EPS's). Bacteria tend to move reversibly on the surface, but the expression of EPSs makes this attachment irreversible. After settlement of bacteria, the synthesis of bacterial flagellum is suppressed and rapid multiplication

*Figure 1. History of antibiotic discovery*

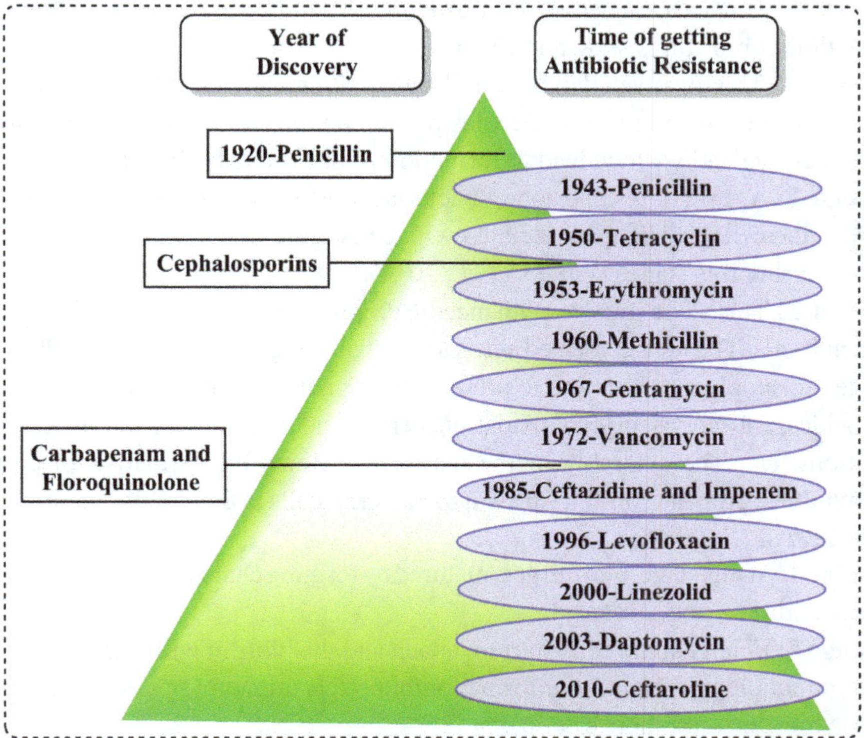


of bacteria results in the formation of a mature biofilm. They all are in aggregates at this stage and forms a barrier for antibiotic action which led to systemic chronic infections. Bacteria within biofilms develop superantigens that tend to evade the immune system. Thus, the formation of biofilms is a serious threat to human health and makes infection control more challenging (Frieri *et al.,* 2017, Von Wintersdorff *et al.*, 2016). These biofilms are embedded into extracellular polymeric substances (EPS) matrix. The EPS is composed of phospholipids, nucleic acids, teichoic acid, exopolysaccharides and extracellular proteins. In addition, mineral crystals, silt, milk residues and blood parts or dirt may be present in EPS matrix, depending upon the location and conditions under which biofilm is formed (Ramos *et al.*, 2018). The biofilm tends to adhere irreversibly to organic phenomenon or biotic surfaces and 80% of pathogenic infections remain persistent due to this biofilm. For example, *P. aeruginosa* (linked to cystic fibrosis) and *Staphylococcus aeruginosa* (accountable for wound infections) are traditional examples of persistent pathogens that lead to formation of biofilm (Algburi *et al.*, 2016). The formation of micron size biofilms occurs around solid surfaces in contact with water, like living tissues and water bodies/systems (Clatworthy *et al.*, 2007). The development of a biofilm involves various steps i.e. adhesion, growth, and production of EPS matrix. A cycle indicating the 5 stages is discussed below (Ramos *et al.*, 2018) (Figure 2) and strategies for prevention of biofilm is shown in Figure 3.

**Stage 1**: This involves deposition of bacterial cells when the microbes arrive to site of adhesion and macromolecules present on that site act as the substrate for these microbial cells which lead to formation of biofilm (Myszka *et al.*, 2011).

**Stage 2**: A reversible adhesion of bacterial cells is followed by communication between the cells which are associated with cells of next stage. The EPS matrix is formed during this stage (Wilkins *et al.*, 2014).

**Stage 3**: This stage involves enhanced EPS production that ends up in a rise of bacterial cells. The matrix formed act as a reservoir for genetic material to alter gene transfer, to supply nutrients and to protect against adverse conditions like exposure to biocides, antibiotics, drying, oxidation, certain metallic cations, ultraviolet (UV) radiations, *etc*. The assembly of the matrix represents the formation of biofilm communities, propagation and survival of bacterial cells in their native environment (O'toole *et al.*, 2000).

**Stage 4:** At this stage, biofilm gets matured by forming characteristic "mushroom" structures due to polysaccharides.

**Stage 5:** Finally some cells start to detach and the biofilm will disperse (https://www.immunology.org/public-information/bitesized-immunology/pathogens-and-disease/biofilms-and-their-role-in).

*Figure 2. Various stages of biofilm formation*

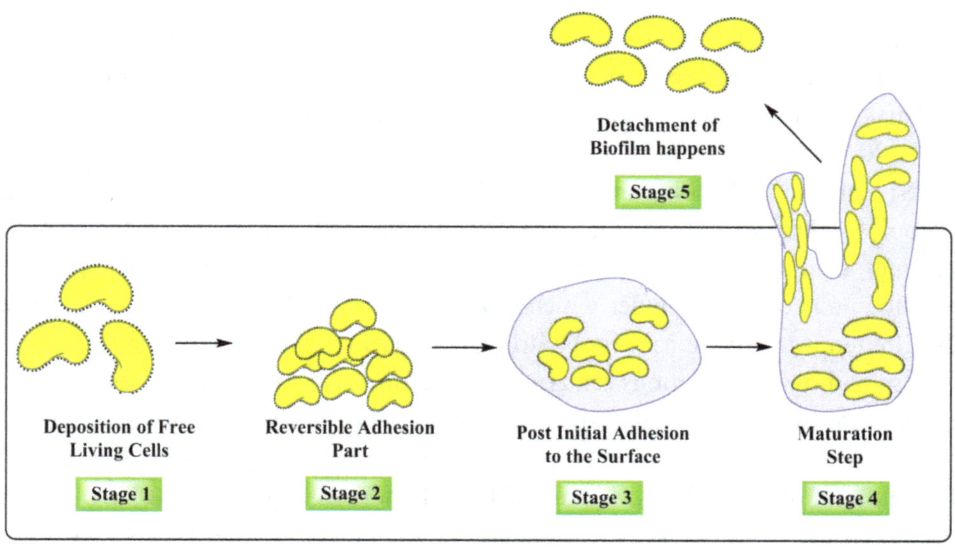

*Figure 3. Strategies for the prevention of biofilm*

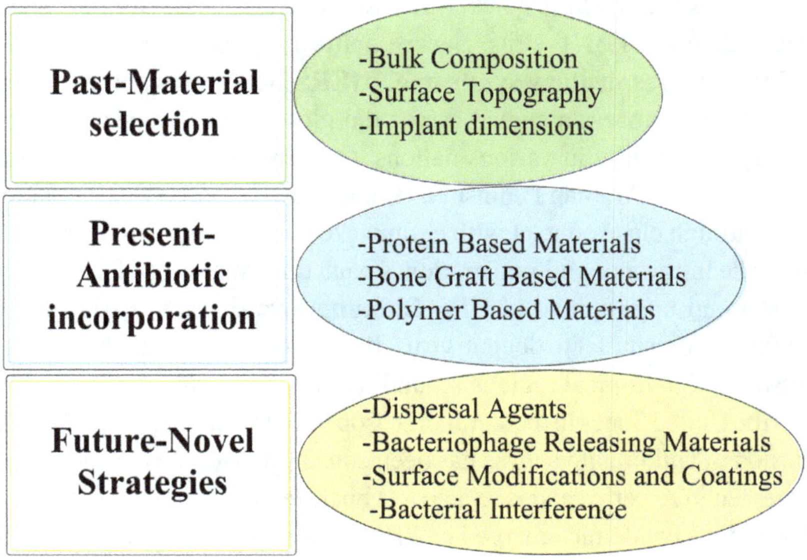


*Table 1. Various Re-emerging diseases with causative agent*

| Disease | Infectious agent |
|---|---|
| Cryptosporidiosis | *Cryptosporidium pavum* (protozoa) |
| Malaria | *Plasmodium* species (protozoon) |
| Diphtheria | *Cornebacterium diptheriae* (bacteria) |

## 1.2 History of Various Infectious Diseases

**Influenza (flu)** is an example of widely spreading disorder that is due to natural and human factor both. The influenza virus is known for its ability to alternate its genetic information. The possibility of huge genetic changes (which are going on and surpassed into people) is increased when human beings are in close contact with agricultural animals such as chickens, pigs and ducks. These animals act as permanent host to influenza virus and can create a new form of influenza virus that has no existence before. Large modifications in influenza virus can be the reason for pandemics since human immune system is not expected to recognize and protect against the new variant strain. **Avian H5N1 influenza (or chicken flu)**, has been confined to rare situations of contamination in humans who came into direct contact with contaminated birds. It is a deadly virus and may cause more than half deaths from the number of infected people. In contrast, **H1N1 influenza virus** (Swine flu) spreads from swine (pigs) and gets transmitted easily. The spread of H1N1 virus in human across the world is quicker than any other viruses. But fortunately, it is less deadly than the H5N1 virus. **Severe acute respiratory syndrome (SARS)** and **Middle East respiratory syndrome (MERS)** represents another example of infectious diseases that can spread across the globe. SARS appeared in China in 2002 and spread rapidly in various nations. Occurrence of MERS in 2012 is due to a related strain in Arabian Peninsula. Humans are supposed to get contaminated with **HIV** through close contact with chimpanzees during hunting in isolated areas of Africa. The intravenous administration, sexual transmission and transfer of blood led to a fast and huge spread of HIV. **Chikungunya disease** is triggered via the chikungunya virus, similar to dengue virus. It was transmitted *via* the tiger mosquito. It was restrained to tropical regions around the Indian Ocean in the past. The virus arrived in the United States in the summer season of 2014, although therefore, the local transmission of chikungunya virus has been constrained to Florida and Texas. **Zika virus** emerged in America and is related to a birth defect identified as microcephaly. The **Ebola virus** pandemic emerged in 2014 in West Africa region. At initial stage, the virus affected only small population (hundreds) but fortunately there was a rapid increase of virus which affected a large population (thousands). Worst outbreak of Ebola in world which had never seen before was due to several factors including high population, closer contact with wild animals, increased travel and weak health care systems. (https://www.bcm.edu/departments/molecularvirologyandmicrobiology/emerginginfectionsandbiodefense/emerginginfectious-diseases). Coronavirus disease (COVID-19) is an infectious disease caused by a newly discovered coronavirus. People infected with COVID-19 virus will experience mild to moderate respiratory illness and recover without requiring special treatment. Older people and those who are suffering from medical problems such as cardiovascular disease, diabetes, chronic respiratory disease and cancer are more likely to develop serious illness. This virus spreads primarily through droplets of saliva or discharge from the nose when an infected person coughs or sneezes (https://www.who.int/health-topics/coronavirus#tab=tab_1). Table 2 summarizes various emerging infectious diseases with their causative agents.

## 1.3 Challenges for Antibacterial Therapy

Superbugs are the biggest threat these days due to development of antibiotic resistance caused by formation of biofilm. Globally, the situation is getting worse as approximately seven lakh deaths occur annually from these drug-resistant infections (Zaman *et al.*, 2017). Therefore, there is an urgent need to control the emergence and re-emergence of superbugs. The alarming situation can be improved by use of antibiotics/anti-microbial, early diagnosis of superbugs, improved sanitation and

*Table 2. Various emerging infectious diseases with infectious agent*

| Year recognized | Disease | Infectious agent |
|---|---|---|
| 1967 | Murburg hemorrhagic | Marburg virus |
| Prior to1976 | Salmonellosis | *S. entertidis* |
| 1976 | Ebola hemorrhagic fever | Ebola virus |
| 1983 | AIDS, gastric ulcers | HIV, *Helicobacter pylori* |
| 1989 | Hepatitis C | Hepatitis C virus(HCV) |
| 1998 | Nipah encephalitis | *Nipah encephalitis* |
| 2002 | VRSA infection | Vancomycin resistant *S. aureus* |
| 2003 | SARS (severe acute respiratory syndrome) | SARS-associated coronavirus |
| 2015 | Zika | *Zika* virus |

hygiene, reduce pollution and increase global surveillance of superbugs (Chen, 2013). The proper use of vaccines and the development of new antibiotics can also reduce antibiotic resistance by preventing infections in its first stage. Some regulations and policies must be practised in every country to stop unnecessary drug promotions. Some measures need to be taken to follow WHO guidelines for minimizing the dual use of antimicrobials related to human beings and food products (Padhy *et al.*, 2016). There are some diagnostic techniques and devices that can identify the presence of superbugs at its earlier stage (Frieri, 2017). Development of nanomaterials for diagnostic and therapeutic application in antibacterial therapy has come out with a wide area of research. Various nanomaterials for diagnostic and preventive applications are discussed in the following sections.

### 1.3.1 Nanomaterials for Diagnosing Superbugs

Nanomaterials are the new tools to fight against superbugs. Various nanosystems have been worked out to control and prevent various infections (Blecher *et al.*, 2011). Diagnosis of infections *via* economical nanosensors or probes fastens the treatment at early stages (Okeke *et al.*, 2011, Zhu *et al.*, 2014). Researchers have developed diagnostic techniques to examine bacterial infections and even other superbugs such as tissue culture, microscopy, lateral flora immunoassays (called immune chromatographic test (ICT), enzyme-linked immunosorbent assay (ELISA) and various biochemical tests (Qasim *et al.*, 2014, Zhu *et al.*, 2014). These techniques are costly and time- consuming. These are mainly utilized in developed countries but are often poorly fitted to developing countries in which infectious diseases are chief causes of death (Cars *et al.*, 2011, Torres *et al.*, 2016). Figure 3 depicts various nanotechnology-based diagnostic techniques used for the early detection of superbugs. A faster diagnostic test *i.e.* In-Dex has been developed to detect dangerous bacteria's like *E. coli, Staphylococcus* and other superbugs. Results may be produced in two hours using a blood sample, spit, wound, urine, stool samples or cerebral spine fluid samples etc. (https://www.hindustantimes.com/health-and-fitness/scientists-develop-faster-method-to-detect-harmful-bacteria-likesuperbugs/storyILaxy8D14lhzuxKt0COcEP.html). Solution- based circuit chip (SSC) for synchronous detection of resistant and non- resistant *E.coli* strains responsible for urinary tract infections has been designed as an another approach. These SSCs were designed by applying a sequence of lithographic steps followed by electrochemical deposition that further create 3-D nanoscopic morphology on microsensors. Further, these microsensors were functionalized via peptide nucleic acid (PNA) probe to target a specific region of pathogen. This multiplexed technique is able to distinguish between bacterial strains such as *E.coli* and *S. aureus* (Lam *et al.*, 2013). Similarly, for the diagnosis of MDR-TB, a novel NanoELIwell device has been developed. The principle of the technique is a combination of *mycobacteria* antigen immunoassay and microwell technology. It is capable of performing quick identification and detection of antigens released by resistant *mycobacteria* (Nguyen *et al.*, 2012). Also, the nanoparticles with better fluorescent property led to improved sensitivity in diagnostic molecular bioimaging technique. Fluorescent silica nanoparticles (FSNPs) are developed to detect *Mycobacterium tuberculosis* (MTB). FSNPs entrap two metallic fluorescent organic dyes *i.e.* Tris (2,2-bipyridyl) osmium bis(hexafluorophosphate) (OsBpy) and Tris (bipyridine) ruthenium(II) dichloride (RuBpy), that are excited by a single wavelength. These FSNPs displayed photostability and high signal amplification. MTB is detected using an anti-MTB primary antibody and a second antibody labelled with FSNPs for better detection of anti-MTB antibody. Using this system, MTB can be detected in a mixture of bacteria and sputum with high sensitivity and in a short time of four hours (Qin *et al.*, 2007). Europium [Eu (III)] polymeric nanoparticles are designed for easy detection of anthrax antibodies by applying fluorescence enzyme- linked immunosorbent assay (ELISA) (Qin *et al.*, 2007). Table 4 depicts various diagnostic tests to detect infectious diseases along with their limitations.

### 1.3.2 Use of Nanomaterials for Prevention Against Superbugs

After diagnosing the specific superbug, the next approach is to reduce the effects of the superbug in the body. Various nanoparticulate carriers of anti-microbial/anti-bacterial compounds have been developed and being used to fight against superbugs or to overcome antibacterial- resistance. The nanoparticulate system has proved it more fatal for microbes and has gained interest due to enhanced bioavailability, penetration ability, effectiveness and site-specific targeting.

*Figure 4. Nanotechnology based approaches for early detection of superbugs*

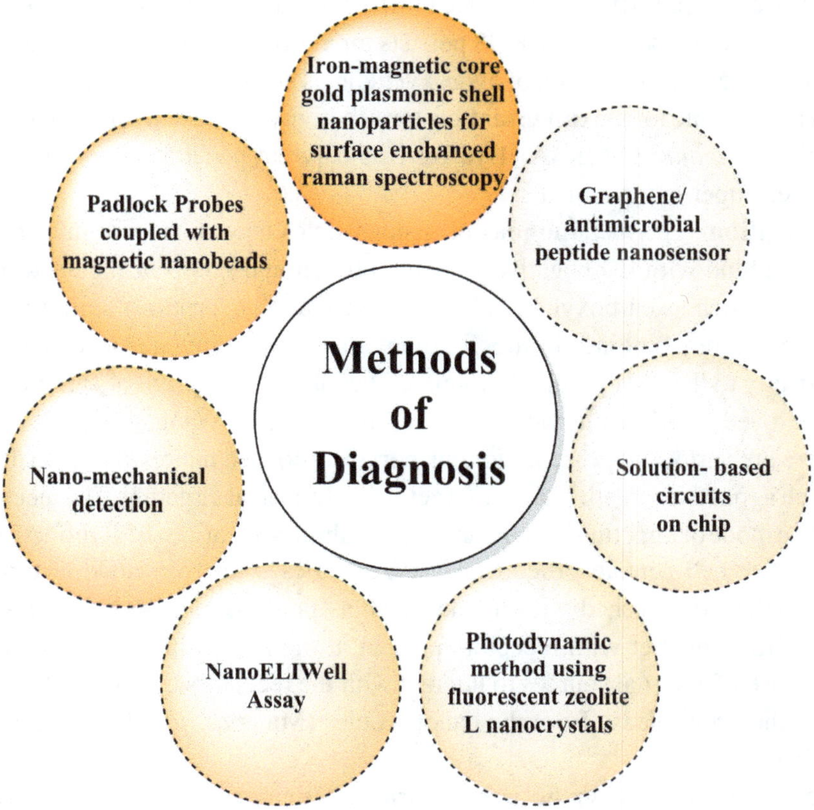


*Table 3. Available diagnostic tests for infectious diseases and their limitations*

| Disease | Diagnostic test | Limitations |
|---|---|---|
| Acute respiratory infection (ARIs) | Blood and sputum culture | • High cost and time consuming method, dependence of results on stringent transport conditions to maintain specimen.<br>• Require sufficient reagents, equipment and well trained staff. |
| Acquired immunodeficiency syndrome (AIDS) | Serology; detection by PCR amplification of LTR region of gag gene, env gene and pol gene | • High rate of false positives and negatives.<br>• PCR cannot be done during HIV incubation period.<br>• Nucleic acid amplification technologies (NAAT) such as PCR are costly and lead to false positives because of contamination. |
| Diarrheal diseases | Microscopy stool culture | • Requirement of trained technologists.<br>• Expensive and time consuming method. |
| Malaria | Blood film, detection of antigens (dipstick) and detection of antibodies (ELISA) | • High false positive rates. |
| Tuberculosis (TB) | Sputum microscopy; tuber skin test ; pathogen solid and liquid culture ; PCR- based Gene Xpert | • Need of expert staff<br>• Occurrence of false positives during antibody tests, tuber skin test (tuberculin) because of allergic reactions. |
| Visceral leishmaniasis | Serological field test; Direct agglutination test (DAT); microscopy / culture of spleen or bone marrow cells. | • Bone marrow sampling is painful and invasive. |

The Penetrating mechanism of nanomaterials is divided into two parts-

**Diffusion-** Nanomaterials generate reactive oxygen species (ROS) into bacteria by using diffusion mechanism. ROS persists for a sufficient and long time to diffuse into bacterial cells. Generation of hydroxyl radicals and their diffusion into bacterial cells is responsible for maximum antibacterial activity against methicillin resistance *Staphylococcus aureus* (MRSA). Hydroxyl radicals may penetrate the cell membrane easily then super oxide radicals.

**Absorption-** The metallic ions of nanoparticles tend to release into the outer media and bind with the negatively charged functional group of the bacterial cell membrane, such as carboxyl and phosphate groups (biosorption). Different metal ions have different binding abilities for their targets for example, zinc ions bind with high affinity to the –SH group of proteins. The inherent structure of cell membrane gets destroyed which further lead to bacterial cell death (Wang *et al.*, 2017).

There are various approaches to transport drug-loaded nanoparticles to diseases sites such as passive targeting, active targeting and physical targeting. The mechanism of passive mode of targeting is to enhance permeability-retention (EPR) effect through which tumor cell gets absorbed to nanosize bodies. In active mode of targeting, nanoparticles are attached with ligands such as proteins, peptides and antibodies which further interact with receptors present at the target site. Physical targeting includes use of external sources to interact with the receptors on target site and also controls the drug release from the nanoparticles (Ma *et al.*, 2013; Egusquiaguirre *et al.*, 2012).

The nanocarriers are synthesized by various techniques which produce sufficient consecutive antimicrobial actions. These can be organic, inorganic, hybrid NPs. The nanomaterials having highest antimicrobial potential are inorganic nanoparticles like gold, silver and alone/ mixed used with a variety of natural polymers. NPs are regularly developed and used as drug carriers and they can directly supply chemotherapeutics to the tumor tissues by reducing the toxicities associated with other organs (Beyth *et al.*, 2015). Several metal oxides, metals, bimetals and metal halides in nanoparticulate forms have been reported for antimicrobial activity because the microbes are less prone to build up resistance for nanomaterials. These comprises of Gold (Au), Copper (Cu), Silver (Ag), Zinc (Zn), Magnesium (Mg), Titanium (Ti), Cerium (Ce), Nickel (Ni), Aluminium (Al), Selenium (Se), Yttrium (Y), Cadmium (Cd), Palladium (Pd) and super-paramagnetic Iron (Fe) (Ma *et al.*,2013). The nano carriers should be stable, non-immunogenic, biodegradable, easy to manufacture, cost efficient and should release their drugs solely at the target site. Regular nano-platform carriers are liposomes, polymeric nanoparticles, solid lipid nanoparticles, metal/metal oxide nanoparticles and dendrimers *etc.* (Chan *et al.*, 2010). Nanocarriers may directly penetrate into biofilms preventing their formation and they also possess some properties such as higher solubility, bioavailability, efficacy and maximum drug entrapment in comparison to traditional delivery systems (Ong *et al.*, 2017).

Silver is widely used in nanocarriers due to its antimicrobial activity. Silver nanocarriers systems may hinder the development of microbial biofilms on medical devices which may be life- threatening. Silver nanoparticles against biofilms can be detected by acid-base indicator, scanning electron microscopy (SEM) and double fluorescent staining and Conofocal laser scanning microscopy (CLSM) (Ansari *et al.*, 2014). Bacteriophages containing drug molecules in its protein coat and antibody on its tip tends to release drug molecules to its complementary receptors. The released drug molecule attack pathogenic bacteria and destroy them. OxiTitan has proved its antibacterial action against *Clostridium difficile* spores and other resistant bacterial strains. It is durable, cost effective and environment friendly (Sadekuzzaman *et al.*, 2015).

## 1.4 Nanocarriers Based Drug Delivery System

### 1.4.1 Polymer Based Nanoparticles

Polymer-based nanoparticulate systems are one of the potential carriers that are used in previous years. Two types of polymer based nanoparticles based on their method of preparation are nanocapsules and nanospheres (Figure 5). Nanocapsules are vesicular carriers in which the drug is entrapped into the cavity surrounded by the polymer membrane and nanospheres are matrix carrier systems where the drug is uniformly dispersed (Han *et al.*, 2018). The utilization of polymeric nanoparticles results in optimized therapeutic outcomes by reducing toxic effects. The ideal polymers for nanoparticles should be easy to synthesize, biocompatible, inexpensive, non-immunogenic, biodegradable, non-toxic and water soluble. Polymers used in fabrication of nanoparticles may be classified as either biodegradable or non-biodegradable. Some of natural biocompatible and biodegradable polymers are gelatin, cellulose, chitosan, pullulan, gliadin and alginate. Various synthetic biodegradable polymers are poly-(lactide-co-glycolide) (PLGA), polylactic acid, poly-ε-caprolactone, polyanhydrides polyphosphazene and poly-alkylcyanoacrylates (Han *et al.*, 2018). One of the study reported a formation of antimicrobial peptides (AMPs), which can act as the first line antibiotics to control resistance and better alternative to antibiotics. Sustained release (SR) of four potent ultra-short lipopeptides conjugated with an aliphatic acid chain (16-C) encapsulated into poly (lactic acid-co-castor oil) and ricinoleic acid-based poly (ester-anhydride) was studied. Better antibacterial efficacy along with anti-biofilm effect and membrane disruption was observed with these biodegradable polymer incorporated peptides (Eckhard *et al.*, 2016). Various polymeric nanoparticles based antibacterial formulations have been developed to overcome antibacterial resistance. For example **Qasim *et al.*, 2018** developed poly-*N*-isopropylacrylamide (pNIPAM)-based polymeric nanoparticles encapsulating silver nanoparticles (AgNPs). The prepared formulation (pNIPAM- and pNIPAM-$NH_2$-based) showed enhanced bacteriostatic action against Gram-negative (*Escherichia coli*) and Gram-positive (*Staphylococcus aureus*) bacteria. **Namasivayam *et al.*, 2017** formulated biocompatible ovalbumin nanoparticles encapsulating cefpirome using coacervation technique. The results showed good biocompatibility along with effective inhibition of pathogenic bacteria and biofilm formation. **Suchomel *et al.*, 2015** reported silver bromide (AgBr) nanoparticles using different polymers (PVP, PEG, PVA, and HEC) and studied the antimicrobial action of Ag and AgBr nanoparticles against both gram positive and negative bacteria along with several strains of *Candida*. PEG and HEC showed less interaction with silver ions. The AgBr NPs were more effective against gram-negative bacteria and tested yeast strains but Ag NPs were more effective against gram-positive bacteria. **Azhdarzadeh *et al.*, 2012** suggested enhanced efficacy of azithromycin (AZI) nanoparticles prepared by modified Quasi-Emulsion Solvent Diffusion to combat *Haemophilus influenzae (PTCC 1623), S. pneumoniae (PTCC 1240) Escherichia coli (PTCC 1330)* using agar well diffusion. The increased potency of AZI NPs

*Figure 5. Different types of polymer based Nanoparticles*

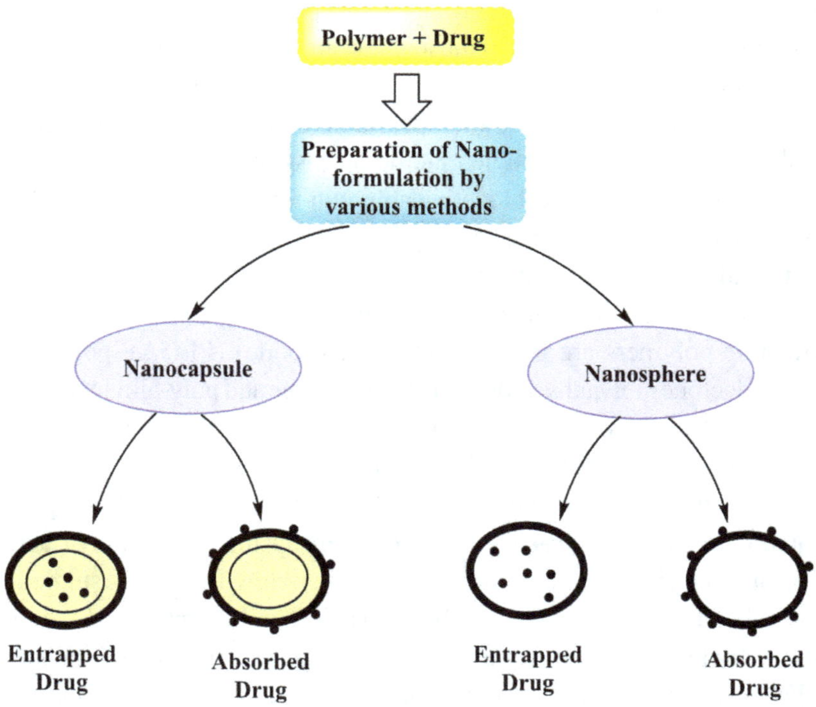

may be recognized to modify surface features as well as increased uptake of drug and adsorption. **Raval *et al.*, 2014** suggested poly (e-caprolactone) microspheres loaded with doxycyclines formulated by double emulsion solvent evaporation for controlled-release drug therapy for longer periods (76 h).

The classification, applications, advantages and disadvantages of polymer-based nanomaterials are given in Table 4.

*Table 4. Polymer based nanoparticles with advantages, disadvantages and applications*

| Classification | Materials | Advantages | Disadvantages | Application |
|---|---|---|---|---|
| Natural polymeric material | Chitosan | Biocompatibilty, easily degradable film formation, Antimicrobial Innocuous, | Poor spinnability, low water solubility, poor strength, | Hemostasis materials, medical dressings, drug delivery carrier, hydrogel, gene transfer |
| | Starch | Cost effective, non-degradable, non-toxic, anti-antigenic | Poor mechanical properties, water-resistant | Hemostasis material, bone repair material, tissue-engineered scaffold |
| | Alginate | Hypotoxicity, suppresses the tumor growth, immunity booster | Low biodegradability, poor cell attachment. | Pharmaceutical excipient, medical dressing. |
| | Cellulose | Extensive sources, cost effective | Rare Adverse reactions | Pharmaceutical adjuvant |
| Biosynthetic Material | Poly β-hydroxybutyrate (PHB) | Biodegradable, non toxic, safe, good physicochemical properties | Poor thermal stability, High Crystallinity | tissue engineering material, Drug delivery carrier |
| Chemosynthetic polymer(Copolymer) | Poly lactic acid | Biocompatible, safe, good mechanical properties, non toxic | Hydrophobicity, Poor toughness, | Anti-adhesion materials, drug delivery carrier, sutures, bone fixing device. |
| | Polyurethane | Low cost, good mechanical properties, rich resource, | Slow speed of degradation | Excipients, medical bandage |
| | Poly(lactic-glycolic acid) (PLGA) | Biocompatible, biodegradable | High cost | Tissue repair, sutures |
| | Polymethyl methacrylate resin | Easy to operate, biocompatible. | Cytotoxic, ease of oxidation. | Dental materials, artificial crystals. |

## 1.4.2 Metal and Metallic Oxides Nanoparticles

Metal nanoparticles (NPs) are promising candidate to fight against growing multi drug resistant strains. These nanoparticles help to create difficulty in building resistance to microbes *via* various mode of action (Ramos *et al.*, 2018). Amongst metallic NPs, the silver NPs have been widely used in distinctive study as they are able to kill gram-positive and gram-negative bacteria. They are efficient against a wide range of antibiotic resistant bacteria such as methicillin resistant *S.epidermis* (MRSE), MRSA, Erythromycin-resistant *S.* pyogenes, Ampicillin-resistant *E. coli*, and Multidrug-resistant *P. aeruginosa*. Multiple mechanisms can also be involved, which includes direct connections between Ag and DNA, Ag and bacterial cell membrane, Ag and enzymes/proteins, and indirect connections *via* development of reactive oxygen species (ROS) (Ramos *et al.*, 2018). The nanohybrids of AgNPs were developed in combination with poly-methyl-methacrylate (PMMA) polymer to improve surface properties of AgNPs (found to adhere to 1nm-thick inorganic silicate clay platelets) and to decrease toxicity on mammalian cells (Zhu *et al.*, 2014). Further, mixture of chitosan acetate polymer and AgNPs exhibited potential activity towards MRSA during its use in burn dressings (Rai *et al.*, 2012, Huang *et al.*, 2011). Gold nanoclusters have been developed with sufficient antimicrobial activity against resistant bacterial strains. Functional gold nanoclusters (AuNCs) have been formulated using lysozyme as sequestering and reducing agent. Titanium dioxide nanoparticles ($TiO_2$-NPs) were observed to improve the action of various antibiotics such as cephalosporins, beta-lactams, aminoglycosides and macrolides against MRSA (Waheed *et al.*, 2015; Bonifacio *et al.*, 2014).

Various studies on metal/metal oxide nanoparticles for antibacterial activity have been evidenced. For example, **Abdellatif *et al.*, 2021** investigated antioxidant and antibacterial activity of formulated silver nanoparticles prepared by different cellulosic polymers. Ethyl cellulose and hydroxypropyl methylcellulose polymers were used due to its own antibacterial activity. The polymers showed its antibacterial activity against *E.coli*. **Gupta *et al.*, 2020** synthesized metal oxide nanoparticles of copper, iron, lead and zinc. The synthesized nanoparticles were found in nano range and were found to have antibacterial activity against *E.coli* and *S. aureus*. **Holubnycha *et al.*, 2018** synthesized chitosan-silver nanoparticles (AgNPs) solution against MRSA. Cetrimonium bromide (CTAB) was used for surface modification of silver nanoparticles. Chitosan-AgNPs solution showed enhanced antimicrobial activity over its pure form. **Eshghi *et al.*, 2018** synthesized silver nanoparticles (Ag NPs) through microwave irradiation using extract of Juglans regia (J. regia) leaves, which act as reducing and stabilizing agent. The developed NPs showed enhanced antimicrobial activity against *Escherichia coli* and *Staphylococcus aureus*. **El-Rashidy *et al.*, 2018** developed zein scaffolds containing silver-doped bioactive glass and investigated antibacterial activity, biocompatibility, and compressive potency. Zein scaffolds with silver-doped sol-gel bioactive glass were effective against *E. coli* and *S. aureus*. **Holubnycha *et al.*, 2017** developed copper nanoparticles (CuNPs) and Cu NPs/chitosan solution against *MRSA* and multidrug resistant *E. coli* clinical isolates.

### 1.4.3 Liposomes

Liposomes are the main class of lipid-based nanocarriers for drug delivery (Waheed *et al.*, 2015). The liposomes are synthesized by combining phospholipids (Emulsifiers) with oil phase. The core material (oil) can be used to encapsulate lipid miscible drugs such as hematoporphyrin, paclitaxel and lipid conjugated prodrug. The phospholipids used in fabrication of liposomes should be biodegradable and nontoxic (Kumar *et al.*, 2012). Antigens and viral enveloped glycoproteins can be incorporated into the core to develop virosomes for treatment of influenza. Combination of 1,2-dioleoyl-3-trimethyl- ammonium propane (DOTAP) modified cationic liposome and a cationic polymer (usually protamine) condensed DNA are known as liposome polycation-DNA nanoparticles (LPD), an often used adjuvant delivery approach in DNA vaccine studies (Zhao *et al.*,2014). Liposome based antibacterial systems have been reported for enhanced antibacterial efficacy. For example, **Moya *et al.*, 2019** prepared and characterized liposomes encapsulating cefepime. Thus, it was concluded that cefepime liposomes were biocompatible nanocarriers for antibacterial activity. The formulation was effective against *E.coli* and showed better encapsulation efficiency than free drug. **Jung *et al.*, 2015** formulated linolenic acid (Lipo LLA) liposomes and concluded an effective antibacterial activity for clinically isolated antibiotic-resistant strains of *H. pylori* and lead to destruction of bacteria within 5 minutes. **Solleti *et al.*, 2014** reported liposomal carriers of azithromycin and examined its antimicrobial effects against *Pseudomonas aeruginosa*. The developed liposomes reduced the biofilm and suppressed the formation of various virulence factors along with a reduction of different patterns of bacterial motility. **Shafaa *et al.*, 2008** developed cephalexin loaded neutral, negative and positive liposomes by phase transition measurements. Negative liposomes encapsulating cephalexin were superior in comparison to neutral and positive liposomes.

### 1.4.4 Dendrimers

Dendrimers are outlined as 3-D structures of nanorange about 1-100nm (Martins *et al.*, 2013). Dendrimers are indicated as tremendous carriers for delivery of antibiotics because of its unique properties (Silva *et al.*, 2012). The extremely branched structure of dendrimers (Figure 6) offers huge surface to size ratio which facilitate greater interaction with microorganism and it can incorporate hydrophobic and hydrophilic drug moieties into it (Medina *et al.*, 2009).

*Figure 6. Diagrammatic presentation of dendrimer*

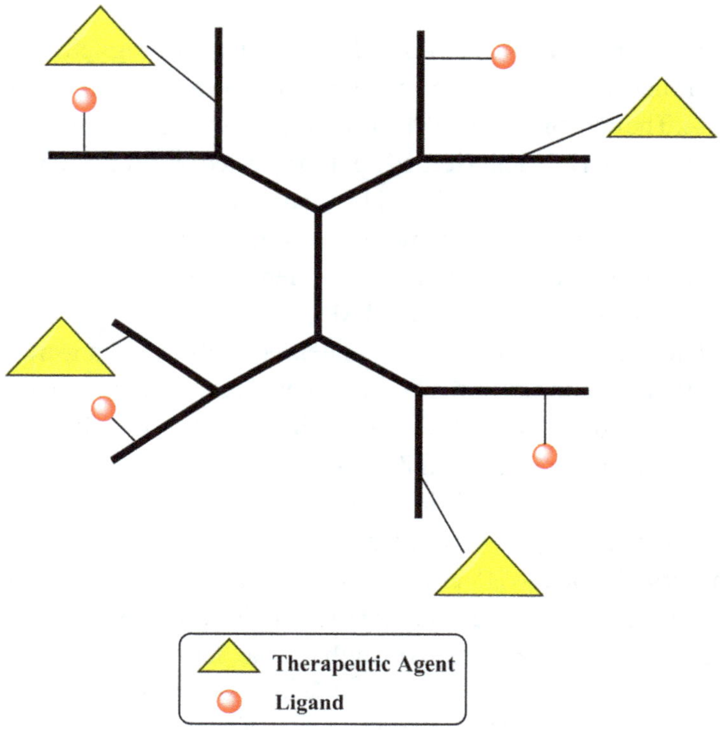


Dendrimer-based nanosystems hold three distinct units central core, polymeric branches and a terminal group at the dendrimers surface. The terminal group is PEGylated to improve biocompatibility and thereby, reducing toxicity. PAMAM [Poly (amidoamine)] was the primary synthesized and commercialized dendrimer and along with poly(propyleneimine) (PPI) that is extensively used as drug delivery agent (Martins *et al.*, 2013; Kumar *et al.*, 2015). The marketed formulation based on this carrier is VivaGel® (http://www.starpharma.com) to prevent and cure bacterial vaginosis and sexually transmissible infections (Puri *et al.*, 2009). VivaGel® is made from SPL7013 which is L-lysine derived dendrimer with methane series at the surface creating a polyanionic exterior. Dendrimer based antibacterial formulations with enhanced activity has been evidenced. For example, **Kannan *et al.*, 2019** formulated poly(aryl ether) based amphiphilic dendrimers. The detailed mechanistic study reveals that optimal tuning of hydrophobicity of dendrimers play a vital role in membrane disruption of bacterial membrane. **Wronska *et al.*, 2019** evaluated antibacterial activity of modified polycationic and polyanionic dendrimers in combination with levofloxacin against *E.coli, Proteus hauseri* and *Staphylococcus aureus.* The formulation showed satisfactory results due to a synergistic effect in antibacterial activity. **Gholami, 2017** reported antibacterial activity of poly(amidoamine) (PAMAM-G7) dendrimers was evaluated against gram-negative and gram-positive bacteria such as *P. Aeruginosa, E. coli, Acinetobacter baumanni, Shigella dysenteriae, Klebsiella pneumoniae, Proteus mirabilis, Bacillus subtilis* and *Staphylococcus aureus*. **Vembu *et al.*, 2015** achieved sustained release of ciprofloxacin by formulating piperazine core (1,3,5-triazine) dendrimer. the developed formulation exhibited 5 times higher activity for tested microbes in comparison to pure ciprofloxacin. **Maleki *et al.*, 2015** evaluated antibacterial activity of different generation (2 and 4) poly (amidoamine) dendrimer (PAMAM) using bacterial strains isolated from water resources like *E. coli, P. aeruginosa, k. oxytoca, B. subtilis,* and *S. aureus*. PAMAM dendrimer was found to be more effective against gram positive bacteria. Amino-terminated generation 2 poly(amidoamine) dendrimer showed positive effect on removal of main bacterial strains proving it as a safe and effective material for water disinfection in future. **Tang *et al.*, 2013** synthesized dendrimers encapsulating silver nanocomposites. These were used as antibacterial agents (against *E. coli* and *S. aureus*) during manufacture of antibacterial cotton fabrics. Increased antimicrobial activity was observed with a decrease in the silver nanoparticle size. **Mazumder *et al.*, 2013** fabricated antibacterial dendrimer-encapsulated zero-valent nickel nanoparticles (NPs) using fourth generation (G4)$NH_2$-terminated poly(amido) amine (PAMAM) dendrimer having antibacterial activity. **Winnicka *et al.*, 2013** investigated the effect of PAMAM-$NH_2$ and PAMAM-OH dendrimers of second generation and third generation incorporating antibiotics such as erythromycin and tobramycin. Erythromycin loaded dendrimers exhibited 8 times increase in water solubility with a minor change in bactericidal action of erythromycin.

### 1.4.5 Microemulsion and Nanoemulsion

Microemulsions (MEs) act as potential drug carrier system in the pharmaceutical drug development (Shah *et al.*, 2010). Drugs with distinct physicochemical properties can be delivered effectively to the tissues/organs with the help of microemulsion and nanoemulsion based nanocarriers. MEs are transparent emulsions or phase-transition systems (water microdroplets dispersed in oil or oil microdroplets dispersed in water) (Peek *et al.*, 2008; O'Hagan, 2007). MEs or NEs are thermodynamically stable which consist of nanosize droplets in an inner phase and are enclosed within an amphiphilic compound (surfactant) or amphiphile along with an appropriate co-surfactant. MEs /NEs are able to carry antigens in the internal core for effective vaccine delivery. An oil-in-water emulsion of MF59TM was proved for influenza in more than 20 countries (Waheed *et al.*, 2015). Microemulsions are of versatile nature having low surface tension, enhanced physicochemical properties such as stability, solubility resulting in enhanced permeation and absorption (Mishra *et al.*, 2014; Hung *et al.*, 2007). Figure 8 showed a diagrammatic view of micro and nanoemulsion.

*Figure 7. Diagrammatic representation of micro and nano-emulsions*

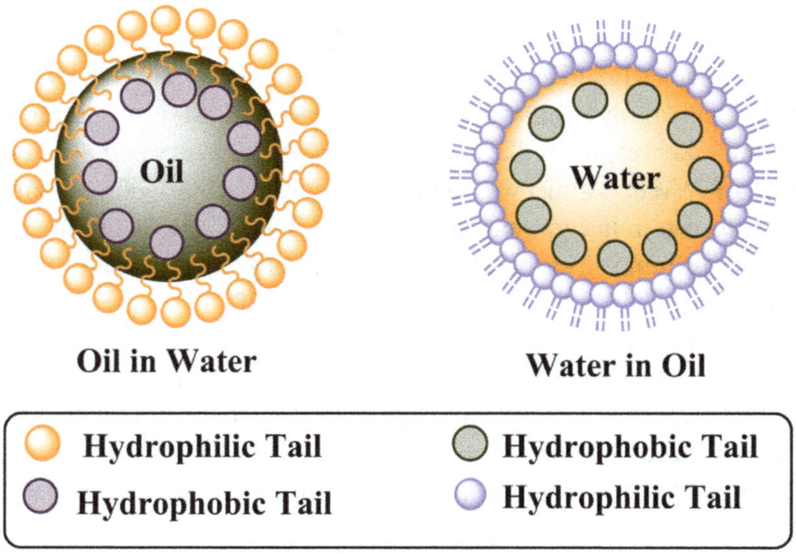


Montanide ™ is an oil-in-water emulsion and water-in-oil emulsion (ISA 50V, 51, 201, 206 and 720). Montanide ISA 201 and 206 have been used in various infections associated with foot and mouth. Montanide ISA 51 and 720 was used in treatment of malaria (Aucouturier *et al.*, 2002; Kumar *et al.*, 2004). An oil-in-water nanoemulsion was developed using biosurfactant. Using this self assembling, the PEG moiety and a receptor specific antibody was encapsulated on the aqueous interface of nanoemulsion (Zhang *et al.*, 2008). Researchers have synthesized various micro/nanoemulsion systems with enhanced bactericidal effect. For example, **Prakash *et al.*, 2019** investigated coriander, cumin, lemongrass, pepper and fennel nanoemulsions for evaluation of antibacterial and anti-biofilm activities against *Salmonella enteric* Typhimurium. The formulation showed that particle size was in nanorange. The citral nanoemulsion can be further explored as natural antibacterial and anti-biofilm agent for food preservation applications. **Kumari *et al.*, 2018** prepared an antibacterial nanoemulsion by using thymol, an essential oil component of plant and *Quillaja* saponin, a glycoside surfactant of *Quillaja* tree. The nano scale formulation could be a potential antimicrobial and plant growth promoting agent for agriculture. **Lu *et al.*, 2018** reported Citral-in water nanoemulsion and evaluated its antimicrobial activity. Citral can lose its bactericidal activity under normal storage. The antimicrobial activities of nanoemulsions were significantly large and therefore can be used as effective antimicrobials in cosmetics, food products and agrochemical industries. **Yildirim *et al.*, 2017** prepared cinnamon oil based nanoemulsions /micro emulsions, which has good antimicrobial activity and also used for improving the quality, shelf life of foods. **Ma Q *et al.*, 2016** prepared microemulsions with a mixture of soybean oil with other oils such as cinnamon bark oil, eugenol/thyme oil. The antimicrobial activity of eugenol or thyme oil microemulsions was significantly greater than un-encapsulated drug. **Hu *et al.*, 2016** developed a microemulsion to enhance the water solubility of the antimicrobials. This mixture consisted of N-methyl-2-pyrrolidone as solvent and dimethyl sulfoxide as co-solvent, polyoxyethylated castor oil, polyalkylene glycol, and polyoxyethylene tridecyl ether phosphate as surfactants. This formulation significantly enhanced the solubility of SecA inhibitors. The SecA inhibitors posses equal bactericidal activity as that of streptomycin. **Tian *et al.*, 2016** reported stable self emulsifying nanoemulsion using cinnamaldehyde. The formulation provided a long-term inhibition towards *E. coli* as compared with pure cinnamaldehyde. **Shaaban and Edris, 2015** prepared carvacrol microemulsions to be used as preservatives and disinfectants. The carvacrol microemulsion exhibited significantly higher antibacterial activity. It was concluded that cationic carvacrol microemulsions can also be used as disinfectants and preservatives. **Jerobin *et al.*, 2015** formulated neem oil based nanoemulsions (O/W) using high energy ultrasonication method. The developed systems proved antibacterial against *Vibrio vulnificus* by disrupting their cell membrane. It was nontoxic at lower concentration to human lymphocytes. **Hwang *et al.*, 2013** reported cetylpyridinium chloride (CPC) nanoemulsion for antimicrobial action against *A. baumannii.* The formulation contains 10% Triton X-100, 1% and 25% soybean oil which indicated the best efficacy against *A. baumannii.* A specific amount of CPC is responsible for death of planktonic forms of *A. baumannii* while the destruction of biofilm was obtained by emulsified oil and its detergent fractions. **Anjali *et al.*, 2010** developed refined sunflower oil and tween 20 based o/w microemulsion. A decline in bacterial growth was observed with all the formulations prepared.

### 1.4.6 Solid Lipid Nanoparticles

Solid lipid nanoparticles are colloidal system alternative to emulsions, polymeric nanoparticles, liposomes *etc*. The lipid-lipid (oil) has been substituted by a solid-lipid in these submicron sized lipid emulsions. Solid lipid nanoparticles have high surface area, small in size, high drug loading capacity and potential approach to target the drugs. They exhibited several advantages such as low toxicity, compatibility and high stability (Ekambaram *et al.*, 2012). Solid Lipid Nanoparticles of various antibiotics have been formulated and studied. For example, **Anjum *et al.*, 2020** developed anacardic acid loaded solid lipid nanoparticles by homogenization method. The results suggested the enhanced efficacy of developed formulation to overcome the biofilm-mediated antimicrobial resistance. **Bolla *et al.*, 2019** formulated furosemide-silver complex for antibacterial activity. The particles were spherical and in nanorange. The formulation could be considered as promising topical antibacterial agent against bacterial infections. **Pignatello *et al.*, 2018** synthesized ciprofloxacin loaded solid lipid nanoparticles using quasi-emulsion solvent diffusion and solvent injection method. Positively charged and stable formulation was formed by incorporation of cationic lipid (didecyl dimethylammonium bromide; DDAB). The formulation showed effective antibacterial effect and could be promising nanocarrier for delivery of antibiotics by reducing toxicities associated with pure drug. **Gaspar *et al.*, 2017** reported the anti-mycobacterial activity of solid lipid nanoparticle containing rifabutin (RFB) *via* pulmonary administration. SLN's were loaded into uniform sized microspheres to facilitate their pulmonary administration. The particle diameter indicated a uniform distribution of drug in the alveolar region. The *in vivo* bio-distribution studies of microencapsulated RFB-SLNs suggested that it could be used for delivery of antibiotic. **Shazly *et al.*, 2017** formulated SLNs loaded with ciprofloxacin *via* ultrasonic melt-emulsification method for controlled release and improved bactericidal effect.

### 1.4.7 Immunostimulating Complex (ISCOM)

ISCOMs are nanorange spherical, hollow, cage like assembled particles consisting of saponin Quil A adjuvant, phospholipids, cholesterol and antigen protein. The particle without antigen is known as an ISCOM matrix (Zhao *et al.*, 2014). Quil A possesses affinity for cholesterol which induces stability in the ISCOM matrix. Hydrophobic molecules can be only loaded within these complexes. The assembly of ISCOM is facilitated by hydrophobic interaction between these components and the antigen. **Kostetsky *et al.*, 2011** developed tubular immunostimulating complex which was able to increase the immunogenicity (three to four folds) for protein antigens from *Yersinia pseudotuberculosis*.

Various nanocarriers incorporating antibacterial drugs have been summarized in Table 5.

*Table 5. Nanocarriers incorporating antibacterial drugs*

| S.No. | Drug Carrier | Drug | Micro-organism | References |
|---|---|---|---|---|
| 1 | Chitosan-AgNPs | Cetrimonium bromide | Methicillin-resistant *Staphylococcus aureus* (MRSA) | Holubnycha *et al.*, 2018 |
| 2 | AgNPs | Juglans regia leaf extract | *E.coli and S.aureus* | Eshghi *et al.*, 2018 |
| 3 | Silver doped bioactive glass | Zein scaffolds | *E.coli and S.aureus* | El-Rashidy *et al.*, 2018 |
| 5 | CuNPs/Chitosan solution | Ginger and ascorbic acid | MRSA and *E.coli* | Holubnycha *et al.*, 2017 |
| 6 | PAMAM-G7 dendrimers | Trimethoprim-sulfamethaxozole, Ampicillin, Tetracycline, Chloramphenicol | *Pseudomonas aeruginosa, E.coli, Acinetoacter baumanni, Shigella dysenteria, Klebsiella pneumonia, proteus mirabilis, S.aureus, Bacillus subtilis* | Gholami, 2017 |
| 8 | PAMAM(G2-G4) dendrimer | Quaterinary ammonium salts using halogen groups(Cl, Br, I) | *E.coli, P.aeruginosa, klebsiella oxytoca, Bacillus subtilis* and *staphylococcus aureus* | Maleki *et al.*, 2015 |
| 9 | PAMAM dendrimer(low generation) | Silver nanocomposites | *E.coli* and *S.aureus* | Tang *et al.*, 2013 |
| 10 | PAMAM dendrimer(G4) | Zero-valent Nickel "Ni(O)" NPs | *Gram positive* and *Gram negative* | Mazumder *et al.*, 2013 |
| 11 | PAMAM dendrimer NH2 and OH (G2 and G3) | Erythromycin and Tobramycin | *S.aureus, Bacillus Subtilis, E.coli, Pseudomonas aeruginosa* and *Proteus mirabilis* | Winnicka *et al.*, 2013 |
| 12 | 1,3,5-Triazine dendrimer | Ciprofloxacin | *Gram positive* and *Gram negative bacteria like (S.aureus, Bacillus Subtilis, E.coli, Pseudomonas aeruginosa, Proteus mirabilis)* | Vembu *et al.*, 2015 |
| 13 | Liposomal formulation | Azithromycin | *P.aeruginosa* | Solleti *et al.*, 2014 |
| 14 | Liposomal formulation | Linolenic acid | *H.pylori* | Jung *et al.*, 2015 |
| 15 | Liposomal formulation | Cephalexin | *S.aureus* | Shafaa *et al.*, 2008 |
| 16 | Ovalbumin NPs | Cefpirome | *Vero cells* and *blood cells* | Namasivayam *et al.*, 2017 |

*continues on following page*

*Table 5. Continued*

| S.No. | Drug Carrier | Drug | Micro-organism | References |
|---|---|---|---|---|
| 17 | PLGA polymer | Azithromycin | *E.coli, Haemophilus influenza, streptococcus pneumonia* | Azhdarzadeh *et al.*, 2012 |
| 18 | Poly(lactic acid co Castor oil) {PLACO} and Ricinoleic acid based poly(ester anhydride){PSA-RA} | Ultra short lipopeptides conjugated to an aliphatic acid chain | *E.coli, S.aureus* | Eckhard *et al.*, 2016 |
| 19 | Poly(e-caprolactone) | Doxycycline | *E.coli, Haemophilus influenza, streptococcus pneumonia* | Raval *et al.*, 2014 |
| 20 | Poly-N-isopropylacryl amide (pNIPAM) | AgNPs | *E.coli, S.aureus* | Qasim *et al.*, 2018 |
| 21 | PEG, PVP, PVA and HEC | AgBr NPs | Gram positive and Gram negative bacteria | Suchomel *et al.*, 2015 |
| 22 | ISCOM | Curcumarinoside A2-2-cholestrol-MGalDG | *Yersinia pseudotuberculosis* | Kostetsky *et al.*, 2011 |
| 24 | Micro emulsion | Soyabean oil with Cinnamon bark oil, eugenol or thyme oil, tween 80 | *Listeria monocytogenes, Salmonella entrica, E.coli* | Ma Q *et al.*, 2016 |
| 25 | Micro emulsion | Refined sunflower oil, tween 20 | *E. coli* | Anjali *et al.*, 2010 |
| 26 | Micro emulsion | SecA of Candidatus Liberibacter asiaticus | *Agrobacterium tumefaceins, liberibacter crescens, Rhizobium etli, Buadyrhizobium japonium, mesorhizobium loti, sinorhizobium melikoti* | Hu *et al.*, 2016 |
| 27 | Nanoemulsion and microemulsion | Cinnamon oil | *Campylobacter jejuni, Salmonella enteridis, E.coli, S. aureus* and *Listeria monocytogens* | Yildirim *et al.*, 2017 |
| 28 | Microemulsion | Triton, Soyabean oil, Cetylpyridinium chloride | *A.baumanii* | Hwang *et al.*, 2013 |
| 29 | Microemulsion | Carvacrol | *E.coli, S. aureus* and *Listeria monocytogens* | Shaaban and Edris, 2015 |
| 31 | Nanoemulsion | Citral | *Bacillus cereus, B. subtilis, Haemophilis influenza, Neisseria gonorrheae, S. pneumonia, Vibrio cholera* | Lu *et al.*,2018 |
| 33 | Nanoemulsion | Thymol and Quillaja | *Xanthomonas axonopodis* | Kumari *et al.*, 2018 |

*continues on following page*

*Table 5. Continued*

| S.No. | Drug Carrier | Drug | Micro-organism | References |
|---|---|---|---|---|
| 34 | Nanoemulsion | Cinnamaldehyde | *E. coli* | Tian *et al.*, 2016 |
| 35 | Nanoemulsion | Neem | *Vibrio vulnificus* | Jerobin, *et al.*, 2015 |
| 36 | SLNs | Rifabutin | *Mycobacterium tuberculosis* | Gaspar, *et al.*, 2017 |
| 38 | SLNs | Ciprofloxacin | *E.coli* | Pignatello, *et al.*, 2018 |
| 40 | SLNs | Ciprofloxacin | *Gram positive, Gram negative bacteria* | Shazly, 2017 |
| 42 | SLNs | Minocycline and ciprofloxacin | *P.aerugiosa, S. aureus* | Valdes *et al.*, 2018 |

## 1.4.8 Bacteriocins

Bacteriocins are a kind of ribosomal synthesized antimicrobial peptides produced by bacteria, which can kill or inhibit bacterial strains (Yang *et al.*, 2014). In other words, antimicrobial peptides or proteins produced by bacteria are categorized as Bacteriocins. Bacteriocins can be classified on the basis of origin. Bacteriocins from gram-negative bacteria include colicins and microcins and Bacteriocins from gram-positive bacteria are generally divided into class I (modified peptides, lantibiotics), class II (unmodified peptides, non-lanthionine) and class III (large proteins, heat unstable) (Cotter *et al.*, 2013). The bacteriocins come under GRAS status (generally recognized as safe) and have no side effects. The use of various nanocarriers such as liposomes, nanofibers, chitosan, and metallic nanoparticles can be used to shield bacteriocins from degradation (Dimov *et al.*, 2005). Bacteriocins are safe for human use as they are acted upon by gastrointestinal enzymes (proteases). Bacteriocins exhibit antibacterial activity against various Gram positive bacteria including *B. cereus, S. aureus, C. botulinum,* and *L. monocytogenes* (Saavedra *et al.*, 2004). Bacteriocins are associated with several disadvantages such as restricted antimicrobial spectrum, high doses required for inhibiting multi-drug resistant strains, high sensitivity to gastrointestinal proteases, high production cost and low production yield due to incomplete recovery during purification. The commonly used bacteriocins include pediocin and nisin. Nisin is a polypeptide, synthesized as pre-pro-peptide and carries 57 amino acid residues and it undergoes post-translational modifications (Alishahi, 2014). It is found to have activity against various pathogenic food-spoiling bacteria such as *S. aureus* and *L. monocytogenes* (Zohri *et al.*, 2013). However, it has been observed that it is less effective against gram negative bacteria, yeasts and moulds (Zacharof and Lovitt, 2012). Nisin is not poisonous for human consumption as per toxicological studies and it is restrained in acidic food materials due to its less stability at neutral pH (Jeevaratnam *et al.*, 2005). Pediocin polypeptide is composed of 44 amino acids. Unlike nisin, it does not show any post-translational changes (Ovchinnikov *et al.*, 2016; Cotter *et al.*, 2005). Nanoformulations of bacteriocins include conjugation of bacteriocins with nanoparticles (Fahim *et al.*, 2016); encapsulation of bacteriocins in nanoliposomes, chitosan nanoparticles and metallic nanoparticles (Sidhu and Nehra, 2017) etc. Bacteriocin nisin is widely used and it does not develop resistance due to its narrow antimicrobial spectrum (Zendo, 2013; Rea *et al.*, 2013; Sidhu and Nehra, 2017). Table 6 depicts various nanocarriers developed by bacteriocins along with targeting bacteria.

*Table 6. Various nanocarriers systems incorporating bacteriocins*

| Approaches | Bacteriocins | Drug delivery | Results | Test organisms | References |
|---|---|---|---|---|---|
| Encapsulation of bacteriocins in nanoliposomes | Nisin Z produced from bacillus licheniformis P40 | Nanovesicles of phosphatidylcholine | Inhibitory action | *Listeria monocytogenes* | Teixeira *et al.*, 2008 |
| Conjugation with chitosan NPs | Nisin | Dipalmitoyl phosphatidylcholine/ Dicetylphosphate/ cholesterol chitosan NPs | Antimicrobial activity | *Bacillus subtilis* | Colas *et al.*, 2007 |
| | Nisin | Nanoparticles of Chitosan | Antimicrobial activity two folds | *E.coli* | Alishahi, 2014 |
| | Nisin | Chitosan-alginate nanoparticles | Greater antimicrobial activity | *S. aureus* | Zohri *et al.*, 2010 |
| | Bacteriocin | Chitosan nanoconjugates | More antimicrobial potential | *S. aureus* | Namasivayam *et al.*, 2015 |
| Bacteriocin Conjugation with metallic nanoparticles | Enterocin | Silver NPs | Broad antimicrobial activity against food spoiling bacteria | *Listeria monocytogenes*, *S. aureus* | Sharma *et al.*, 2012 |
| | Nisin | Silver NPs | High antimicrobial spectrum | *Listeria monocytogenes* | Kumari, 2012 |

## 1.4.9 Carbon Nanotubes (CNTs)

Carbon nanotubes are cylindrical molecules made of hexagonal arrangements of $sp^2$ hybridized carbon atoms which are classified as the members of fullerene and belongs to family of carbon allotropes. The size varies from 0.5 to 1 nm of diameter to several hundred in length. When the CNTs walls are made of single graphene sheets then these are referred as single-walled carbon nanotubes (SWCNTs) and when these are made of several graphene sheets then these are called multi-walled carbon nanotubes (MWCNTs). These possess unique mechanical, electronic and structural properties which make them potential nanocarriers for delivery of antibacterial drugs. The organic functionalization of CNTs can improve its biocompatibility profile with low cytotoxicity. CNTs can be used as carriers for anticancer molecules, immunoactive compounds, proteins and genetic materials (Tostes *et al.*, 2013). Table 7 comprises patents on CNTs (Martins *et al.*, 2013).

*Table 7. Various patents on carbon nanotubes*

| Patent no. | Title | Nanocarriers | Therapeutic indications | References |
|---|---|---|---|---|
| US20100324315 A1 | Poly (citric acid) functionalized CNTs DDS. | CNTs-g-PCA | Therapeutic agents and cancer | Atyabi *et al.*, 2010 |
| WO2012031164A2 | Drug delivery via CNT arrays | CNTs | Therapeutic agents and cancer | Aria *et al.*, 2012 |
| US20120220921 A1 | Immunologically modified CNT for cancer treatment | SWCNTs | Photodynamic therapy and cancer | Chen *et al.*, 2014 |
| WO 2010123989 A1 | Chitosan/CNT composite scaffolds for drug delivery | Chitosan coated SWCNTs/ MWCNTs | Therapeutic agents and antibiotic delivery | Jennings *et al.*, 2010 |
| WO 2009070380 A3 | Water-soluble CNT compositions for drug delivery and medical applications | SWCNTs | Therapeutic agents and medical applications | Tour *et al.*, 2014 |
| WO 2013101983 A3 | Targeted self assembly of functionalized CNT on tumors | Oligo-morpholino coated SWCNTs | Therapeutic agents and cancer | Scheinberg *et al.*, 2013 |
| US 20130034610 A1 | Hydrophobic nanotubes and NPs as transporters for delivery of drugs into cells | SWCNTs/ MWCNTs | Therapeutic agents | Dai *et al.*, 2012 |
| US 20130158377 A1 | Drug delivery and substance transfer facilitated by nano enchanced device having aligned CNTs protruding from device surface | Coated CNTs | Therapeutic agents and medical applications | Gharib *et al.*, 2012 |

## 1.5. Synergistic Effect of Antibodies with Nanomaterials to Combat Superbugs

Antibodies are used since 19th century; the first antibody (immune-boosting serums) was developed to treat tetanus and diphtheria patients. The first noble prize for medicinal drug was awarded to Emil Von Behring in 1901 for his efforts in field of antibiotics (Cal *et al.*, 2017). Researchers (New York City) developed an antibody serum that captured the interest of the state health commissioner for treatment of pneumonia in 1910. Thomas Parran Jr conducted a pneumonia control program in which the serum was dispensed to human beings all over the country. Monoclonal antibodies (MAbs) are the targeting molecules with high specificity, affinity and versatility. Antibody-conjugated nanoparticles were prepared to reach directly at a target site by overcoming physiological barriers (Hu *et al.*, 2010). The delivery of targeted nanomaterials to the infection site can be achieved by surface modification with ligands such as polyethyleneimines, chitosan and glucosamine which may also improve the therapeutic efficacy and reduce the side effects of antimicrobial drugs (M Cardoso *et al.*, 2012). A study reported an antibody-antibiotic conjugate consisting of an anti-*Staphylococcus aureus* antibody conjugated to an exceptionally efficacious antibiotic that was activated solely after its release in the proteolytic environment of the phagolysosome. The antibody-antibiotic conjugate proved to be effective to vancomycin for treatment of bacteraemia and provided direct evidence that intracellular *Staphylococcus aureus* represents the necessary component of invasive infections (Azevedo *et al.*, 2014; Peng *et al.*, 2017).

*Table 8. Various antibodies with their target organism*

| Target organism | Antibody | References |
|---|---|---|
| *Bacillus anthracis* | Raxibacumab | Subramanian *et al.*, 2005 |
| *Bacillus anthracis* | Anthim(ETI-204) | www.elusys.com |
| *Clostridium difficile* | CDA1/CDB1 | Lowy *et al.*, 2010 |
| *Bacillus anthracts* | Valortim (MDX- 1303) | www.pharmathene.com |
| *Shiga toxin- Producing E.coli* | Urtoxazumab | Lopez *et al.*, 2010 |
| *Shiga toxin- Producing E.coli* | ShigamAbs | Bitzan *et al.*, 2009 |
| *Pseudomonas aeruginosa* | KB001 | www.Kalobios.com |
| *Staphylococcus aureus* | Pagibaximab | Weisman *et al.*, 2009 |
| *Pseudomonas aeruginosa* | Panobacumab(KBPA101) | Lazar *et al.*, 2009 |
| *Pseudomonas aeruginosa* | Anti- Pseudomonas IgY | Nilsson *et al.*, 2008 |

### 1.5.1 Emergence of Natural Antibodies in Form of Nanomaterials

Anti-bacterial antibodies attach directly with pathogen or intend to deactivate toxins or other virulence factors (Subramanian *et al.*, 2005). The bacteria bind to antibodies by opsonizing process. Some antibody-antibacterial drugs are presently being evaluated in clinical trials and are shown in Table 8.

## 1.6 CONCLUSION AND FUTURE PERSPECTIVE

Nanotechnology has become a revolutionary approach to improve and investigate novel pharmaceutical drug products with antibacterial activity. Nanoparticles present promising choices in the plan of next generation therapeutics towards fungal, bacterial and viral threats. The development of new nano antimicrobial drugs with multiple functions will revolutionize medical medicines and it will also play a role in pacifying infectious diseases (Sharma *et al.*, 2012; Stern *et al.*, 2005). Thus, it has become important to fight multi drug resistant (MDR) microorganisms using conventional antibiotics in clinically proved nano formulations (MacRae *et al.*, 2003). The nanoparticles improve the shelf life of drugs by nano-based aspects like decreased lipophilicity, increased solubility, cost-effectiveness, patient compliance, reduced toxicity, high bioavailability and sustained-release (SR) of a drug over a long period of time. Scientists have engineered quantum dot nanoparticles made up of cadmium that makes bacteria more vulnerable to antibiotics telluride (Aruguete *et al.*, 2013). This will hopefully be a step forward in fighting against drug-resistant pathogens, like superbugs and the infections caused by bacteria's. Currently, some nano based antibiotics had been approved clinically for the use of human beings in several infectious diseases. Lipoquin™ and Pulmaquin™ (S.A., Grifols, Spain, Barcelona, and Aradigm Corporation, Hayward, CA, USA) are the liposomal formulations (inhalable) of ciprofloxacin, for curing the diseases such as cystic fibrosis (CF) or in non-CF bronchiectasis. Thus, future research will present a great potential to fight against multidrug-resistant superbugs.

## REFERENCES

Abdellatif, A. A., Alturki, H. N., & Tawfeek, H. M. (2021). Different cellulosic polymers for synthesizing silver nanoparticles with antioxidant and antibacterial activities. *Scientific Reports*, *11*(1), 1–8. doi:10.103841598-020-79834-6 PMID:33420131

Algburi, A., Comito, N., Kashtanov, D., Dicks, L. M., & Chikindas, M. L. (2016). Control of biofilm formation: Antibiotics and beyond. *Applied and Environmental Microbiology*, AEM-02508.

Alishahi, A. (2014). Antibacterial effect of chitosan nanoparticle loaded with nisin for the prolonged effect. *Journal of Food Safety*, *34*(2), 111–118. doi:10.1111/jfs.12103

Anjali, C. H., Dash, M., Chandrasekaran, N., & Mukherjee, A. (2010). Antibacterial activity of sunflower oil microemulsion. *International Journal of Pharmacy and Pharmaceutical Sciences*, *2*(1), 123–128.

Anjum, M. M., Patel, K. K., Dehari, D., Pandey, N., Tilak, R., Agrawal, A. K., & Singh, S. (2020). Anacardic acid encapsulated solid lipid nanoparticles for Staphylococcus aureus biofilm therapy: Chitosan and DNase coating improves antimicrobial activity. *Drug Delivery and Translational Research*, •••, 1–13. PMID:32519201

Ansari, M. A., Khan, H. M., Khan, A. A., Cameotra, S. S., & Pal, R. (2014). Antibiofilm efficacy of silver nanoparticles against biofilm of extended spectrum β-lactamase isolates of Escherichia coli and Klebsiella pneumoniae. *Applied Nanoscience*, *4*(7), 859–868. doi:10.100713204-013-0266-1

Aria, Beizai, & Gharib. (2012). *Drug delivery by carbon nanotube arrays.* WO2012031164A2.

Aruguete, D. M., Kim, B., Hochella, M. F., Ma, Y., Cheng, Y., Hoegh, A., Liu, J., & Pruden, A. (2013). Antimicrobial nanotechnology: Its potential for the effective management of microbial drug resistance and implications for research needs in microbial nanotoxicology. *Environmental Science. Processes & Impacts*, *15*(1), 93–102. doi:10.1039/C2EM30692A PMID:24592430

Atyabi, F., Adeli, M., Sobhani, Z., Dinarvand, R., & Ghahremani, M. H. (2013). *U.S. Patent No. 8,460,711*. Washington, DC: U.S. Patent and Trademark Office.

Aucouturier, J., Dupuis, L., Deville, S., Ascarateil, S., & Ganne, V. (2002). Montanide ISA 720 and 51: A new generation of water in oil emulsions as adjuvants for human vaccines. *Expert Review of Vaccines*, *1*(1), 111–118. doi:10.1586/14760584.1.1.111 PMID:12908518

Azevedo, M. M., Ramalho, P., Silva, A. P., Teixeira-Santos, R., Pina-Vaz, C., & Rodrigues, A. G. (2014). Polyethyleneimine and polyethyleneimine-based nanoparticles: Novel bacterial and yeast biofilm inhibitors. *Journal of Medical Microbiology*, *63*(9), 1167–1173. doi:10.1099/jmm.0.069609-0 PMID:24913563

Azhdarzadeh, M., Lotfipour, F., Zakeri-Milani, P., Mohammadi, G., & Valizadeh, H. (2012). Anti-bacterial performance of azithromycin nanoparticles as colloidal drug delivery system against different gram-negative and gram-positive bacteria. *Advanced Pharmaceutical Bulletin*, *2*(1), 17. PMID:24312766

Beyth, N., Houri-Haddad, Y., Domb, A., Khan, W., & Hazan, R. (2015). Alternative antimicrobial approach: Nano-antimicrobial materials. *Evidence-Based Complementary and Alternative Medicine*, 2015. PMID:25861355

Bitzan, M., Poole, R., Mehran, M., Sicard, E., Brockus, C., Thuning-Roberson, C., & Rivière, M. (2009). Safety and pharmacokinetics of chimeric anti-Shiga toxin 1 and anti-Shiga toxin 2 monoclonal antibodies in healthy volunteers. *Antimicrobial Agents and Chemotherapy*, *53*(7), 3081–3087. doi:10.1128/AAC.01661-08 PMID:19414580

Blecher, K., Nasir, A., & Friedman, A. (2011). The growing role of nanotechnology in combating infectious disease. *Virulence*, *2*(5), 395–401. doi:10.4161/viru.2.5.17035 PMID:21921677

Bloom, D. E., Black, S., & Rappuoli, R. (2017). Emerging infectious diseases: a proactive approach. *Proceedings of the National Academy of Sciences.*

Bolla, P. K., Kalhapure, R. S., Rodriguez, V. A., Ramos, D. V., Dahl, A., & Renukuntla, J. (2019). Preparation of solid lipid nanoparticles of furosemide-silver complex and evaluation of antibacterial activity. *Journal of Drug Delivery Science and Technology*, *49*, 6–13. doi:10.1016/j.jddst.2018.10.035

Bonifácio, B. V., da Silva, P. B., dos Santos Ramos, M. A., Negri, K. M. S., Bauab, T. M., & Chorilli, M. (2014). Nanotechnology-based drug delivery systems and herbal medicines: A review. *International Journal of Nanomedicine*, *9*, 1. PMID:24363556

Cal, P. M., Matos, M. J., & Bernardes, G. J. (2017). Trends in therapeutic drug conjugates for bacterial diseases: A patent review. *Expert Opinion on Therapeutic Patents*, *27*(2), 179–189. doi:10.1080/13543776.2017.1259411 PMID:27828733

Cardoso, M, M., N Peca, I., & CA Roque, A. (. (2012). Antibody-conjugated nanoparticles for therapeutic applications. *Current Medicinal Chemistry*, *19*(19), 3103–3127. doi:10.2174/092986712800784667 PMID:22612698

Cars, O., Hedin, A., & Heddini, A. (2011). The global need for effective antibiotics—Moving towards concerted action. *Drug Resistance Updates*, *14*(2), 68–69. doi:10.1016/j.drup.2011.02.006 PMID:21444235

Chan, J. M., Zhang, L., Tong, R., Ghosh, D., Gao, W., Liao, G., & Golomb, G. (2010). Spatiotemporal controlled delivery of nanoparticles to injured vasculature. *Proceedings of the National Academy of Sciences*. 10.1073/pnas.0914585107

Chen, L. F. (2013). The changing epidemiology of methicillin-resistant Staphylococcus aureus: 50 years of a superbug. *American Journal of Infection Control*, *41*(5), 448–451. doi:10.1016/j.ajic.2012.06.013 PMID:23219674

Chen, W. R. (2014). *U.S. Patent No. 8,664,198*. Washington, DC: U.S. Patent and Trademark Office.

Clatworthy, A. E., Pierson, E., & Hung, D. T. (2007). Targeting virulence: A new paradigm for antimicrobial therapy. *Nature Chemical Biology*, *3*(9), 541–548. doi:10.1038/nchembio.2007.24 PMID:17710100

Colas, J. C., Shi, W., Rao, V. M., Omri, A., Mozafari, M. R., & Singh, H. (2007). Microscopical investigations of nisin-loaded nanoliposomes prepared by Mozafari method and their bacterial targeting. *Micron (Oxford, England)*, *38*(8), 841–847. doi:10.1016/j.micron.2007.06.013 PMID:17689087

Cotter, P. D., Hill, C., & Ross, R. P. (2005). Food microbiology: bacteriocins: developing innate immunity for food. *Nature Reviews. Microbiology*, *3*(10), 777–788. doi:10.1038/nrmicro1273 PMID:16205711

Cotter, P. D., Ross, R. P., & Hill, C. (2013). Bacteriocins—A viable alternative to antibiotics? *Nature Reviews. Microbiology*, *11*(2), 95–105. doi:10.1038/nrmicro2937 PMID:23268227

Dai, H., Kam, N. W. S., Wender, P. A., & Liu, Z. (2012). *U.S. Patent No. 8,246,995*. Washington, DC: U.S. Patent and Trademark Office.

Department of Molecular Virology and Microbiology. (n.d.). *Mosquito-Borne Diseases*. Available on: https://www.bcm.edu/departments/molecular-virology-and-microbiology/emerging-infections-and-biodefense/emerging-infectious-diseases

Dimov, S., Ivanova, P., & Harizanova, N. (2005). Genetics of bacteriocins biosynthesis by lactic acid bacteria. *Biotechnology & Biotechnological Equipment, 19*(sup2), 4-10.

Eckhard, L. H., Houri-Haddad, Y., Sol, A., Zeharia, R., Shai, Y., Beyth, S., & Beyth, N. (2016). Sustained release of antibacterial lipopeptides from biodegradable polymers against oral pathogens. *PLoS One*, *11*(9), e0162537. doi:10.1371/journal.pone.0162537 PMID:27606830

Egusquiaguirre, S. P., Igartua, M., Hernández, R. M., & Pedraz, J. L. (2012). Nanoparticle delivery systems for cancer therapy: Advances in clinical and preclinical research. *Clinical & Translational Oncology*, *14*(2), 83–93. doi:10.100712094-012-0766-6 PMID:22301396

Ekambaram, P., Sathali, A. A. H., & Priyanka, K. (2012). Solid lipid nanoparticles: A review. *Sci Rev Chem Commun*, *2*(1), 80–102.

El-Rashidy, A., Waly, G., Gad, A., Roether, J. A., Hum, J., Yang, Y., & Boccaccini, A. R. (2018). Antibacterial activity and biocompatibility of zein scaffolds containing silver-doped bioactive glass. *Biomedical Materials (Bristol, England)*, *13*(6), 065006. doi:10.1088/1748-605X/aad8cf PMID:30088480

Eshghi, M., Vaghari, H., Najian, Y., Najian, M., Jafarizadeh-Malmiri, H., & Berenjian, A. (2018). Microwave-Assisted Green Synthesis of Silver Nanoparticles Using Juglans regia Leaf Extract and Evaluation of Their Physico-Chemical and Antibacterial Properties. *Antibiotics (Basel, Switzerland)*, *7*(3), 68. doi:10.3390/antibiotics7030068 PMID:30061517

Fahim, H. A., Khairalla, A. S., & El-Gendy, A. O. (2016). Nanotechnology: A valuable strategy to improve bacteriocin formulations. *Frontiers in Microbiology*, *7*, 1385. doi:10.3389/fmicb.2016.01385 PMID:27695440

Frieri, M., Kumar, K., & Boutin, A. (2017). Antibiotic resistance. *Journal of Infection and Public Health*, *10*(4), 369–378. doi:10.1016/j.jiph.2016.08.007 PMID:27616769

Gaspar, D. P., Gaspar, M. M., Eleutério, C. V., Grenha, A., Blanco, M., Gonçalves, L. M., & Remuñán-López, C. (2017). Microencapsulated solid lipid nanoparticles as a hybrid platform for pulmonary antibiotic delivery. *Molecular Pharmaceutics*, *14*(9), 2977–2990. doi:10.1021/acs.molpharmaceut.7b00169 PMID:28809501

Gharib, M., Sansom, E. B., & Aria, A. I. (2010). *U.S. Patent Application No. 12/657,862*. US Patent Office.

Gholami, M., Mohammadi, R., Arzanlou, M., Dourbash, F. A., Kouhsari, E., Majidi, G., & Nazari, S. (2017). In vitro antibacterial activity of poly (amidoamine)-G7 dendrimer. *BMC Infectious Diseases*, *17*(1), 395. doi:10.118612879-017-2513-7 PMID:28583153

Global Antibiotic Research & Development Partnership. (n.d.). *Thailand activates first site in global trial of zoliflodacin as new gonorrhea treatment*. Available on: https://www.hindustantimes.com/health-and-fitness/scientists-develop-faster-method-to-detect-harmful-bacteria-like-superbugs-/story-ILaxy-8D14l-hzuxK-t0COcEP.html

Gu, D., Dong, N., Zheng, Z., Lin, D., Huang, M., Wang, L., & Chen, S. (2018). A fatal outbreak of ST11 carbapenem-resistant hypervirulent Klebsiella pneumoniae in a Chinese hospital: A molecular epidemiological study. *The Lancet. Infectious Diseases*, *18*(1), 37–46. doi:10.1016/S1473-3099(17)30489-9 PMID:28864030

Gupta, V., & Chandra, N. (2020). Biosynthesis and Antibacterial Activity of Metal Oxide Nanoparticles Using Brassica oleracea subsp. botrytis (L.) leaves, an Agricultural Waste. *Proceedings of the National Academy of Sciences. India. Section B, Biological Sciences*, *90*(5), 1093–1100. doi:10.100740011-020-01184-0

Han, J., Zhao, D., Li, D., Wang, X., Jin, Z., & Zhao, K. (2018). Polymer-based nanomaterials and applications for vaccines and drugs. *Polymers*, *10*(1), 31. doi:10.3390/polym10010031 PMID:30966075

Hollmann, B., Perkins, M., & Walsh, D. (n.d.). *Biofilms and their role in pathogenesis*. British Society for Immunology. https://www.immunology.org/public-information/bitesized-immunology/patho gens -and-disease/biofilms-and-their-role-in

Holubnycha, V., Kalinkevich, O., Ivashchenko, O., & Pogorielov, M. (2018). Antibacterial Activity of *In Situ* Prepared Chitosan/Silver Nanoparticles Solution Against Methicillin-Resistant Strains of Staphylococcus aureus. *Nanoscale Research Letters*, *13*(1), 71. doi:10.118611671-018-2482-9 PMID:29500654

Holubnycha, V., Pogorielov, M., Korniienko, V., Kalinkevych, O., Ivashchenko, O., Peplinska, B., & Jarek, M. (2017). Antibacterial activity of the new copper nanoparticles and Cu NPs/chitosan solution. In *Nanomaterials: Application & Properties (NAP), 2017 IEEE 7th International Conference* (pp. 04NB10-1). IEEE.

Hu, C. M. J., Kaushal, S., Cao, H. S. T., Aryal, S., Sartor, M., Esener, S., & Zhang, L. (2010). Half-antibody functionalized lipid– polymer hybrid nanoparticles for targeted drug delivery to carcinoembryonic antigen presenting pancreatic cancer cells. *Molecular Pharmaceutics*, *7*(3), 914–920. doi:10.1021/mp900316a PMID:20394436

Hu, J., Akula, N., & Wang, N. (2016). Development of a microemulsion formulation for antimicrobial SecA inhibitors. *PLoS One*, *11*(3), e0150433. doi:10.1371/journal.pone.0150433 PMID:26963811

Huang, L., Dai, T., Xuan, Y., Tegos, G. P., & Hamblin, M. R. (2011). Synergistic combination of chitosan acetate with nanoparticle silver as a topical antimicrobial: Efficacy against bacterial burn infections. *Antimicrobial Agents and Chemotherapy*, *55*(7), AAC-01803. doi:10.1128/AAC.01803-10 PMID:21502618

Hung, C. F., Fang, C. L., Liao, M. H., & Fang, J. Y. (2007). The effect of oil components on the physicochemical properties and drug delivery of emulsions: Tocol emulsion versus lipid emulsion. *International Journal of Pharmaceutics*, *335*(1-2), 193–202. doi:10.1016/j.ijpharm.2006.11.016 PMID:17129692

Hwang, Y. Y., Ramalingam, K., Bienek, D. R., Lee, V., You, T., & Alvarez, R. (2013). Antimicrobial activity of nanoemulsion in combination with cetylpyridinium chloride on multi-drug resistant Acinetobacter baumannii. *Antimicrobial Agents and Chemotherapy*, •••, AAC-02109. PMID:23669390

Jeevaratnam, K., Jamuna, M., & Bawa, A. S. (2005). *Biological preservation of foods–Bacteriocins of lactic acid bacteria*. Academic Press.

Jennings, J. A., Haggard, W. O., & Bumgardner, J. D. (2010). *U.S. Patent Application No. 12/427,546*. US Patent Office.

Jerobin, J., Makwana, P., Kumar, R. S., Sundaramoorthy, R., Mukherjee, A., & Chandrasekaran, N. (2015). Antibacterial activity of neem nanoemulsion and its toxicity assessment on human lymphocytes in vitro. *International Journal of Nanomedicine*, *10*(Suppl 1), 77. PMID:26491309

Jung, S. W., Thamphiwatana, S., Zhang, L., & Obonyo, M. (2015). Mechanism of antibacterial activity of liposomal linolenic acid against Helicobacter pylori. *PLoS One*, *10*(3), e0116519. doi:10.1371/journal.pone.0116519 PMID:25793403

Kannan, R., Prabakaran, P., Basu, R., Pindi, C., Senapati, S., Muthuvijayan, V., & Prasad, E. (2019). Mechanistic study on the antibacterial activity of self-assembled poly (aryl ether)-based amphiphilic dendrimers. *ACS Applied Bio Materials*, *2*(8), 3212–3224. doi:10.1021/acsabm.9b00140

Kostetsky, E. Y., Sanina, N. M., Mazeika, A. N., Tsybulsky, A. V., Vorobyeva, N. S., & Shnyrov, V. L. (2011). Tubular immunostimulating complex based on cucumarioside A 2-2 and monogalactosyldiacylglycerol from marine macrophytes. *Journal of Nanobiotechnology*, *9*(1), 35. doi:10.1186/1477-3155-9-35 PMID:21888630

Kumar, P. D., Kumar, P. V., Anneer Selvam, T. P., & Rao, K. S. (2015). Prolonged drug delivery system of PEGylated PAMAM dendrimers with a anti-HIV drug. *Research in Pharmacy, 3*(2).

Kumar, S., Dilbaghi, N., Saharan, R., & Bhanjana, G. (2012). Nanotechnology as emerging tool for enhancing solubility of poorly water-soluble drugs. *BioNanoScience*, *2*(4), 227–250. doi:10.100712668-012-0060-7

Kumar, S., Jones, T. R., Oakley, M. S., Zheng, H., Kuppusamy, S. P., Taye, A., Krieg, A. M., Stowers, A. W., Kaslow, D. C., & Hoffman, S. L. (2004). CpG oligodeoxynucleotide and Montanide ISA 51 adjuvant combination enhanced the protective efficacy of a subunit malaria vaccine. *Infection and Immunity*, *72*(2), 949–957. doi:10.1128/IAI.72.2.949-957.2004 PMID:14742540

Kumari, S. (2012). Enhancing the Antimicrobial Activity of Nisin by Encapsulating on Silver Nanoparticle Synthesized by Bacillus sp. *International Journal of Pharmaceutical and Biological Archives*, *3*(2).

Kumari, S., Kumaraswamy, R. V., Choudhary, R. C., Sharma, S. S., Pal, A., Raliya, R., Biswas, P., & Saharan, V. (2018). Thymol nanoemulsion exhibits potential antibacterial activity against bacterial pustule disease and growth promotory effect on soybean. *Scientific Reports*, *8*(1), 8. doi:10.103841598-018-24871-5 PMID:29703904

Lam, B., Das, J., Holmes, R. D., Live, L., Sage, A., Sargent, E. H., & Kelley, S. O. (2013). Solution-based circuits enable rapid and multiplexed pathogen detection. *Nature Communications*, *4*(1), 2001. doi:10.1038/ncomms3001 PMID:23756447

Lazar, H., Horn, M. P., Zuercher, A. W., Imboden, M. A., Durrer, P., Seiberling, M., ... Lang, A. B. (2009). Pharmacokinetics and safety profile of the human anti P. aeruginosa serotype O11 IgM monoclonal antibody KBPA-101 in healthy volunteers. *Antimicrobial Agents and Chemotherapy*. Advance online publication. doi:10.1128/AAC.01699-08

López, E. L., Contrini, M. M., Glatstein, E., Ayala, S. G., Santoro, R., Allende, D., & .... (2010). Safety and pharmacokinetics of urtoxazumab, a humanized monoclonal antibody, against Shiga-like toxin 2 in healthy adults and in pediatric patients infected with Shiga-like toxin-producing Escherichia coli. *Antimicrobial Agents and Chemotherapy*, *54*(1), 239–243. doi:10.1128/AAC.00343-09 PMID:19822704

Lowy, I., Molrine, D. C., Leav, B. A., Blair, B. M., Baxter, R., Gerding, D. N., Nichol, G., Thomas, W. D. Jr, Leney, M., Sloan, S., Hay, C. A., & Ambrosino, D. M. (2010). Treatment with monoclonal antibodies against Clostridium difficile toxins. *The New England Journal of Medicine*, *362*(3), 197–205. doi:10.1056/NEJMoa0907635 PMID:20089970

Lu, W. C., Huang, D. W., Wang, C. C., Yeh, C. H., Tsai, J. C., Huang, Y. T., & Li, P. H. (2018). Preparation, characterization, and antimicrobial activity of nanoemulsions incorporating citral essential oil. *Yao Wu Shi Pin Fen Xi*, *26*(1), 82–89. PMID:29389592

Ma, Q., Davidson, P. M., & Zhong, Q. (2016). Antimicrobial properties of microemulsions formulated with essential oils, soybean oil, and Tween 80. *International Journal of Food Microbiology*, *226*, 20–25. doi:10.1016/j.ijfoodmicro.2016.03.011 PMID:27016636

MacRae, C. A., & Peterson, R. T. (2003). Zebrafish-based small molecule discovery. *Chemistry & Biology*, *10*(10), 901–908. doi:10.1016/j.chembiol.2003.10.003 PMID:14583256

Maleki, A., Ahmadi Jebelli, M., Hayati, B., Daraei, H., & Gharibi, F. (2015). Antimicrobial Effect of Poly (amidoamine)-G2 and G4 Dendrimers on Some Bacteria in water resources. *Majallah-i Danishgah-i Ulum-i Pizishki-i Mazandaran*, *25*(124), 56–68.

Martins, P., Rosa, D. R., Fernandes, A., & Baptista, P. V. (2013). Nanoparticle drug delivery systems: Recent patents and applications in nanomedicine. *Recent Patents on Nanomedicine*, *3*(2), 105–118. doi:10.2174/1877912304666140304000133

Mazumder, A., Davis, J., Rangari, V., & Curry, M. (2013). Synthesis, Characterization, and Applications of Dendrimer-Encapsulated Zero-Valent Ni Nanoparticles as Antimicrobial Agents. *ISRN Nanomaterials*, *2013*, 2013. doi:10.1155/2013/843709

Medina, S. H., & El-Sayed, M. E. (2009). Dendrimers as carriers for delivery of chemotherapeutic agents. *Chemical Reviews*, *109*(7), 3141–3157. doi:10.1021/cr900174j PMID:19534493

Mishra, A., Panola, R., & Rana, A. C. (2014). Microemulsions: As drug delivery system. *J Sci Innov Res*, *3*(4), 467–474.

Moyá, M. L., López-López, M., Lebrón, J. A., Ostos, F. J., Pérez, D., Camacho, V., Beck, I., Merino-Bohórquez, V., Camean, M., Madinabeitia, N., & López-Cornejo, P.Bactericidal Activity of Cefepime Encapsulated into Cationic Liposomes. (2019). Preparation and characterization of new liposomes. Bactericidal activity of cefepime encapsulated into cationic liposomes. *Pharmaceutics*, *11*(2), 69. doi:10.3390/pharmaceutics11020069 PMID:30736367

Myszka, K., & Czaczyk, K. (2011). Bacterial biofilms on food contact surfaces-a review. *Polish Journal of Food and Nutrition Sciences*, *61*(3), 173–180. doi:10.2478/v10222-011-0018-4

Namasivayam, K. R. S., Samrat, K., Debnath, S., & Jayaprakash, C. (2015). Biocompatible chitosan nanoparticles incorporated bacteriocin (CSNps-B) preparation for the controlled release and improved anti-bacterial activity against food borne pathogenic bacteria Listeria monocytogenes. *Research Journal of Pharmaceutical, Biological and Chemical Sciences*, *6*(5), 625–631.

Namasivayam, S. K. R. (2017). Nanoformulation of Antibacterial Antibiotics Cefpirome with Biocompatible Polymeric Nanoparticles and Evaluation for the Improved Antibacterial Activity and Nontarget Toxicity Studies. *Asian Journal of Pharmaceutics*, *11*(02).

Newly emerged superbug discovered. (2017). *The Hong Kong Polytechnic University*. https://www.sciencedaily.com/releases/2017/08/170831101508.html

Nguyen, Y. H., Ma, X., & Qin, L. (2012). Rapid identification and drug susceptibility screening of ESAT-6 secreting Mycobacteria by a NanoELIwell assay. *Scientific Reports*, *2*(1), 635. doi:10.1038rep00635 PMID:22957139

Nilsson, E., Larsson, A., Olesen, H. V., Wejåker, P. E., & Kollberg, H. (2008). Good effect of IgY against Pseudomonas aeruginosa infections in cystic fibrosis patients. *Pediatric Pulmonology*, *43*(9), 892–899. doi:10.1002/ppul.20875 PMID:18680179

O'Hagan, D. T. (2007). MF59 is a safe and potent vaccine adjuvant that enhances protection against influenza virus infection. *Expert Review of Vaccines*, *6*(5), 699–710. doi:10.1586/14760584.6.5.699 PMID:17931151

O'Toole, G., Kaplan, H. B., & Kolter, R. (2000). Biofilm formation as microbial development. *Annual Review of Microbiology*, *54*(1), 49–79. doi:10.1146/annurev.micro.54.1.49 PMID:11018124

Okeke, I. N., Peeling, R. W., Goossens, H., Auckenthaler, R., Olmsted, S. S., de Lavison, J. F., & Nordqvist, K. (2011). Diagnostics as essential tools for containing antibacterial resistance. *Drug Resistance Updates*, *14*(2), 95–106. doi:10.1016/j.drup.2011.02.002 PMID:21398170

Ong, T. H., Chitra, E., Ramamurthy, S., Siddalingam, R. P., Yuen, K. H., Ambu, S. P., & Davamani, F. (2017). Chitosan-propolis nanoparticle formulation demonstrates anti-bacterial activity against Enterococcus faecalis biofilms. *PLoS One*, *12*(3), e0174888. doi:10.1371/journal.pone.0174888 PMID:28362873

Ovchinnikov, K. V., Chi, H., Mehmeti, I., Holo, H., Nes, I. F., & Diep, D. B. (2016). A novel group of leaderless and multi-peptide bacteriocins from Gram-positive bacteria. *Applied and Environmental Microbiology*, AEM-01094. PMID:27316965

Padhy, A., Sahu, A. R., Ranabijuli, S., & Ganguly, S. (2016). *Superbug, an Emerging Global Threat in Current Scenario*. Academic Press.

Peek, L. J., Middaugh, C. R., & Berkland, C. (2008). Nanotechnology in vaccine delivery. *Advanced Drug Delivery Reviews, 60*(8), 915–928. doi:10.1016/j.addr.2007.05.017 PMID:18325628

Peng, Y., Song, C., Yang, C., Guo, Q., & Yao, M. (2017). Low molecular weight chitosan-coated silver nanoparticles are effective for the treatment of MRSA-infected wounds. *International Journal of Nanomedicine, 12*, 295–304. doi:10.2147/IJN.S122357 PMID:28115847

Pignatello, R., Leonardi, A., Fuochi, V., Petronio Petronio, G., Greco, A. S., & Furneri, P. M. (2018). A Method for Efficient Loading of Ciprofloxacin Hydrochloride in Cationic Solid Lipid Nanoparticles: Formulation and Microbiological Evaluation. *Nanomaterials (Basel, Switzerland), 8*(5), 304. doi:10.3390/nano8050304 PMID:29734771

Prakash, A., Baskaran, R., Nithyanand, P., & Vadivel, V. (2019). Effect of Nanoemulsification on the Antibacterial and Anti-biofilm Activities of Selected Spice Essential Oils and Their Major Constituents against Salmonella enterica Typhimurium. *Journal of Cluster Science*, 1–13. doi:10.100710876-019-01720-7

Puri, A., Loomis, K., Smith, B., Lee, J. H., Yavlovich, A., Heldman, E., & Blumenthal, R. (2009). Lipid-based nanoparticles as pharmaceutical drug carriers: from concepts to clinic. *Critical Reviews™ in Therapeutic Drug Carrier Systems, 26*(6).

Qasim, M., Lim, D. J., Park, H., & Na, D. (2014). Nanotechnology for diagnosis and treatment of infectious diseases. *Journal of Nanoscience and Nanotechnology, 14*(10), 7374–7387. doi:10.1166/jnn.2014.9578 PMID:25942798

Qasim, M., Udomluck, N., Chang, J., Park, H., & Kim, K. (2018). Antimicrobial activity of silver nanoparticles encapsulated in poly-N-isopropylacrylamide-based polymeric nanoparticles. *International Journal of Nanomedicine, 13*, 235–249. doi:10.2147/IJN.S153485 PMID:29379284

Qin, D., He, X., Wang, K., Zhao, X. J., Tan, W., & Chen, J. (2007). Fluorescent nanoparticle-based indirect immunofluorescence microscopy for detection of Mycobacterium tuberculosis. *BioMed Research International*. PMID:18273415

Rai, M. K., Deshmukh, S. D., Ingle, A. P., & Gade, A. K. (2012). Silver nanoparticles: The powerful nanoweapon against multidrug-resistant bacteria. *Journal of Applied Microbiology, 112*(5), 841–852. doi:10.1111/j.1365-2672.2012.05253.x PMID:22324439

Ramos, M. A. D. S., Da Silva, P. B., Spósito, L., De Toledo, L. G., Bonifácio, B. V., Rodero, C. F., ... Bauab, T. M. (2018). Nanotechnology-based drug delivery systems for control of microbial biofilms: A review. *International Journal of Nanomedicine*, *13*, 1179–1213. doi:10.2147/IJN.S146195 PMID:29520143

Raval, J. P., Naik, D. R., Amin, K. A., & Patel, P. S. (2014). Controlled-release and antibacterial studies of doxycycline-loaded poly (ε-caprolactone) microspheres. *Journal of Saudi Chemical Society*, *18*(5), 566–573. doi:10.1016/j.jscs.2011.11.004

Rea, M. C., Alemayehu, D., Ross, R. P., & Hill, C. (2013). Gut solutions to a gut problem: Bacteriocins, probiotics and bacteriophage for control of Clostridium difficile infection. *Journal of Medical Microbiology*, *62*(9), 1369–1378. doi:10.1099/jmm.0.058933-0 PMID:23699066

Saavedra, L., Minahk, C., de Ruiz Holgado, A. P., & Sesma, F. (2004). Enhancement of the enterocin CRL35 activity by a synthetic peptide derived from the NH2-terminal sequence. *Antimicrobial Agents and Chemotherapy*, *48*(7), 2778–2781. doi:10.1128/AAC.48.7.2778-2781.2004 PMID:15215149

Sadekuzzaman, M., Yang, S., Mizan, M. F. R., & Ha, S. D. (2015). Current and recent advanced strategies for combating biofilms. *Comprehensive Reviews in Food Science and Food Safety*, *14*(4), 491–509. doi:10.1111/1541-4337.12144

Scheinberg, D. A., McDevitt, M. R., Villa, C. H., & Mulvey, J. J. (2018). *U.S. Patent No. 9,976,137*. Washington, DC: U.S. Patent and Trademark Office.

Shaaban, H. A., & Edris, A. E. (2015). Factors affecting the phase behavior and antimicrobial activity of carvacrol microemulsions. *Journal of Oleo Science*, *64*(4), 393–404. doi:10.5650/jos.ess14194 PMID:25766929

Shah, P., Bhalodia, D., & Shelat, P. (2010). Nanoemulsion: a pharmaceutical review. *Systematic Reviews in Pharmacy, 1*(1).

Sharma, A., Kumar Arya, D., Dua, M., Chhatwal, G. S., & Johri, A. K. (2012). *Nanotechnology for targeted drug delivery to combat antibiotic resistance*. Academic Press.

Sharma, T. K., Sapra, M., Chopra, A., Sharma, R., Patil, S. D., Malik, R. K., Pathania, R., & Navani, N. K. (2012). Interaction of bacteriocin-capped silver nanoparticles with food pathogens and their antibacterial effect. *International Journal of Green Nanotechnology*, *4*(2), 93–110. doi:10.1080/19430892.2012.678757

Shazly, G. A. (2017). Ciprofloxacin controlled-solid lipid nanoparticles: Characterization, in vitro release, and antibacterial activity assessment. *BioMed Research International*.

Sidhu, P. K., & Nehra, K. (2017). Bacteriocin-Nanoconjugates as Emerging Compounds for Enhancing Antimicrobial Activity of Bacteriocins. *Journal of King Saud University. Science*.

Silva, N. P., Menacho, F. P., & Chorilli, M. (2012). Dendrimers as potential platform in nanotechnology-based drug delivery systems. *IOSR Journal of Pharmacy, 2*, 23–30. doi:10.9790/3013-25602330

Solleti, V. S., Alhariri, M., Halwani, M., & Omri, A. (2014). Antimicrobial properties of liposomal azithromycin for Pseudomonas infections in cystic fibrosis patients. *The Journal of Antimicrobial Chemotherapy, 70*(3), 784–796. doi:10.1093/jac/dku452 PMID:25416744

StarPharma. (n.d.). *Vivagel®*. Availabe on: http://www.starpharma.com

Stern, H. M., Murphey, R. D., Shepard, J. L., Amatruda, J. F., Straub, C. T., Pfaff, K. L., Weber, G., Tallarico, J. A., King, R. W., & Zon, L. I. (2005). Small molecules that delay S phase suppress a zebrafish bmyb mutant. *Nature Chemical Biology, 1*(7), 366–370. doi:10.1038/nchembio749 PMID:16372403

Subramanian, G. M., Cronin, P. W., Poley, G., Weinstein, A., Stoughton, S. M., Zhong, J., & Freimuth, W. W. (2005). A phase 1 study of PAmAb, a fully human monoclonal antibody against Bacillus anthracis protective antigen, in healthy volunteers. *Clinical Infectious Diseases, 41*(1), 12–20. doi:10.1086/430708 PMID:15937757

Suchomel, P., Kvitek, L., Panacek, A., Prucek, R., Hrbac, J., Vecerova, R., & Zboril, R. (2015). Comparative study of antimicrobial activity of AgBr and Ag nanoparticles (NPs). *PloS one, 10*(3).

Tang, J., Chen, W., Su, W., Li, W., & Deng, J. (2013). Dendrimer-encapsulated silver nanoparticles and antibacterial activity on cotton fabric. *Journal of Nanoscience and Nanotechnology, 13*(3), 2128–2135. doi:10.1166/jnn.2013.6883 PMID:23755656

Teixeira, M. L., dos Santos, J., Silveira, N. P., & Brandelli, A. (2008). Phospholipid nanovesicles containing a bacteriocin-like substance for control of Listeria monocytogenes. *Innovative Food Science & Emerging Technologies, 9*(1), 49–53. doi:10.1016/j.ifset.2007.05.001

Tian, W. L., Lei, L. L., Zhang, Q., & Li, Y. (2016). Physical Stability and Antimicrobial Activity of Encapsulated Cinnamaldehyde by Self-Emulsifying Nanoemulsion. *Journal of Food Process Engineering, 39*(5), 462–471. doi:10.1111/jfpe.12237

Torres-Sangiao, E., Holban, A. M., & Gestal, M. C. (2016). Advanced nanobiomaterials: Vaccines, diagnosis and treatment of infectious diseases. *Molecules (Basel, Switzerland)*, *21*(7), 867. doi:10.3390/molecules21070867 PMID:27376260

Tostes Gazzinelli, Ladeira, Furtado, & Batista Faria Gontijo. (2013). *Anti-tumour vaccine formulation based on carbon nanotubes and the use thereof.* WO2013110150A1.

Tour, J. M., Lucente-Schultz, R., Leonard, A., Kosynkin, D. V., Price, B. K., Hudson, J. L., & Milas, Z. L. (2014). *U.S. Patent No. 8,784,866*. Washington, DC: U.S. Patent and Trademark Office.

Valdes, C., Bustos, G., Martinez, J. L., & Laurido, C. (2018). Antinociceptive antibiotics-loaded into solid lipid nanoparticles of prolonged release: Measuring pharmacological efficiency and time span on chronic monoarthritis rats. *PLoS One*, *13*(4), e0187473. doi:10.1371/journal.pone.0187473 PMID:29649262

Vembu, S., Pazhamalai, S., & Gopalakrishnan, M. (2015). Potential antibacterial activity of triazine dendrimer: Synthesis and controllable drug release properties. *Bioorganic & Medicinal Chemistry*, *23*(15), 4561–4566. doi:10.1016/j.bmc.2015.06.009 PMID:26113186

Ventola, C. L. (2015). The antibiotic resistance crisis: part 1: causes and threats. *P&T*, *40*(4), 277–283. PMID:25859123

Von Wintersdorff, C. J., Penders, J., van Niekerk, J. M., Mills, N. D., Majumder, S., van Alphen, L. B., Savelkoul, P. H. M., & Wolffs, P. F. G. (2016). Dissemination of antimicrobial resistance in microbial ecosystems through horizontal gene transfer. *Frontiers in Microbiology*, *7*, 173. doi:10.3389/fmicb.2016.00173 PMID:26925045

Waheed, A., Gupta, A., & Patel, P. (2015). Targeted Drug Delivery Systems for Lung Cancer. *PharmaTutor*, *3*(2), 38–42.

Wang, L., Hu, C., & Shao, L. (2017). The antimicrobial activity of nanoparticles: Present situation and prospects for the future. *International Journal of Nanomedicine*, *12*, 1227–1249. doi:10.2147/IJN.S121956 PMID:28243086

Weisman, L. E., Thackray, H. M., Garcia-Prats, J. A., Nesin, M., Schneider, J. H., Fretz, J., Kokai-Kun, J. F., Mond, J. J., Kramer, W. G., & Fischer, G. W. (2009). Phase 1/2 double-blind, placebo-controlled, dose escalation, safety, and pharmacokinetic study of pagibaximab (BSYX-A110), an antistaphylococcal monoclonal antibody for the prevention of staphylococcal bloodstream infections, in very-low-birth-weight neonates. *Antimicrobial Agents and Chemotherapy*, *53*(7), 2879–2886. doi:10.1128/AAC.01565-08 PMID:19380597

Wilkins, M., Hall-Stoodley, L., Allan, R. N., & Faust, S. N. (2014). New approaches to the treatment of biofilm-related infections. *The Journal of Infection*, *69*, S47–S52. doi:10.1016/j.jinf.2014.07.014 PMID:25240819

Winnicka, K., Wroblewska, M., Wieczorek, P., Sacha, P. T., & Tryniszewska, E. A. (2013). The effect of PAMAM dendrimers on the antibacterial activity of antibiotics with different water solubility. *Molecules (Basel, Switzerland)*, *18*(7), 8607–8617. doi:10.3390/molecules18078607 PMID:23881050

World Health Organisation. (n.d.). *Coronavirus*. https://www.who.int/health-topics/coronavirus#tab=tab_1

Wrońska, N., Majoral, J. P., Appelhans, D., Bryszewska, M., & Lisowska, K. (2019). Synergistic effects of anionic/cationic dendrimers and levofloxacin on antibacterial activities. *Molecules (Basel, Switzerland)*, *24*(16), 2894. doi:10.3390/molecules24162894 PMID:31395831

Yang, S. C., Lin, C. H., Sung, C. T., & Fang, J. Y. (2014). Antibacterial activities of bacteriocins: Application in foods and pharmaceuticals. *Frontiers in Microbiology*, *5*, 241.

Yildirim, S. T., Oztop, M. H., & Soyer, Y. (2017). Cinnamon oil nanoemulsions by spontaneous emulsification: Formulation, characterization and antimicrobial activity. *Lebensmittel-Wissenschaft + Technologie*, *84*, 122–128.

Zacharof, M. P., & Lovitt, R. W. (2012). Bacteriocins produced by lactic acid bacteria a review article. *APCBEE Procedia*, *2*, 50–56.

Zaman, S. B., Hussain, M. A., Nye, R., Mehta, V., Mamun, K. T., & Hossain, N. (2017). A review on antibiotic resistance: Alarm bells are ringing. *Cureus*, *9*(6).

Zendo, T. (2013). Screening and characterization of novel bacteriocins from lactic acid bacteria. *Bioscience, Biotechnology, and Biochemistry*, *77*(5), 893–899.

Zhang, L., Chan, J. M., Gu, F. X., Rhee, J. W., Wang, A. Z., Radovic-Moreno, A. F., & ... , & Farokhzad, O. C. (2008). Self-assembled lipid– polymer hybrid nanoparticles: A robust drug delivery platform. *ACS Nano*, *2*(8), 1696–1702.

Zhao, L., Seth, A., Wibowo, N., Zhao, C. X., Mitter, N., Yu, C., & Middelberg, A. P. (2014). Nanoparticle vaccines. *Vaccine*, *32*(3), 327–337.

Zhu, X., Radovic-Moreno, A. F., Wu, J., Langer, R., & Shi, J. (2014). Nanomedicine in the management of microbial infection–overview and perspectives. *Nano Today*, *9*(4), 478–498.

Zohri, M., Alavidjeh, M. S., Haririan, I., Ardestani, M. S., Ebrahimi, S. E. S., Sani, H. T., & Sadjadi, S. K. (2010). A comparative study between the antibacterial effect of nisin and nisin-loaded chitosan/alginate nanoparticles on the growth of Staphylococcus aureus in raw and pasteurized milk samples. *Probiotics and Antimicrobial Proteins*, *2*(4), 258–266.

Zohri, M., Shafiee Alavidjeh, M., Mirdamadi, S. S., Behmadi, H., Hossaini Nasr, S. M., Eshghi Gonbaki, S., & Jabbari Arabzadeh, A. (2013). Nisin-Loaded Chitosan/Alginate Nanoparticles: A Hopeful Hybrid Biopreservative. *Journal of Food Safety*, *33*(1), 40–49.

# Chapter 11
# Harnessing the Capability of CADD Methods in the Prediction of Anti-COVID Drug Likeliness

**Shubhra Chaturvedi**
*Institute of Nuclear Medicine and Allied Sciences, India*

**Vishaka Chaudhary**
*Institute of Nuclear Medicine and Allied Sciences, India*

**Tina Klauss**
*Université de Bordeaux, France*

**Philippe Barthélémy**
*ChemBioPharm, France*

**Anil Kumar Mishra**
*Institute of Nuclear Medicine and Allied Sciences, India*

## ABSTRACT

*The COVID-19 pandemic has claimed many lives and added to the social, economic, and psychological distress. The contagious disease has quickly spread to almost 200 countries following the regional outbreak in China. As the number of infected populations increases exponentially, there is a pressing demand for anti-COVID drugs and vaccines. Virtual screening provides possible leads while extensively cutting down the time and resources required for ab-initio drug design. The chapter aims to highlight the various computer-aided drug design methods to predict an anti-COVID drug molecule.*

DOI: 10.4018/978-1-7998-0307-2.ch011

## 1. INTRODUCTION

Nearly the entire year 2020 was marred by the deleterious effect of the pandemic, COVID 19 infection. The infection claimed many lives and added to the social, economic, and psychological distress. The contagious disease that had an initial regional outbreak in Wuhan, China, quickly spread to almost all countries. The World Health Organization declared the disease as a pandemic in March 2020. As the number of infected populations increased exponentially, the medical research fraternity faced pressing demand for anti-COVID therapeutic solutions like drugs and vaccines. Despite some recent success claims in vaccine development, the search for an ideal drug candidate continues. Predominantly, three drugs that have ambiguous success include hydroxyquinoline- an antimalarial drug with quinoline as the pharmacophore, remedesivir- a nucleoside analog with an interfering role in RNA replication, and dexamethasone- a steroid that modulates the inflammatory response. However, as none of the above candidates are "the drug"against COVID-19, the scientists continue searching for drugs. With limited workforce resources due to the infected personnel or preventive lockdown and time-pressures for early therapeutic solutions, computer-aided drug design emerged as a preferred tool with many scientific reports that focussed on drug repurposing using screening methods.

Researchers applied computational screening methods to the FDA approved drugs to cut down on cost and time as the already tested and approved drugs were assumed to be safe for immediate clinical application. Media reports included mega-projects of Scripps Research, US announcing compound library screening of over 14000 drugs, the screening of drugs for augmenting remedisivir by Calibr scientist, and many more.

The quantum of the reported work in a short span is unparallel. However, in the absence of uniform protocols, though the protocol may be complete and standardized in itself, there have been contradictory and ambiguous findings. The reliability of the screening data hence needs to be ascertained. Further, as the predicted drugs, especially the established antivirals, did not elicit desired results in clinical settings, the scientists were forced to include varied categories of the molecules while screening. While mining extensive databases with varied scaffolds, newer screening protocols based extensively on machine learning methods were also developed. The chapter aims to highlight the various computer-aided drug design methods to predict an anti-COVID drug molecule. (Figure 1)

*Figure 1. Highlights of the chapter*

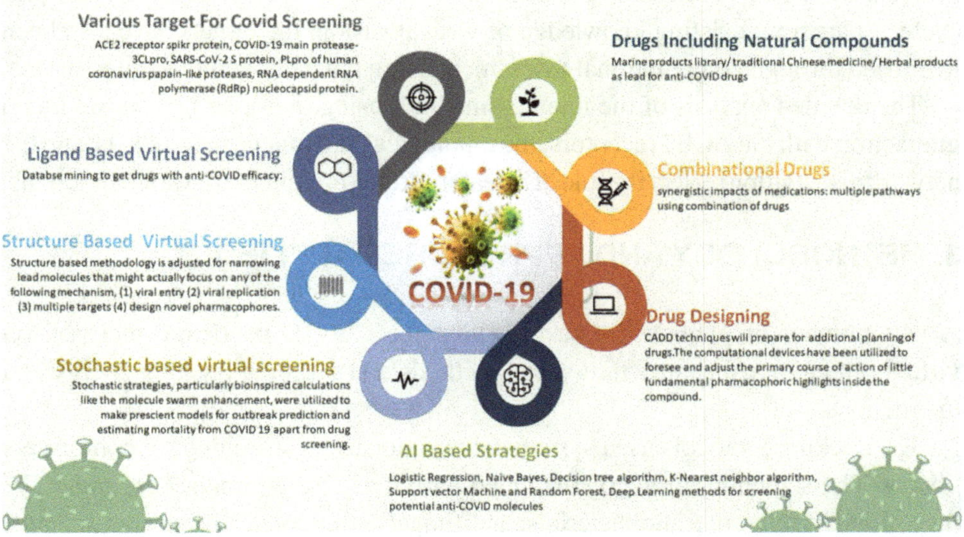


## 2. TARGETS USED FOR COVID SCREENING

The targets for anti-COVID drug development play vital role in virus pathogenesis and are classified as targets during (1) viral entry, (2) viral replication, (3) viral repackaging and release, and (4) immune response modulators of the host (human). The initial and widely reported targets for the anti-COVID drug screening have been the proteins, particularly proteases involved in the various COVID life cycle stages (Hu et al., 2020)

A detailed literature covering the biochemical role of these targets already exists. For computer aided drug design (CADD) applications, the targets and associated relevant information is summarized in Table 1. Among the various targets, the spike protein has been of the particular interest, inhibition of which can potentially impede virus entry into the host cell. The 3D structure of the variants of coronavirus spike protein are available in RCSB PDB repository, for example PDB 6ACD as a complex with the ACE2 receptor (Song et al., 2018), PDBs 6CRV, 6CRW, 6CRX, 6CRZ, 6CS0, 6CS1 and 6CS2 as cryo-EM analyses of ACE-2 bound trimeric SARS-CoV spike protein, and various trypsin-cleaved, stabilized SARS-CoV spike protein variants (Kirchdoerfer et al., 2018). Homology predictive model of the spike protein using the SARS-CoV-2 sequence (GenBank accession number: MT159721.1, length: 29882 bp) was used as a target when the structure was not known (Bharath et al., 2020). In March 2020, SARS-CoV-2 RBD–ACE2 complex cryo-structure was deposited as PDB 6M0J, that has been used extensively for CADD applications (Lan et al., 2020).

As listed in the Table 1, structured and non-structured proteins and accessory proteins have been indicated as potential targets at different stages of the viral cycle. Apart from existing knowledge of virus invasion, the targets have also been identified through computational assessment using the pathogen-host interactome.

The detailed analysis of the interactome built using the spike protein's domain interaction with the ACE2 receptors, the topological and functional analysis, and the network distribution reflect the role of at least fourteen proteins with varied functions.

## 3. METHODOLOGY AND APPROACHES APPLIED

A vast number of published data following the COVID pandemic focussed on virtual screening using structure-based methods and explored docking energy and interactions.

The docking protocols used systematic and stochastic searching for conformers library. The screening protocols included antivirals like antimalarials, anti-HCV (hepatitis-Cvirus, *etc.*, antibacterials, antifungals, diuretics, immune-stimulants, and naturally occurring compounds. However, due to the absence of well-defined protocols, the results have varied key molecules. Few compounds also showed the ability for multi-targeting. Such compounds have been predicted to have better pharmacological utility. The results of published data have been summarized in the review article by Mohamed et al., 2020. Reports used structure-based (SBVS) and ligand-based virtual screening (LBVS) methods alone or in combination, and frequently, the results were also supplemented using molecular dynamics. In screening, multiple databases and softwares have been used.

### 3.1 Databases for Anti-Covid Screening

Databases are the backbone for *in silico* drug design. The databases list not only the compounds, but the pharmacological data, activities, potential targets, mechanism of action and various other important parameters. Conventionally and for COVID -19 repurposing of drugs screening has been carried using FDA approved drug databases, DrugCentral, ZINC, ChEMBL, Chemical Abstracts Service (CAS) that offers ≤50,000 compounds with tested or predicted antiviral activities, Natural compounds databases listing compounds in marine and traditional medicines, PubChem - a convenient access to antiviral chemicals and related information, SuperDRUG2 that is a repository of more than 4000 drugs, etc. In addition dedicated databases came into effect to save the time and resources, for example COVID-19 Drug-Repurposing Database (COVIDDRD) that provides data for COVID-19 drug-repurposing, DrugBank designed "COVID-19 Information Dashboard" that enables the direct access to pharmacological data of ~40 repurposed drugs, IUPHAR/BPS

Guide to PHARMACOLOGY (GtoPdb) having expert-curated SAR relationships on investigative COVID-19 drugs, and Therapeutic Target Database (TTD) provides data of clinical/preclinical COVID-19 drugs. (Wang et al., 2020)

*Table 1. Various COVID Targets reported in the studies referred in the chapter and their representative available PDBs*

| Targeted pathway in viral life cycle | Protein Target | Role | PDBs |
|---|---|---|---|
| Viral Entry and structural protein | ACE2 receptor-spike protein interface | restrict the binding of SARS-CoV2 spike protein to the ACE2 receptor | Templates reported for predictive homology modelling: 6ACD, PDBs 6CRV, 6CRW, 6CRX, 6CRZ, 6CS0, 6CS1, 6CS2 |
| | | | Crystallographic PDB:6M0J, 6VXX (Walls et al., 2020) |
| Cleaving RNA genome | EndoRNAse (nsp15) | Cleave both single-stranded and double-stranded RNA. | 6W01 (RNA uridylate-specific endoribonuclease (NendoU) Kim et al., 2020) |
| viral replication and transcription | COVID-19 main protease-$M^{pro}$/C30 endopeptidase/ 3CLpro | Cysteine proteases | 6LU7 (Jin et al., 2020) |
| Targeting viral replication | PLpro of human coronavirus papain-like proteases/ ORF3a protein | 1. responsible for the N-terminus cleavage of the replicate poly-protein to nsp1-3<br>2. Essential for correcting virus replication.<br>3. significant in modulation of the innate immunity of the host | 6W9C (Osipiuk,et al., to be published) |
| Replicating Viral Genome | RNA dependent RNA polymerase (RdRp)/ nsp 12 | Viral replication | 7BTF, 6M71 (Gao et al., 2020) |
| | helicase (nsp 13) | Blocks the ATPase activity to inhibit the replication of viral genome | 6JYT (Jia et al., 2020) |
| | 3′-to-5′ exonuclease (nsp 14) | Critically involved in SARS-COV-2 RNA synthesis | Models reported using predictive modelling |
| transcription and assembly/ structural proteins | nucleocapsid protein | forms complexes with genomic RNA of the virus | 6M3M (Kang et al., 2020) |
| RNA synthesis and processing | 2'-o- methyltransferase/ nsp 16 | involved in the capping of coronaviral mRNAs | 6W75 (Rosas-Lemus et al., 2020) |
| structural coronaviral protein | small envelope (E) glycoprotein (ORF 4) / membrane (M) glycoprotein (ORF 5) | Involved in several aspects of virus life cycle such as assembly, envelop formation and pathogenesis | Homology modelling: Apo form (PDB-ID: 6M03) and Holo form (PDB-ID: 6LU7, bound with N3 inhibitor) forms |
| Non structural proteins (nsp) | Various classes: nsp1, nsp 2, nsp4 | Multiple functions in replication, packaging and release | 6YWL for nsp3 (Schroeder et al., to be published, RCSB) |
| Host target | transmembrane protease serine 2 (TMPRSS2) | Modulates Viral entry | Predictive modelling using 2OQ5 as reported Elmezayen et al., 2020 |

## 3.2 Ligand based Virtual Screening (LBVS)

Ligand based virtual screening has been reported using representative softwares like FLAP (Fingerprints for Ligands and Proteins: Baroni et al., 2007), JChem fingerprint along with in-house developed softwares like D3Similarity (Yang et al, 2021).Purely based on ligand-based approach, Bocci et al. narrowed down on nine hits while screening a library of 4000 approved drugs. The database molecules were compared and ranked based on Glob-Prod (GP) score evaluated using different molecular interaction fields along with other structure-activity parameters. Hydroxyquinoline was regarded as the cut-off molecule and the potential molecules were then assessed *in vitro* on a cell line for cytotoxicity on SARS-CoV-2.

## 3.3 Structure Based Virtual Screening (SBVS)

Schrödinger's Glide, Autodock Vina, rDOCK, UCSF Chimera platform are among the different softwares that were used to carry SBVS for COVID 19 potential drug molecules. Structure based methodology was adapted for narrowing drugs that could potentially target, though not limited to, any of the following mechanisms: (1) inhibit entry of the virus by binding with the viral S-protein (Choudhary et al., 2020), (2) inhibit replicase machinery of the virus like disrupting nsp 9 (Chandra et al., 2021), (3) screening inhibitors for multiple implicated COVID 19 targets (Hosseini, et al., 2020), (4) to study interactions with COVID 19 targets like interaction studies of Mpro with therapeutically indicated drugs for COVID (Marinho, et al., 2020).

Dedicated servers to carry docking studies were also reported like the DINC-COVID that account for the flexibility in protein through ensemble docking algorithm. The server was validated through inhibitor prediction for Mpro (Hall-Swan et al., 2021).

The poses obtained from the docking studies were also subjected to dynamic studies that indicate the stability of the protein-ligand complex. For instance, 50 to 100 ns simulations using different platforms like AMBER/ GROMACS (GROningenMAchine for Chemical Simulations) / YASARA have been performed on docked poses of synthetic or naturally derived chemical compounds with COVID target proteins to understand the dynamic behavior of ligands (Quimque, et al., 2020: Sen Gupta et al., 2020).

Various parameters like the surface areas (solvent–accessible surface area (SASA) and molecular surface area (MolSA)), bond parameters, radius of gyration (Rg), binding energy per-frame, root-mean-square deviation (RMSD), hydrogen-bonding (HB), root-mean square fluctuation (RMSF), center-of-mass (CoM) distance and principal component analysis (PCA) can be studied to arrive at stable conformation and interactions of the ligand with protein. In addition, for a more accurate analysis binding energies ($\Delta$Gbinding) can also be calculated using the Molecular mechanics (MM), MM-GBSA (generalized Born surface area) and MM/PBSA (Poisson–Boltzmann surface area) approaches.

## 3.4 Stochastic Methods Based Virtual Screening

Stochastic methods, especially bioinspired algorithms like the particle swarm optimization, were used to make predictive models for outbreak prediction and estimating mortality from COVID- 19. Probabilistic models were also reported by Czuppon et al., 2020, to predict the prophylactic activity of antivirals when used alone or in combination. The model included within-host viral dynamics based on viral kinetics in epithelial cell along with parameters for humoral immune response and virion release. The prophylactic prediction of the antiviral therapy accounted for (i) viral infectivity reduction $\beta$, (ii) enhanced viral clearance c, (iii) decrease in viral production p, and (iv) enhanced infected cell death $\delta$. The critical efficacy and time scales for viral load and extinction were also reported using the model. Using mathematical modeling, viral dynamics has been studied after developing suitable within- host viral models using single and dual (upper and lower respiratory tract) physiological compartments. The models were used to predict viral load, infectiousness of an individual, infection to symptom onset, probability of transmission etc. (Ke et al., 2020).

Genetic algorithms using the GOLD software was used for docking and predicting the inhibitors for Mpro protein. The results indicated that Apixabin, an anticoagulant can be a potential candidate molecule (da Silva Hage-Melim et al., 2020).

## 3.5 AI Based Methods

Apart from the conventional docking programs, artificial intelligence and machine learning methods found wide application, not only in the utilization to analyze the clinical data of chest X-rays and CT data but also for virtual screening. As the AI methods are designed to analyze, categorize and predict with some degree of in-built decision making, AI methods distinguish from the conventional algorithms for screening. Few algorithms used for AI methods are listed in Table 2.

The AI methods were applied to predict the antibody variants, prediction of targets apart from the conventional proteases, and screening of massive drug/ compound datasets for the novel specific scaffold. The methodology typically followed the following pattern: (1) Compound database with approved or drug-like molecules including natural products, (2) in silico representation of the compounds through descriptors, (3) conformational searching through extensive sampling, or activity screening through docking or other parameters or both and (4) parallel application of

AI based methods. For a rigorous validation of results, combined scores of different algorithm programs were also used. The concept can be appreciated through the research published by Onawole et al., wherein scoring obtained from machine learning method based virtual screening and molecular docking were considered in parallel to evaluate surface glycoprotein inhibitors. The approach is termed as vote rank method.

When different methods for example docking and molecular vector- / structure based screening using deep learning methods gave different lead molecules while screening TargetMol-Approved-Drug-Library, molecular dynamics studies were carried to ascertain the results. Thus, hybrid screening methods can help to identify potential lead molecules. (Zhang et al., 2020(b))

Because of several AI algorithms, Xu et al., 2020 evaluated the algorithms for their capability in virtual screening. A combination of LB and SBVS dataset screening of 2030 Chinese natural medicine was carried using six commonly used ML algorithms. The six algorithms used were Logistic Regression (LR), Naïve Bayes, Decision tree algorithm (DT), K-Nearest neighbor algorithm (kNN), Support vector Machine (SVM) and Random Forest (RF). When compared on the basis of the statistical parameters like area under the curve (AUC), the LR outperformed the other five ML algorithms. Similarly, Zhang et al., 2020(a), evaluated eight AI algorithms (DT, kNN, SVM, RF, ERT, AdaBoost, GBT and XGBoost) using the various statistical parameters. Random forest (RF) regression models combined with Vina docking scores were used to screen library of 19000 molecules that can have potential to disrupt virus entry by binding with the S protein. Important molecular fragments in the compounds were oxolanes, benzene sulfonate groups and imidazoles that aid in binding. The results also indicated the importance of certain descriptors like the topological surface area and ring count. (Batra et al., 2020)

The protein-protein interactions are vital for the entry of the viral particles into the host. Learning vector Quantization in conjugation with multiple supervised learning algorithms was used for predicting as many 1326 potential human proteins as a target for anti-COVID strategies due to viral protein and human protein interactions (Dey et al., 2020). Deep docking protocols are machine learning methods that utilize QSAR deep models trained on docking parameters for screening billions of compounds with a certain level of accuracy. In the study (Ton et al., 2020), Deep docking protocols were developed for mining as many as 1.3 billion compounds from the ZINC dataset. The QSAR model utilized Morgan fingerprinting for screening, along with a neural network that works on the feed-forward back-propagation principle.

*Table 2. Examples of few algorithms for AI based methods used in references and cited in the chapter*

| AI methods | support vector machine | Random Forest | Naive Bayes | K-nearest neighbors' algorithm | Decision Tree | Logistic Regression |
|---|---|---|---|---|---|---|
| | (SVM) | (RF) | (NB) | (k-NN) | (DT) | (LR) |
| **Type of learning method** | Supervised machine learning technique (instance based method) | Supervised machine learning technique (Ensemble learning methods) | Supervised learning methods (Probabilistic classifier, non-linear model) | Supervised learning methods (non-parametric instance-based learning, /lazy learning/ memory-based method) | Supervised learning method (Regression analysis methods, non-linear model). | Supervised learning algorithm (classification algorithm, statistical model, linear model) |
| **Primary tasks**<br>classification and regression includes pattern recognition and outlier's detection | Classification, and regression | Classification, and regression | Used for constructing classifiers | Classification, and regression and making predictions | Classify, or predict values for input data | classify, or predict values for input data |
| **Ideal for** | Variable separation | Large datasets with multiple features. | Large datasets with random noise. | Calculating properties with strong locality, e.g. protein function prediction | Dimensionality reduction | solving classification related problems. |
| **Based on** | vectors that help in creating the hyperplane. | Multitude of uncorrelated decision trees as an ensemble | Applies bayes' theorem with the "naive" assumption of conditional independence between features and the class variable. | Local approximation for classification | Flowchart-like /Tree-shaped structure having representations of chains of decisions. | Utilizes a sigmoid function and works best on binary classification problems. |
| **Aides in** | Attributing missing data, working with outliers, and estimating characteristics for classification. | Attributing missing data, working with outliers, and estimating characteristics for classification | Predicting biological properties. E.g.. Toxicity, bioactivity classification for drug-like molecules | Predict the class, property, or rank of a molecule based on nearest training examples. | Multiclass problems, and aides l in handling non linear data sets | Hypothesis testing, model coefficient and classification data interpretation. |
| **Error frequency/ Limitation** | Least number of errors | Least number of errors | Least number of errors | Irrelevant attributes in high-dimensional space may mislead and may not handle large volume of data | Poor handling of noisy or incomplete data | Target variables classification important |
| **Applications (*though not restricted to)** | Categorization, classification of images, classification of data like SAR | Regression and classification task | Medical data classification and real-time predictions | Detecting outliers. | Classification | Prediction. |
| | Drug-target interactions: distinguishing between active and inactive compounds: ranking compounds<br>Regression model | | target interactions, novel targets classification tools for non-related information biomedical data, and<br>Predicting ligand-target interactions | Chemical similarity search, Predict other bioactivities of drug-like molecules. | For predicting QSAR relationships,<br>Fast conformational search<br>Promising new compounds | |

(Wang et al. 2005; Segal et al. 2004; Madhukar et al. 2019: Lavecchia, et al 2015: Zhang et al. 2017: The features listed are in general though exceptions may exist.)

## 4. APPLICATIONS OF CADD METHODS FOR DRUG SCREENING

### 4.1 Combinational Drugs

Researchers also explored not only multi-targeting but synergistic effects of drugs. Zhou et al. used network proximity analysis and complementary exposure patterns to identify drugs that target the various aspects of the viral cycle and other potential host cellular pathways. Three combinations were proposed; (1) Sirolimus plus Dactinomycin can be used to target mTOR signaling, and RNA synthesis pathway, (2) structural interference in viral glycoproteins and disruption of interaction of spike protein with ACE2 could potentially be effected using Toremifene plus Emodin combination, and (3) Inhibition of PLP and modulation of the inflammatory response through a combination of Mercaptopurine plus Melatonin (Zhou et al., 2020). Synergistic effects of drugs (A) lopinavir, (B) oseltamivir and (C) ritonavir was studied through sequential docking (Muralidharan et al., 2020).

### 4.2 Designing of Drugs

CADD methods will pave the way for further designing of drugs. Zhang et al. proposed alpha-keto amides as peptidomimetic compounds that can potentially serve as broad-spectrum antiviral agents for alphacoronaviruses, beta coronaviruses, and enteroviruses. The genesis is based on docking results with crystal structures of the main protease of different viruses. The *in silico* results were validated by wet-lab synthesis and *in vitro* validation (Zhang et al., 2020). The computational tools have also been used to predict and fine-tune the structural arrangement of small essential pharmacophoric features within the compounds. For instance, in the report by Kumar et al., the authors using computational structure-activity relationship predicted the role and structural arrangement of hydroxyethyl amine (HEA) within antiviral compounds. Similarly, an effective inhibitor for $3CL^{Pro}$ was speculated to have a preference for an aromatic bicyclic scaffold bearing N and O atoms and flexible linkers with NH groups to facilitate interactions with the substrate site. The presence of carboxyl or trifluoromethyl groups in the rings was also noted among the scorers in the virtual screening of compounds (Olubiyi et al., 2020). Heterocyclic bearing scaffolds like the oxadiazole and isatin moieties were screened as potential inhibitors for the main protease $M^{pro}$ (Badavath et al., 2020)

For example, *in silico* and machine learning methods, an unsupervised Markov Cluster algorithm was also applied to identify aptamers as recognition molecules against COVID. The 51- and 67- base containing hairpin structured aptamers were screened as potential candidates using the SMART-Aptamer algorithm. Unique scoring methods viz., motif enrichment score (K), aptamer abundance (F), stability score (S), and penalty score for families to counter opposing changes (P) were used. After optimizing the aptamers, MD simulations offered an insight into the binding of the aptamers with the spike protein (Song et al., 2020).

## 1.3 Novel Classes of Drugs Including Natural Compounds

Other classes of drugs that have been explored include antipsychotics, aptamers, and drugs from other natural sources. Antipsychotic drug molecules were screened *in silico* and *in vitro* for their inhibitory effect via binding with the ACE2 receptor. Nineteen drugs were docked using the ACE2 PDB (6MOJ), of which five antipsychotics, especially phenothiazines, were realized to block SARS-CoV-2 interaction with ACE2 (Lu et al., 2020). Histamine-2 receptor antagonist, Famotidine was also investigated computationally for exact mechanism of action as an anti-viral, after reports of its usage in clinically on COVID-19 patients. Though an indirect effect of Famotidine can be traced in modulation of immune response Sen Gupta et al., 2020 investigated the molecule as potential inhibitor against COVID protein targets using SBVS and molecular dynamics of 100 ns.

Bacterial compounds have for long exhibited antibacterial and antiviral compounds. Gentile et al. narrowed down on phlorotannins (brown alga), flavonoids such as Apigenin-7-*O*-neohesperidoside, Luteolin-7-rutinoside, and Resinoside and Peptidomimetic derivatives like pseudotheonamides based on docking and molecular dynamics results with SARS-CoV-2 $M^{pro}$ after screening the marine natural product library with nearly 14000 compounds. The group initially screened the library using a binding-site-derived pharmacophore model to realize the importance of dimensional pharmacophore that accounts for stearic and electronic arrangement. The objective was to retain compounds that can serve as Michael acceptors covalent inhibitor and, if not non-peptidic molecules, then should have peptidomimetic regions. A combination of the supervised machine learning algorithm, Random forest, and docking protocol was utilized to classify and screen marine natural compounds library with 494 entries that ultimately lead to the identification of five potential inhibitors for $M^{pro}$. (Gaudêncio et al., 2020).

Lianhuaqingwen (LH) is a traditional Chinese medicine (TCM) preparation with broad-spectrum antiviral effects on different influenza virus strains. LH also showed anti-2019nCoV activity. Having honeysuckle and forsythia as active compounds in LH with antiviral potency, the LH components were examined for anti-Covid activity. The reported application of LH includes its usage in the treatment of SARS in 2003. In this study, the binding of chloroquine, remdesivir, ribavirin, and luteolin (the active component of honeysuckle) with the main proteins of 2019-nCoV (3CLpro, PLpro, RdRp, and S) were carried out by computational methods. (Yu et al., 2020)

Xu et al., while screening Chinese medicinal compounds not only evaluated different AI methods (referred above) but also narrowed on active compounds like Rutin as effective inhibitor for 3CLpro·

Natural products belonging to Nigerian and Indian plants were also screened for efficacy against COVID-19. Over 3000 natural products of Nigerian plants origin, majority being flavonoids, along with the ZINC database compounds were screened for inhibitory activities against 3CLpro.Amentoflavone, Glabrolide and zeylanone were some of the lead molecules. The work was based on structure based screening using AutoDock. (Olubiyi, et al., 2020)

Rather than targeting viral proteins, host (human) proteins were also proposed to serve as therapeutic targets against COVID 19. For example, Transmembrane Protease Serine 2 or TMPRSS2, a host protease or cathepsins that plays important processesing step for viral entry. Indian herbs phytochemicals were virtually screened in the study by Vivek-Ananth et al., 2020 through docking interactions using AutoDock Vina and molecular dynamics simulations.

More than 35,000 compounds in the Natural Product Activity and Species Source (NPASS) database were screened through SBVS for anti-COVID activity by targeting multiple targets (endoribonuclease, exoribonuclease, RNA-dependent RNA polymerase, methyltransferase and 3C-like proteinase) (Naik et al., 2020).

With modeling tools, ninety seven compounds that are antiviral secondary metabolites from fungi were screened for potency. The five SARS-CoV2 targets PLpro, 3CLpro, RdRp, nsp15, and the spike binding domain were docked with the compounds followed by molecular dynamics simulation. Three fumiquinazoline alkaloids, the polyketide isochaetochromin D1, and terpenoid based metabolite exhibited high binding affinities. (Quimque et al., 2020)

## 5. CONCLUSION

CADD tools can edge scientists in preliminary screening by cutting short the time and resources investment. However, the lead molecules obtained as results of CADD need extensive testing before any clinical applications. Past evidences exist wherein even when designed for a dedicated application drugs failed during pre-clinical to clinical translation. Repurposed drugs will certainly require testing on actual targets before any conclusive efficacy.

The other aspect is the standardization of CADD tools. With different algorithms and scoring functions, arriving on a single molecule during screening may not be feasible. Validating the CADD workflows will help to address the ambiguity in results.

However, CADD tools have certainly proved their utility, especially in the urgency during the COVID-19 infection.

## REFERENCES

Badavath, V. N., Kumar, A., Samanta, P. K., Maji, S., Das, A., Blum, G., Jha, A., & Sen, A. (2020). Determination of potential inhibitors based on isatin derivatives against SARS-CoV-2 main protease (mpro): A molecular docking, molecular dynamics and structure-activity relationship studies. *Journal of Biomolecular Structure & Dynamics*, 1–19. doi:10.1080/07391102.2020.1845800 PMID:33200681

Baroni, M., Cruciani, G., Sciabola, S., Perruccio, F., & Mason, J. S. (2007). A common reference framework for analyzing/comparing proteins and ligands. Fingerprints for Ligands and Proteins (FLAP): Theory and application. *Journal of Chemical Information and Modeling*, *47*(2), 279–294. doi:10.1021/ci600253e PMID:17381166

Batra, R., Chan, H., Kamath, G., Ramprasad, R., Cherukara, M. J., & Sankaranarayanan, S. K. (2020). Screening of therapeutic agents for COVID-19 using machine learning and ensemble docking studies. *The Journal of Physical Chemistry Letters*, *11*(17), 7058–7065. doi:10.1021/acs.jpclett.0c02278 PMID:32787328

Bharath, B. R., Damle, H., Ganju, S., & Damle, L. (2020). In silico screening of known small molecules to bind ACE2 specific RBD on Spike glycoprotein of SARS-CoV-2 for repurposing against COVID-19. *F1000 Research*, 9. PMID:32765844

Blower, P. E., & Cross, K. P. (2006). Decision tree methods in pharmaceutical research. *Current Topics in Medicinal Chemistry*, *6*(1), 31–39. doi:10.2174/156802606775193301 PMID:16454756

Bocci, G., Bradfute, S. B., Ye, C., Garcia, M. J., Parvathareddy, J., Reichard, W., Surendranathan, S., Bansal, S., Bologa, C. G., Perkins, D. J., Jonsson, C. B., Sklar, L. A., & Oprea, T. I. (2020). Virtual and In Vitro Antiviral Screening Revive Therapeutic Drugs for COVID-19. *ACS Pharmacology & Translational Science*, *3*(6), 1278–1292. doi:10.1021/acsptsci.0c00131 PMID:33330842

Chandra, A., Gurjar, V., Ahmed, M. Z., Alqahtani, A. S., Qamar, I., & Singh, N. (2021). Exploring potential inhibitor of SARS-CoV2 replicase from FDA approved drugs using insilico drug discovery methods. *Journal of Biomolecular Structure & Dynamics*, 1–8. PMID:33491573

Choudhary, S., Malik, Y. S., & Tomar, S. (2020). Identification of SARS-CoV-2 cell entry inhibitors by drug repurposing using in silico structure-based virtual screening approach. *Frontiers in Immunology*, *11*, 1664. doi:10.3389/fimmu.2020.01664 PMID:32754161

Czuppon, P., Débarre, F., Gonçalves, A., Tenaillon, O., Perelson, A. S., Guedj, J., & Blanquart, F. (2020). *Predicted success of prophylactic antiviral therapy to block or delay SARS-CoV-2 infection depends on the targeted mechanism*. MedRxiv.

da Silva Hage-Melim, L. I., Federico, L. B., de Oliveira, N. K. S., Francisco, V. C. C., Correia, L. C., de Lima, H. B., ... Francischini, I. A. G. (2020). Virtual screening, ADME/Tox predictions and the drug repurposing concept for future use of old drugs against the COVID-19. *Life Sciences*, *256*, 117963. doi:10.1016/j.lfs.2020.117963 PMID:32535080

Dey, L., Chakraborty, S., & Mukhopadhyay, A. (2020). Machine learning techniques for sequence-based prediction of viral–host interactions between SARS-CoV-2 and human proteins. *Biomedical Journal*, *43*(5), 438–450. doi:10.1016/j.bj.2020.08.003 PMID:33036956

Elmezayen, A. D., Al-Obaidi, A., Şahin, A. T., & Yelekçi, K. (2020). Drug repurposing for coronavirus (COVID-19): In silico screening of known drugs against coronavirus 3CL hydrolase and protease enzymes. *Journal of Biomolecular Structure & Dynamics*, 1–13. doi:10.1080/07391102.2020.1798812 PMID:32306862

Gao, Y., Yan, L., Huang, Y., Liu, F., Zhao, Y., Cao, L., & Rao, Z. (2020). Structure of the RNA-dependent RNA polymerase from COVID-19 virus. *Science*, *368*(6492), 779–782. doi:10.1126cience.abb7498 PMID:32277040

Gaudêncio, S. P., & Pereira, F. (2020). A Computer-Aided Drug Design Approach to Predict Marine Drug-Like Leads for SARS-CoV-2 Main Protease Inhibition. *Marine Drugs*, *18*(12), 633. doi:10.3390/md18120633 PMID:33322052

Gentile, D., Patamia, V., Scala, A., Sciortino, M. T., Piperno, A., & Rescifina, A. (2020). Putative inhibitors of SARS-CoV-2 main protease from a library of marine natural products: A virtual screening and molecular modeling study. *Marine Drugs*, *18*(4), 225. doi:10.3390/md18040225 PMID:32340389

Hall-Swan, S., Antunes, D. A., Devaurs, D., Rigo, M. M., Kavraki, L. E., & Zanatta, G. (2021). DINC-COVID: A webserver for ensemble docking with flexible SARS-CoV-2 proteins. *bioRxiv*.

Hosseini, M., Chen, W., Xiao, D., & Wang, C. (2020). Computational molecular docking and virtual screening revealed promising SARS-CoV-2 drugs. *Precision Clinical Medicine*. PMID:33842834

Hu, B., Guo, H., Zhou, P., & Shi, Z. L. (2020). Characteristics of SARS-CoV-2 and COVID-19. *Nature Reviews. Microbiology*, 1–14. PMID:33024307

Jia, Z., Yan, L., Ren, Z., Wu, L., Wang, J., Guo, J., Zheng, L., Ming, Z., Zhang, L., Lou, Z., & Rao, Z. (2019). Delicate structural coordination of the Severe Acute Respiratory Syndrome coronavirus Nsp13 upon ATP hydrolysis. *Nucleic Acids Research*, *47*(12), 6538–6550. doi:10.1093/nar/gkz409 PMID:31131400

Jin, Z., Du, X., Xu, Y., Deng, Y., Liu, M., Zhao, Y., Zhang, B., Li, X., Zhang, L., Peng, C., Duan, Y., Yu, J., Wang, L., Yang, K., Liu, F., Jiang, R., Yang, X., You, T., Liu, X., ... Yang, H. (2020). Structure of M pro from SARS-CoV-2 and discovery of its inhibitors. *Nature*, *582*(7811), 289–293. doi:10.103841586-020-2223-y PMID:32272481

Kang, S., Yang, M., Hong, Z., Zhang, L., Huang, Z., Chen, X., He, S., Zhou, Z., Zhou, Z., Chen, Q., Yan, Y., Zhang, C., Shan, H., & Chen, S. (2020). Crystal structure of SARS-CoV-2 nucleocapsid protein RNA binding domain reveals potential unique drug targeting sites. *Acta Pharmaceutica Sinica. B*, *10*(7), 1228–1238. doi:10.1016/j.apsb.2020.04.009 PMID:32363136

Ke, R., Zitzmann, C., Ribeiro, R. M., & Perelson, A. S. (2020). Kinetics of SARS-CoV-2 infection in the human upper and lower respiratory tracts and their relationship with infectiousness. medRxiv. doi:10.1101/2020.09.25.20201772

Kim, Y., Jedrzejczak, R., Maltseva, N. I., Wilamowski, M., Endres, M., Godzik, A., Michalska, K., & Joachimiak, A. (2020). Crystal structure of Nsp15 endoribonuclease NendoU from SARS-CoV-2. *Protein Science*, *29*(7), 1596–1605. doi:10.1002/pro.3873 PMID:32304108

Kirchdoerfer, R. N., Wang, N., Pallesen, J., Wrapp, D., Turner, H. L., Cottrell, C. A., Corbett, K. S., Graham, B. S., McLellan, J. S., & Ward, A. B. (2018). Stabilized coronavirus spikes are resistant to conformational changes induced by receptor recognition or proteolysis. *Scientific Reports*, *8*(1), 1–11. doi:10.103841598-018-34171-7 PMID:29311619

Kumar, S., Sharma, P. P., Shankar, U., Kumar, D., Joshi, S. K., Pena, L., & Kempaiah, P. (2020). Poonam,; Rathi, B. Discovery of New HydroxyethylamineAnalogs against 3CLpro Protein Target of SARS-CoV-2: Molecular Docking, Molecular Dynamics Simulation, and Structure–Activity Relationship Studies. *J. Chem. Inf. Model.*

Lan, J., Ge, J., Yu, J., Shan, S., Zhou, H., Fan, S., Zhang, Q., Shi, X., Wang, Q., Zhang, L., & Wang, X. (2020). Structure of the SARS-CoV-2 spike receptor-binding domain bound to the ACE2 receptor. *Nature*, *581*(7807), 215–220. doi:10.103841586-020-2180-5 PMID:32225176

Lavecchia, A. (2015). Machine-learning approaches in drug discovery: Methods and applications. *Drug Discovery Today*, *20*(3), 318–331. doi:10.1016/j.drudis.2014.10.012 PMID:25448759

Lo, Y. C., Rensi, S. E., Torng, W., & Altman, R. B. (2018). Machine learning in chemoinformatics and drug discovery. *Drug Discovery Today*, *23*(8), 1538–1546. doi:10.1016/j.drudis.2018.05.010 PMID:29750902

Lu, J., Hou, Y., Ge, S., Wang, X., Wang, J., Hu, T., & Wang, C. (2021). Screened antipsychotic drugs inhibit SARS-CoV-2 binding with ACE2 in vitro. *Life Sciences*, *266*, 118889. doi:10.1016/j.lfs.2020.118889 PMID:33310043

Madhukar, N. S., Khade, P. K., Huang, L., Gayvert, K., Galletti, G., Stogniew, M., & Elemento, O. (2019). A Bayesian machine learning approach for drug target identification using diverse data types. *Nature Communications*, *10*(1), 1–14. doi:10.103841467-019-12928-6 PMID:31745082

Marinho, E. M., de Andrade Neto, J. B., Silva, J., da Silva, C. R., Cavalcanti, B. C., Marinho, E. S., & Júnior, H. V. N. (2020). Virtual screening based on molecular docking of possible inhibitors of Covid-19 main protease. *Microbial Pathogenesis*, *148*, 104365. doi:10.1016/j.micpath.2020.104365 PMID:32619669

Mohamed, K., Yazdanpanah, N., Saghazadeh, A., & Rezaei, N. (2021). Computational drug discovery and repurposing for the treatment of Covid-19: A systematic review. *Bioorganic Chemistry*, *106*, 104490. doi:10.1016/j.bioorg.2020.104490 PMID:33261845

Muralidharan, N., Sakthivel, R., Velmurugan, D., & Gromiha, M. M. (2020). Computational studies of drug repurposing and synergism of lopinavir, oseltamivir and ritonavir binding with SARS-CoV-2 protease against COVID-19. *Journal of Biomolecular Structure & Dynamics*, 1–6. PMID:32248766

Naik, B., Gupta, N., Ojha, R., Singh, S., Prajapati, V. K., & Prusty, D. (2020). High throughput virtual screening reveals SARS-CoV-2 multi-target binding natural compounds to lead instant therapy for COVID-19 treatment. *International Journal of Biological Macromolecules*, *160*, 1–17. doi:10.1016/j.ijbiomac.2020.05.184 PMID:32470577

Olubiyi, O. O., Olagunju, M., Keutmann, M., Loschwitz, J., & Strodel, B. (2020). High throughput virtual screening to discover inhibitors of the main protease of the coronavirus SARS-CoV-2. *Molecules (Basel, Switzerland)*, *25*(14), 3193. doi:10.3390/molecules25143193 PMID:32668701

Onawole, A. T., Sulaiman, K. O., Kolapo, T. U., Akinde, F. O., & Adegoke, R. O. (2020). COVID-19: CADD to the rescue. *Virus Research*, *285*, 198022. doi:10.1016/j.virusres.2020.198022 PMID:32417181

Quimque, M. T. J., Notarte, K. I. R., Fernandez, R. A. T., Mendoza, M. A. O., Liman, R. A. D., Lim, J. A. K., Pilapil, L. A. E., Ong, J. K. H., Pastrana, A. M., Khan, A., Wei, D.-Q., & Macabeo, A. P. G. (2020). Virtual screening-driven drug discovery of SARS-CoV2 enzyme inhibitors targeting viral attachment, replication, post-translational modification and host immunity evasion infection mechanisms. *Journal of Biomolecular Structure & Dynamics*, 1–18. doi:10.1080/07391102.2020.1776639 PMID:32476574

Rosas-Lemus, M., Minasov, G., Shuvalova, L., Inniss, N. L., Kiryukhina, O., Brunzelle, J., & Satchell, K. J. (2020). High-resolution structures of the SARS-CoV-2 2′-O-methyltransferase reveal strategies for structure-based inhibitor design. *Science Signaling*, *13*(651), eabe1202. doi:10.1126cisignal.abe1202 PMID:32994211

Segal, M. R. (2004). *Machine learning benchmarks and random forest regression*. Academic Press.

Sen Gupta, P. S., Biswal, S., Singha, D., & Rana, M. K. (2020). Binding insight of clinically oriented drug famotidine with the identified potential target of SARS-CoV-2. *Journal of Biomolecular Structure & Dynamics*, 1–7. PMID:32579065

Song, W., Gui, M., Wang, X., & Xiang, Y. (2018). Cryo-EM structure of the SARS coronavirus spike glycoprotein in complex with its host cell receptor ACE2. *PLoS Pathogens*, *14*(8), e1007236. doi:10.1371/journal.ppat.1007236 PMID:30102747

Song, Y., Song, J., Wei, X., Huang, M., Sun, M., Zhu, L., Lin, B., Shen, H., Zhu, Z., & Yang, C. (2020). Discovery of aptamers targeting the receptor-binding domain of the SARS-CoV-2 spike glycoprotein. *Analytical Chemistry*, *92*(14), 9895–9900. doi:10.1021/acs.analchem.0c01394 PMID:32551560

Ton, A. T., Gentile, F., Hsing, M., Ban, F., & Cherkasov, A. (2020). Rapid identification of potential inhibitors of SARS-CoV-2 main protease by deep docking of 1.3 billion compounds. *Molecular Informatics*, *39*(8), 2000028. doi:10.1002/minf.202000028 PMID:32162456

Vivek-Ananth, R. P., Rana, A., Rajan, N., Biswal, H. S., & Samal, A. (2020). In silico identification of potential natural product inhibitors of human proteases key to SARS-CoV-2 infection. *Molecules (Basel, Switzerland)*, *25*(17), 3822. doi:10.3390/molecules25173822 PMID:32842606

Walls, A. C., Park, Y. J., Tortorici, M. A., Wall, A., McGuire, A. T., & Veesler, D. (2020). Structure, Function, and Antigenicity of the SARS-CoV-2 Spike Glycoprotein. *Cell, 181*(2), 281–292.e6. doi:10.1016/j.cell.2020.02.058 PMID:32155444

Wang, L. (Ed.). (2005). *Support vector machines: theory and applications* (Vol. 177). Springer Science & Business Media. doi:10.1007/b95439

Wang, Y., Li, F., Zhang, Y., Zhou, Y., Tan, Y., Chen, Y., & Zhu, F. (2020). Databases for the targeted COVID-19 therapeutics. *British Journal of Pharmacology, 177*(21), 4999–5001. doi:10.1111/bph.15234 PMID:32845521

Xu, Z., Yang, L., Zhang, X., Zhang, Q., Yang, Z., Liu, Y., Wei, S., & Liu, W. (2020). Discovery of Potential Flavonoid Inhibitors Against COVID-19 3CL Proteinase Based on Virtual Screening Strategy. *Frontiers in Molecular Biosciences, 7*, 556481. doi:10.3389/fmolb.2020.556481 PMID:33134310

Yang, Y., Zhu, Z., Wang, X., Zhang, X., Mu, K., Shi, Y., ... Zhu, W. (2021). Ligand-based approach for predicting drug targets and for virtual screening against COVID-19. *Briefings in Bioinformatics*.

Yu, R., Chen, L., Lan, R., Shen, R., & Li, P. (2020). Computational screening of antagonists against the SARS-CoV-2 (COVID-19) coronavirus by molecular docking. *International Journal of Antimicrobial Agents, 56*(2), 106012. doi:10.1016/j.ijantimicag.2020.106012 PMID:32389723

Zhang, L., Lin, D., Kusov, Y., Nian, Y., Ma, Q., Wang, J., von Brunn, A., Leyssen, P., Lanko, K., Neyts, J., de Wilde, A., Snijder, E. J., Liu, H., & Hilgenfeld, R. (2020a). α-Ketoamides as broad-spectrum inhibitors of coronavirus and enterovirus replication: Structure-based design, synthesis, and activity assessment. *Journal of Medicinal Chemistry, 63*(9), 4562–4578. doi:10.1021/acs.jmedchem.9b01828 PMID:32045235

Zhang, H., Yang, Y., Li, J., Wang, M., Saravanan, K. M., Wei, J., Tze-Yang Ng, J., Tofazzal Hossain, M., Liu, M., Zhang, H., Ren, X., Pan, Y., Peng, Y., Shi, Y., Wan, X., Liu, Y., & Wei, Y. (2020b). A novel virtual screening procedure identifies Pralatrexate as inhibitor of SARS-CoV-2 RdRp and it reduces viral replication in vitro. *PLoS Computational Biology, 16*(12), e1008489. doi:10.1371/journal.pcbi.1008489 PMID:33382685

Zhang, L., Tan, J., Han, D., & Zhu, H. (2017). From machine learning to deep learning: Progress in machine intelligence for rational drug discovery. *Drug Discovery Today, 22*(11), 1680–1685. doi:10.1016/j.drudis.2017.08.010 PMID:28881183

Zhang, Y., Wang, Y., Zhou, W., Fan, Y., Zhao, J., Zhu, L., Lu, S., Lu, T., Chen, Y., & Liu, H. (2019). A Combined Drug Discovery Strategy Based on Machine Learning and Molecular Docking. *Chemical Biology & Drug Design*, *93*(5), 685–699. Advance online publication. doi:10.1111/cbdd.13494 PMID:30688405

Zhou, Y., Hou, Y., Shen, J., Huang, Y., Martin, W., & Cheng, F. (2020). Network-based drug repurposing for novel coronavirus 2019-nCoV/SARS-CoV-2. *Cell Discovery*, *6*(1), 1–18. doi:10.103841421-020-0153-3 PMID:33723226

# Related Readings

To continue IGI Global's long-standing tradition of advancing innovation through emerging research, please find below a compiled list of recommended IGI Global book chapters and journal articles in the areas of microbial infections, drug resistance, and solutions to avoid future epidedemics. These related readings will provide additional information and guidance to further enrich your knowledge and assist you with your own research.

Aasi, P., Rusu, L., Leidner, D., Perjons, E., & Estrada, M. C. (2018). What is the Role of Organizational Culture in IT Governance Performance of Collaborative Virtual Networks? *International Journal of IT/Business Alignment and Governance*, *9*(1), 21–37. doi:10.4018/IJITBAG.2018010102

Abdullah, S., You, K. Y., Chong, C. Y., & Ali, M. S. (2020). Milk Pasteurization and Characterization Using Mono-Mode Microwave Reactor and Slotted Coaxial Antenna. In V. Kharchenko & P. Vasant (Eds.), *Handbook of Research on Energy-Saving Technologies for Environmentally-Friendly Agricultural Development* (pp. 107–138). IGI Global. doi:10.4018/978-1-5225-9420-8.ch005

Adom, D., Sekyere, P. A., & Krishnappa, M. K. (2020). The Chemical Constituents, Anti-Inflammatory, Anti-Oxidant, and Ethnomedicinal Properties of Aloe barbadensis. In Ethnomedicinal Plant Use and Practice in Traditional Medicine (pp. 181-198). IGI Global. http://doi:10.4018/978-1-7998-1320-0.ch009

Ahmad, J., & Majid, K. (2021). Nano-Structured Materials for Efficient Energy Storage and Conversion. In M. A. Bhat, I. A. Wani, & S. Ashraf (Eds.), *Applications of Nanomaterials in Agriculture, Food Science, and Medicine* (pp. 165–187). IGI Global. doi:10.4018/978-1-7998-5563-7.ch010

Ahmed, E., & Matouk, A. E. (2020). Fractional-Order Calculus in Antimicrobial Resistance and Waning Vaccination. In A. E. Matouk (Ed.), *Advanced Applications of Fractional Differential Operators to Science and Technology* (pp. 1–20). IGI Global. doi:10.4018/978-1-7998-3122-8.ch001

Ajaykumar, C. (2021). Efficacy of Herbal Medicine in Treating Metabolic and Endocrine Disorders. In A. Hussain & S. Behl (Eds.), *Treating Endocrine and Metabolic Disorders With Herbal Medicines* (pp. 236–255). IGI Global. doi:10.4018/978-1-7998-4808-0.ch010

Akash, N., Bhandari, B. S., Bisht, S. S., & Mansotra, D. K. (2020). Ethnobotany. In Ethnomedicinal Plant Use and Practice in Traditional Medicine (pp. 38-57). IGI Global. http://doi:10.4018/978-1-7998-1320-0.ch003

Akhoon, S. A., Sofi, A. H., Khan, R. A., Tantray, A. M., & Rubab, S. (2021). Application of LiMn2O4 Nanostructures as Efficient Cathodes for Energy Storage Devices. In M. A. Bhat, I. A. Wani, & S. Ashraf (Eds.), *Applications of Nanomaterials in Agriculture, Food Science, and Medicine* (pp. 204–228). IGI Global. doi:10.4018/978-1-7998-5563-7.ch012

Ambure, P., Ballesteros, A., Huertas, F., Camilleri, P., Barigye, S. J., & Gozalbes, R. (2020). Development of Generalized QSAR Models for Predicting Cytotoxicity and Genotoxicity of Metal Oxides Nanoparticles. *International Journal of Quantitative Structure-Property Relationships*, *5*(4), 83–100. doi:10.4018/IJQSPR.20201001.oa2

Amin, S. A., Adhikari, N., Gayen, S., & Jha, T. (2018). First Report on the Validated Classification-Based Chemometric Modeling of Human Rhinovirus 3C Protease (HRV 3Cpro) Inhibitors. *International Journal of Quantitative Structure-Property Relationships*, *3*(2), 1–20. doi:10.4018/IJQSPR.2018070101

Ashokkumar, K., Murugan, M., Dhanya, M. K., Sathyan, T., Raj, S., & Mathews, N. (2020). Traditional Uses, Phytochemistry, and Pharmacological Properties of Zingiber officinale Essential Oil and Extracts. In N. Mishra (Ed.), *Ethnopharmacological Investigation of Indian Spices* (pp. 62–84). IGI Global. doi:10.4018/978-1-7998-2524-1.ch005

Awasthi, M. D., Pandey, M. K., Chauhan, T., Danish, M. S., Kumar, D., & Faisal, N. (2021). Contemporary Developments in Waste Water Treatment Technologies. In Eco-Friendly Energy Processes and Technologies for Achieving Sustainable Development (pp. 196-219). IGI Global. doi:10.4018/978-1-7998-4915-5.ch011

Babikir, O. M. (2021). Agribusiness and Innovation Systems in Africa. In F. N. Che, K. D. Strang, & N. R. Vajjhala (Eds.), *Opportunities and Strategic Use of Agribusiness Information Systems* (pp. 13–32). IGI Global. doi:10.4018/978-1-7998-4849-3.ch002

Babu, N., & Navneet. (2020). Ethnopharmacological Properties of Family Asteraceae. In *Ethnomedicinal Plant Use and Practice in Traditional Medicine* (pp. 199-219). IGI Global. http://doi:10.4018/978-1-7998-1320-0.ch010

Bajaj, S., & Manchanda, S. (2019). Nutraceutical and Functional Foods in Cancer Management and Therapy. In R. K. Keservani, A. K. Sharma, & R. K. Kesharwani (Eds.), *Nutraceutical and Functional Foods in Disease Prevention* (pp. 341–390). IGI Global. doi:10.4018/978-1-5225-3267-5.ch012

Bano, F., Fakhri, K. U., & Rizvi, M. M. (2021). Targeted Drug Delivery in Cancer Treatment. In S. Kumar, M. A. Rizvi, & S. Verma (Eds.), *Handbook of Research on Advancements in Cancer Therapeutics* (pp. 356–381). IGI Global. doi:10.4018/978-1-7998-6530-8.ch012

Barhoi, D., Giri, S., Upadhaya, P., & Nath Barbhuiya, S. (2021). Chemopreventive and Therapeutic Potential of Phytopharmaceuticals Against Oral Cancer. In S. Kumar, M. A. Rizvi, & S. Verma (Eds.), *Handbook of Research on Advancements in Cancer Therapeutics* (pp. 541–569). IGI Global. doi:10.4018/978-1-7998-6530-8.ch020

Bashir, R., Chisti, H. T., Rangreez, T. A., & Mobin, R. (2021). Role of Nanostructured Materials in Health and Medicine. In M. A. Bhat, I. A. Wani, & S. Ashraf (Eds.), *Applications of Nanomaterials in Agriculture, Food Science, and Medicine* (pp. 310–329). IGI Global. doi:10.4018/978-1-7998-5563-7.ch017

Basu, A., Sarkar, A., & Basak, P. (2018). Nutraceuticals for Human Health and Hypersensitivity Reaction. In A. Verma, K. Srivastava, S. Singh, & H. Singh (Eds.), *Nutraceuticals and Innovative Food Products for Healthy Living and Preventive Care* (pp. 323–341). IGI Global. doi:10.4018/978-1-5225-2970-5.ch015

Belyanina, L. A. (2018). Project-Design Activity as a Condition for the Formation of Research Competence Among Students. In V. Mkrttchian & L. Belyanina (Eds.), *Handbook of Research on Students' Research Competence in Modern Educational Contexts* (pp. 107–131). IGI Global. doi:10.4018/978-1-5225-3485-3.ch006

Bhandary, T. M. M., Vickraman, I., & K. A., P. (2020). Probiotics as an Alternative Food Therapy. In Technological Developments in Food Preservation, Processing, and Storage (pp. 97-127). IGI Global. http://doi:10.4018/978-1-7998-1924-0.ch006

Bhat, A. H., & Nigar, S. (2020). Traditional Medicinal Systems. In A. Singh, P. Singh, & N. Bithel (Eds.), *Advanced Pharmacological Uses of Medicinal Plants and Natural Products* (pp. 1–24). IGI Global. doi:10.4018/978-1-7998-2094-9.ch001

Bhat, A. H., & Sharma, H. G. (2020). Medicinal Plants. In A. Singh, P. Singh, & N. Bithel (Eds.), *Advanced Pharmacological Uses of Medicinal Plants and Natural Products* (pp. 67–90). IGI Global. doi:10.4018/978-1-7998-2094-9.ch004

Bhatt, R., & Gupta, S. (2018). Nano Particles and Their Mode of Action in Environment. In Microbial Biotechnology in Environmental Monitoring and Cleanup (pp. 212-219). IGI Global. doi:10.4018/978-1-5225-3126-5.ch013

Bhukta, A., & Jana, S. K. (2020). Biopiracy. In Intellectual Property Rights and the Protection of Traditional Knowledge (pp. 140-157). IGI Global. http:// doi:10.4018/978-1-7998-1835-9.ch007

Biney, R. P., Mpofana, T., & Kasanga, E. A. (2019). Free Radicals in Oxidative Stress, Aging, and Neurodegenerative Disorders. In M. Uddin & M. Amran (Eds.), *Handbook of Research on Critical Examinations of Neurodegenerative Disorders* (pp. 48–75). IGI Global. doi:10.4018/978-1-5225-5282-6.ch003

Bogueva, D., Marinova, D., & Phau, I. (2018). Is Meat a Luxury? In D. Bogueva, D. Marinova, & T. Raphaely (Eds.), *Handbook of Research on Social Marketing and Its Influence on Animal Origin Food Product Consumption* (pp. 172–186). IGI Global. doi:10.4018/978-1-5225-4757-0.ch012

Bozyel, M. E., Merdamert Bozyel, E., & Canli, K. (2020). Ethnobotanical Uses of Medicinal Plants for the Treatment of Herniated Disc in Turkish Traditional Medicine. In Ethnomedicinal Plant Use and Practice in Traditional Medicine (pp. 221-234). IGI Global. doi:10.4018/978-1-7998-1320-0.ch011

Bozyel, M. E., Merdamert Bozyel, E., & Canli, K. (2020). Turkish Herbal Medicine in the Treatment of Back Pain. In Ethnomedicinal Plant Use and Practice in Traditional Medicine (pp. 235-251). IGI Global. doi:10.4018/978-1-7998-1320-0.ch012

Branquinho, R., Gonçalves, V., Vaz, P. F., Fachada, I., & Aguiar, C. (2020). Integra(-te). In S. Mosquera & P. Jardim da Palma (Eds.), *Multidisciplinary Approach to Entrepreneurship Education for Migrants* (pp. 81–96). IGI Global. doi:10.4018/978-1-7998-2925-6.ch005

Bratovcic, A. (2020). New Solar Photocatalytic Technologies for Water Purification as Support for the Implementation of Industry 4.0. In I. Karabegović, A. Kovačević, L. Banjanović-Mehmedović, & P. Dašić (Eds.), *Handbook of Research on Integrating Industry 4.0 in Business and Manufacturing* (pp. 385–412). IGI Global. doi:10.4018/978-1-7998-2725-2.ch017

Casañola-Martin, G. M., & Pham-The, H. (2019). Machine Learning Applications in Nanomedicine and Nanotoxicology. *International Journal of Applied Nanotechnology Research, 4*(1), 1–7. doi:10.4018/IJANR.2019010101

Cascella, M., Muzio, M. R., Bimonte, S., & Cuomo, A. (2018). Nutraceuticals for Prevention of Chemotherapy-Induced Peripheral Neuropathy. In A. Verma, K. Srivastava, S. Singh, & H. Singh (Eds.), *Nutraceuticals and Innovative Food Products for Healthy Living and Preventive Care* (pp. 236–259). IGI Global. doi:10.4018/978-1-5225-2970-5.ch011

Chakravorty, A., Rather, G. A., Ali, A., Bhat, B. A., Sana, S. S., Abhishek, N., & Nanda, A. (2020). Nano Approach. In N. Mishra (Ed.), *Ethnopharmacological Investigation of Indian Spices* (pp. 205–241). IGI Global. doi:10.4018/978-1-7998-2524-1.ch016

Chang, D. M., & Mohd Suki, N. (2021). Traditional and Complementary Medicine (TCM) Usage vs. Health Technology. In N. Mohd Suki (Ed.), *Handbook of Research on Technology Applications for Effective Customer Engagement* (pp. 315–328). IGI Global. doi:10.4018/978-1-7998-4772-4.ch020

Chetri, B. K., Rinchen, N., & Wangchuck, K. (2020). Ethnobotanical Wealth of Home Gardens. In Ethnomedicinal Plant Use and Practice in Traditional Medicine (pp. 17-37). IGI Global. doi:10.4018/978-1-7998-1320-0.ch002

Chooramani, G., & Singh, P. (2018). Impact of Tuberculosis in Elderly Population. In B. V. Prasad & S. Akbar (Eds.), *Handbook of Research on Geriatric Health, Treatment, and Care* (pp. 326–338). IGI Global. doi:10.4018/978-1-5225-3480-8.ch018

Dangroo, N. A., Malik, A. A., Ara, T., & Ali, G. (2021). Nanotechnology as an Emerging Field in the Arena of Medicine. In M. A. Bhat, I. A. Wani, & S. Ashraf (Eds.), *Applications of Nanomaterials in Agriculture, Food Science, and Medicine* (pp. 106–115). IGI Global. doi:10.4018/978-1-7998-5563-7.ch006

Dar, Z. A., Rifat, B., Bhat, J. I., Bhatti, A. A., Haq, S., Amin, A., & Dar, S. A. (2019). Potential Role of Endophytes for Sustainable Environment. In A. Karmaoui (Ed.), *Climate Change and Its Impact on Ecosystem Services and Biodiversity in Arid and Semi-Arid Zones* (pp. 78–95). IGI Global. doi:10.4018/978-1-5225-7387-6.ch005

Das, P. (2020). Designing a Framework of Ethnomedicinal Plant Knowledge Integration Using OSS. In A. Kaushik, A. Kumar, & P. Biswas (Eds.), *Handbook of Research on Emerging Trends and Technologies in Library and Information Science* (pp. 332–345). IGI Global. doi:10.4018/978-1-5225-9825-1.ch022

Das, U. (2019). Veterinary Public Health. In S. Yasobant & D. Saxena (Eds.), *Global Applications of One Health Practice and Care* (pp. 113–141). IGI Global. doi:10.4018/978-1-5225-6304-4.ch005

David, S. K., Saeb, A. T., Rafiullah, M., & Rubeaan, K. (2019). Classification Techniques and Data Mining Tools Used in Medical Bioinformatics. In S. K. Strydom & M. Strydom (Eds.), *Big Data Governance and Perspectives in Knowledge Management* (pp. 105–126). IGI Global. doi:10.4018/978-1-5225-7077-6.ch005

De la Pena, A. S. (2020). Just a Group of Oil Ladies. In C. L. Wang (Ed.), *Handbook of Research on the Impact of Fandom in Society and Consumerism* (pp. 127–148). IGI Global. doi:10.4018/978-1-7998-1048-3.ch007

Deng, N., & Xiong, C. (2020). Serialized Co-Training-Based Recognition of Medicine Names for Patent Mining and Retrieval. *International Journal of Data Warehousing and Mining*, *16*(3), 87–107. doi:10.4018/IJDWM.2020070105

Dutta, M. M. (2021). Nanomaterials for Food and Agriculture. In M. A. Bhat, I. A. Wani, & S. Ashraf (Eds.), *Applications of Nanomaterials in Agriculture, Food Science, and Medicine* (pp. 75–97). IGI Global. doi:10.4018/978-1-7998-5563-7.ch004

Dutta, M. M., & Goswami, M. (2021). Coating Materials. In S. Roy & G. K. Bose (Eds.), *Advanced Surface Coating Techniques for Modern Industrial Applications* (pp. 1–30). IGI Global. doi:10.4018/978-1-7998-4870-7.ch001

Dwivedi, M. K., Pandey, S. K., & Singh, P. K. (2021). Nanomaterials Useful in Health and Medicine to Improve Public Health. In M. A. Bhat, I. A. Wani, & S. Ashraf (Eds.), *Applications of Nanomaterials in Agriculture, Food Science, and Medicine* (pp. 49–74). IGI Global. doi:10.4018/978-1-7998-5563-7.ch003

Dwivedi, M. K., & Singh, P. K. (2020). Anti-Malarial Drug Resistance. In A. Singh, P. Singh, & N. Bithel (Eds.), *Advanced Pharmacological Uses of Medicinal Plants and Natural Products* (pp. 154–176). IGI Global. doi:10.4018/978-1-7998-2094-9.ch008

Endong, F. P. (2019). Cultural Heritage Preservation in a Modernizing Africa. In C. Inglese & A. Ippolito (Eds.), *Conservation, Restoration, and Analysis of Architectural and Archaeological Heritage* (pp. 28–47). IGI Global. doi:10.4018/978-1-5225-7555-9.ch002

Epshtein, M., & Yushkov, A. (2019). School Week of High Technologies and Technical Entrepreneurship. In M. Epshtein (Ed.), *Business Community Engagement for Educational Initiatives* (pp. 206–234). IGI Global. doi:10.4018/978-1-5225-6951-0.ch011

Escribano, A. J. (2018). The Evolution of the Feed Industry Towards Its "Sustainabilization". In V. Sima (Ed.), *Organizational Culture and Behavioral Shifts in the Green Economy* (pp. 181–209). IGI Global. doi:10.4018/978-1-5225-2965-1.ch008

Fakhri, K. U., Alotaibi, B. S., Kumar, S., Kumar, U., Sonkar, S. C., Mathur, A., Kazim, Z., Perwez, A., Yasin, D., Zeya, B., Ahmad, F. J., & Rizvi, M. M. (2021). Phytopharmaceuticals in Cancer Treatment. In S. Kumar, M. A. Rizvi, & S. Verma (Eds.), *Handbook of Research on Advancements in Cancer Therapeutics* (pp. 639–668). IGI Global. doi:10.4018/978-1-7998-6530-8.ch025

Faris, M. A., & Ghazzawi, H. (2020). Health-Improving and Disease-Preventing Potential of Camel Milk Against Chronic Diseases and Autism. In O. A. Alhaj, B. Faye, & R. P. Agrawal (Eds.), *Handbook of Research on Health and Environmental Benefits of Camel Products* (pp. 155–184). IGI Global. doi:10.4018/978-1-7998-1604-1.ch008

Ferrari, S. (2020). National Parks, Territorial Brands and Co-Branding Initiatives. In Global Opportunities and Challenges for Rural and Mountain Tourism (pp. 91-111). IGI Global. http://doi:10.4018/978-1-7998-1302-6.ch006

Gahtori, R., & Paliwal, A. (2021). Role of Herbal Medicine/Phyto-Therapy in Cancer Prevention by Inhibiting Epithelial-Mesenchymal Transition (EMT) Pathways. In A. Hussain & S. Behl (Eds.), *Treating Endocrine and Metabolic Disorders With Herbal Medicines* (pp. 215–234). IGI Global. doi:10.4018/978-1-7998-4808-0.ch009

Gautam, A. (2020). Therapeutic Properties of Syzygium cumini (Jamun) and Tinospora cordifolia (Giloy) Against Various Lethal Diseases. In A. Singh, P. Singh, & N. Bithel (Eds.), *Advanced Pharmacological Uses of Medicinal Plants and Natural Products* (pp. 316–337). IGI Global. doi:10.4018/978-1-7998-2094-9.ch017

Goswami, S., Mahanta, K., Goswami, S., Jigdung, T., & Devi, T. P. (2018). Ageing and Cancer. In B. V. Prasad & S. Akbar (Eds.), *Handbook of Research on Geriatric Health, Treatment, and Care* (pp. 206–235). IGI Global. doi:10.4018/978-1-5225-3480-8.ch012

Goswami, S., Mitra, S., Paul, P., Dey, D., & Das, S. (2019). Biochemic System of Medicine. In R. K. Keservani, A. K. Sharma, & R. K. Kesharwani (Eds.), *Nutraceutical and Functional Foods in Disease Prevention* (pp. 403–431). IGI Global. doi:10.4018/978-1-5225-3267-5.ch014

Gozalbes, R., & Vicente de Julián-Ortiz, J. (2018). Applications of Chemoinformatics in Predictive Toxicology for Regulatory Purposes, Especially in the Context of the EU REACH Legislation. *International Journal of Quantitative Structure-Property Relationships*, *3*(1), 1–24. doi:10.4018/IJQSPR.2018010101

Gunjal, A. B., Waghmode, M. S., Patil, N. N., & Kapadnis, B. P. (2018). Endophytes and Their Role in Phytoremediation and Biotransformation Process. In Microbial Biotechnology in Environmental Monitoring and Cleanup (pp. 238-251). IGI Global. http://doi:10.4018/978-1-5225-3126-5.ch015

Hafeez, Z. B., Fakhri, K. U., Zafaryab, M., & Rizvi, M. M. (2021). Functional Mechanisms of Green Tea Polyphenols and Their Molecular Targets in Prevention of Multiple Cancers. In S. Kumar, M. A. Rizvi, & S. Verma (Eds.), *Handbook of Research on Advancements in Cancer Therapeutics* (pp. 587–605). IGI Global. doi:10.4018/978-1-7998-6530-8.ch022

Hamid, S., & Jameel, S. T. (2020). A Study of Green Marketing Practices in the Selected Ayurvedic Resorts of Kerala. In S. Paul & S. Kulshreshtha (Eds.), *Global Developments in Healthcare and Medical Tourism* (pp. 176–187). IGI Global. doi:10.4018/978-1-5225-9787-2.ch010

Hamid, S. B., & Hamid, A. F. (2019). Roles of Nutraceuticals and Functional Food in Prevention of Cardiovascular Disease. In R. K. Keservani, A. K. Sharma, & R. K. Kesharwani (Eds.), *Nutraceutical and Functional Foods in Disease Prevention* (pp. 126–165). IGI Global. doi:10.4018/978-1-5225-3267-5.ch005

Harandi, M. F. (2019). Engineering. In The Normative Nature of Social Practices and Ethics in Professional Environments (pp. 84-114). IGI Global. http://doi:10.4018/978-1-5225-8006-5.ch005

Henao, J., Poblano-Salas, C. A., Vargas, F., Giraldo-Betancur, A. L., Corona-Castuera, J., & Sotelo-Mazón, O. (2021). Principles and Applications of Thermal Spray Coatings. In S. Roy & G. K. Bose (Eds.), *Advanced Surface Coating Techniques for Modern Industrial Applications* (pp. 31–70). IGI Global. doi:10.4018/978-1-7998-4870-7.ch002

Ho, Y., Lim, B., & Lim, J. (2020). Agricultural Runoff and Treatment Methods. In A. C. Affam & E. H. Ezechi (Eds.), *Handbook of Research on Resource Management for Pollution and Waste Treatment* (pp. 550–575). IGI Global. doi:10.4018/978-1-7998-0369-0.ch023

Hussain, A., & Bhattacharya, A. (2019). *Advanced Wastewater Treatments*. IGI Global. doi:10.4018/978-1-5225-9441-3.ch005

Inbaraj, S., Sriram, V. K., Thomas, P., Verma, A., & Chaudhuri, P. (2019). Antibiotic Resistance in the Veterinary Perspective. In S. Yasobant & D. Saxena (Eds.), *Global Applications of One Health Practice and Care* (pp. 142–157). IGI Global. doi:10.4018/978-1-5225-6304-4.ch006

Jain, V. (2020). A Review of Traditional Medicinal Plants Used for Treatment of Leucorrhoea in India. In Ethnomedicinal Plant Use and Practice in Traditional Medicine (pp. 252-266). IGI Global. doi:10.4018/978-1-7998-1320-0.ch013

Jangid, N. K., Yadav, A., Jadoun, S., Srivastava, A., & Srivastava, M. (2020). Outdoor Pollution Management by Nanotechnology. In K. A. Wani, N. K. Jangid, & A. R. Bhat (Eds.), *Impact of Textile Dyes on Public Health and the Environment* (pp. 258–277). IGI Global. doi:10.4018/978-1-7998-0311-9.ch012

Jhandai, P. (2019). One Health Study Design. In S. Yasobant & D. Saxena (Eds.), *Global Applications of One Health Practice and Care* (pp. 59–80). IGI Global. doi:10.4018/978-1-5225-6304-4.ch003

Joshi, G. Y., & Nulkar, G. (2018). Green. In P. Ordóñez de Pablos (Ed.), *Management Strategies and Technology Fluidity in the Asian Business Sector* (pp. 99–113). IGI Global. doi:10.4018/978-1-5225-4056-4.ch006

Kalaiselvam, S., Rajan, D., & Ali, I. H. (2019). Thermal Technologies and Systems for Food Preservation. In P. D. Gaspar & P. D. da Silva (Eds.), *Novel Technologies and Systems for Food Preservation* (pp. 140–159). IGI Global. doi:10.4018/978-1-5225-7894-9.ch006

Kalita, D., & Devi, N. (2020). A Short Review on Gynocardia odorata R. Br. In A. K. Rathoure & P. B. Chauhan (Eds.), *Current State and Future Impacts of Climate Change on Biodiversity* (pp. 135–143). IGI Global. doi:10.4018/978-1-7998-1226-5.ch008

Kaloudas, D., & Penchovsky, R. (2018). Plant-Derived Compounds and Their Potential Role in Drug Development. *International Journal of Biomedical and Clinical Engineering*, 7(1), 53–66. doi:10.4018/IJBCE.2018010104

Kanagasubbulakshmi, S., Lakshmi, K., & Kadirvelu, K. (2019). Cost-Effective Methods of Monitoring Pesticide Pollution in Water. In Handbook of Research on the Adverse Effects of Pesticide Pollution in Aquatic Ecosystems (pp. 236-256). IGI Global. doi:10.4018/978-1-5225-6111-8.ch014

Karnena, M. K., Konni, M., & Saritha, V. (2020). Nano-Catalysis Process for Treatment of Industrial Wastewater. In G. Duca & A. Vaseashta (Eds.), *Handbook of Research on Emerging Developments and Environmental Impacts of Ecological Chemistry* (pp. 229–251). IGI Global. doi:10.4018/978-1-7998-1241-8.ch011

Kasemsap, K. (2018). An Overview of Global Health Initiatives and Practices. In V. E. Papalois & M. Theodosopoulou (Eds.), *Optimizing Health Literacy for Improved Clinical Practices* (pp. 198–209). IGI Global. doi:10.4018/978-1-5225-4074-8.ch011

Kaur, N. (2020). Nanoantimicrobials. In S. Yıkmış (Ed.), *Technological Developments in Food Preservation, Processing, and Storage* (pp. 146–165). IGI Global. doi:10.4018/978-1-7998-1924-0.ch008

Keri, I., Gromova, E., & Sinelschikova, N. (2019). Features of Financial Policy Development by Russian Companies in Today's Conditions. In D. Ushakov (Ed.), *Global Trends of Modernization in Budgeting and Finance* (pp. 64–82). IGI Global. doi:10.4018/978-1-5225-7760-7.ch004

Khan, R., Ullah, M., Shafi, B., Orakzai, U. B., & Shafi, S. (2020). Knocking at Alternate Doors. In S. Umair (Ed.), *Mobile Devices and Smart Gadgets in Medical Sciences* (pp. 22–38). IGI Global. doi:10.4018/978-1-7998-2521-0.ch002

Khati, P., Gangola, S., Bhatt, P., Kumar, R., & Sharma, A. (2018). Application of Nanocompounds for Sustainable Agriculture System. In Microbial Biotechnology in Environmental Monitoring and Cleanup (pp. 194-211). IGI Global. doi:10.4018/978-1-5225-3126-5.ch012

Khelfi, A. (2018). Sources of Air Pollution. In Handbook of Research on Microbial Tools for Environmental Waste Management (pp. 220-258). IGI Global. doi:10.4018/978-1-5225-3540-9.ch011

Kosti, S., & Kundu, J. (2020). SWOT Analysis of Literature. In *Creating a Framework for Dissertation Preparation* (pp. 67-80). IGI Global. http://doi:10.4018/978-1-5225-9707-0.ch004

Kreitinger, J., & Corsi, T. F. (2019). Utilizing a National Association to Increase Access to Education Abroad. In G. F. Malveaux & R. L. Raby (Eds.), *Study Abroad Opportunities for Community College Students and Strategies for Global Learning* (pp. 255–264). IGI Global. doi:10.4018/978-1-5225-6252-8.ch018

Kumar, B., Singh, S., Srivastava, K., & Wani, K. A. (2018). New Product Development and Regulatory Challenges in Nutraceutical Industry. In A. Verma, K. Srivastava, S. Singh, & H. Singh (Eds.), *Nutraceuticals and Innovative Food Products for Healthy Living and Preventive Care* (pp. 280–303). IGI Global. doi:10.4018/978-1-5225-2970-5.ch013

Kumar, D. (2020). Health and Well-Being Among the Jats of Western Uttar Pradesh. In S. Padmanaban & C. Subudhi (Eds.), *Psycho-Social Perspectives on Mental Health and Well-Being* (pp. 136–160). IGI Global. doi:10.4018/978-1-7998-1185-5.ch007

Kumar, I. (2019). Simulation and Modeling of Nanotechnology Aircraft Using MATLAB. In N. Ramdani (Ed.), *Nanotechnology in Aerospace and Structural Mechanics* (pp. 257–290). IGI Global. doi:10.4018/978-1-5225-7921-2.ch008

Kumar, P., & Sharma, A. (2020). Analysis of International Patent Applications for Inventions Like Traditional Herbal Medicines. In *Handbook of Research on Emerging Trends and Technologies in Library and Information Science* (pp. 346-356). IGI Global. http://doi:10.4018/978-1-5225-9825-1.ch023

Kumar, R., Bala, S., & Kumar, A. (2021). Comparative Performance Analysis of Nanowire and Nanotube Field Effect Transistors. In B. Raj (Ed.), *Innovative Applications of Nanowires for Circuit Design* (pp. 54–70). IGI Global. doi:10.4018/978-1-7998-6467-7.ch003

Kurissery, S., Shaw, L. K., & Kanavillil, N. (2019). A Historic Perspective of Endophytes in Vascular Plants and Their Role in Environmental Sustainability. In R. A. Turvey & S. Kurissery (Eds.), *Intellectual, Scientific, and Educational Influences on Sustainability Research* (pp. 14–45). IGI Global. doi:10.4018/978-1-5225-7302-9.ch002

Kurtar, H., Kapusuzoglu, A., & Ceylan, N. B. (2019). Investment Decision Process in Oil Economy for Major Oil Exporting and Importing Countries. In H. Dinçer & S. Yüksel (Eds.), *Handbook of Research on Global Issues in Financial Communication and Investment Decision Making* (pp. 236–255). IGI Global. doi:10.4018/978-1-5225-9265-5.ch012

Largacha-Martinez, C. (2021). Five Global Purpose-Driven Business Platforms. In Handbook of Research on International Business and Models for Global Purpose-Driven Companies (pp. 1-21). IGI Global. doi:10.4018/978-1-7998-4909-4.ch001

Loukanov, A. R., Gagov, H. S., Mishonova, M. Y., & Nakabayashi, S. (2018). Biocompatible Carbon Nanodots for Functional Imaging and Cancer Therapy. *International Journal of Biomedical and Clinical Engineering*, *7*(2), 31–45. doi:10.4018/IJBCE.2018070103

Luo, D., Ma, W., Wu, L., & Nakata, K. (2018). CO2 Reduction by Photocatalysis on TiO2. In S. González-Cortés & F. E. Imbert (Eds.), *Advanced Solid Catalysts for Renewable Energy Production* (pp. 268–299). IGI Global. doi:10.4018/978-1-5225-3903-2.ch010

Luong, H. T. (2020). Healthcare Service, Food Security, and Sustainable Development. In L. Amadi & F. Allen (Eds.), *Global Food Politics and Approaches to Sustainable Consumption* (pp. 119–141). IGI Global. doi:10.4018/978-1-7998-0125-2.ch006

Lutomia, A. N., & Wanzala, W. O. (2019). Intersecting Gender and Culture. In D. P. Peltz & A. C. Clemons (Eds.), *Multicultural Andragogy for Transformative Learning* (pp. 154–170). IGI Global. doi:10.4018/978-1-5225-3474-7.ch009

Lydekaityte, J., & Tambo, T. (2021). Internet of Packaging and the Bridge Between Digital Marketing and Physical Retailing. In Encyclopedia of Organizational Knowledge, Administration, and Technology (pp. 2057-2070). IGI Global. doi:10.4018/978-1-7998-3473-1.ch141

Majumdar, D. (2019). Polyaniline as Proficient Electrode Material for Supercapacitor Applications. In N. Ramdani (Ed.), *Polymer Nanocomposites for Advanced Engineering and Military Applications* (pp. 190–219). IGI Global. doi:10.4018/978-1-5225-7838-3.ch007

Malik, A. A., Dangroo, N. A., Amin, R., Khazir, Z. U., & Ara, T. (2021). Applications of Organic Chemistry in Nano-Medicine. In M. A. Bhat, I. A. Wani, & S. Ashraf (Eds.), *Applications of Nanomaterials in Agriculture, Food Science, and Medicine* (pp. 116–123). IGI Global. doi:10.4018/978-1-7998-5563-7.ch007

Malik, S. (2021). Implication of Nanoparticles in Drosophila Toxicology Research. In M. A. Bhat, I. A. Wani, & S. Ashraf (Eds.), *Applications of Nanomaterials in Agriculture, Food Science, and Medicine* (pp. 330–340). IGI Global. doi:10.4018/978-1-7998-5563-7.ch018

Malveaux, G. F. (2019). How to Survive and Thrive as a Community College Consortium. In G. F. Malveaux & R. L. Raby (Eds.), *Study Abroad Opportunities for Community College Students and Strategies for Global Learning* (pp. 265–283). IGI Global. doi:10.4018/978-1-5225-6252-8.ch019

Maravilhas-Lopes, S. (2020). Sustainable Innovation Projects From Patent Information to Leverage Economic Development. In G. L. Jamil, F. Ribeiro, A. Malheiro da Silva, & S. Maravilhas Lopes (Eds.), *Handbook of Research on Emerging Technologies for Effective Project Management* (pp. 169–184). IGI Global. doi:10.4018/978-1-5225-9993-7.ch010

Marinova, D., & Raphaely, T. (2018). Taxing Meat and Animal Food Products. In D. Bogueva, D. Marinova, & T. Raphaely (Eds.), *Handbook of Research on Social Marketing and Its Influence on Animal Origin Food Product Consumption* (pp. 121–134). IGI Global. doi:10.4018/978-1-5225-4757-0.ch008

Marquez, O., & Jairo Márquez, P. (2018). Synthesis of Electrocatalysts for Electrochemistry in Energy. In S. González-Cortés & F. E. Imbert (Eds.), *Advanced Solid Catalysts for Renewable Energy Production* (pp. 300–385). IGI Global. doi:10.4018/978-1-5225-3903-2.ch011

McGrowder, D. A., Miller, F. G., Nwokocha, C., Wilson-Clarke, C. F., Anderson, M., Anderson-Jackson, L., Williams, L., & Alexander-Lindo, R. (2021). Medical Herbs and the Treatment of Diabetes Mellitus. In A. Hussain & S. Behl (Eds.), *Treating Endocrine and Metabolic Disorders With Herbal Medicines* (pp. 48–73). IGI Global. doi:10.4018/978-1-7998-4808-0.ch003

Medhekar, A. (2020). Emerging Trends of Wellness and Medical Tourism in India. In S. Paul & S. Kulshreshtha (Eds.), *Global Developments in Healthcare and Medical Tourism* (pp. 1–22). IGI Global. doi:10.4018/978-1-5225-9787-2.ch001

Mir, M. A., Bhat, B. A., Sheikh, B. A., Rather, G. A., Mehraj, S., & Mir, W. R. (2021). Nanomedicine in Human Health Therapeutics and Drug Delivery. In M. A. Bhat, I. A. Wani, & S. Ashraf (Eds.), *Applications of Nanomaterials in Agriculture, Food Science, and Medicine* (pp. 229–251). IGI Global. doi:10.4018/978-1-7998-5563-7.ch013

Mishra, A., Chaudhari, D. D., Patel, H. K., & Patel, B. D. (2018). Nutraceutical Properties of Important Weeds in India. In A. Verma, K. Srivastava, S. Singh, & H. Singh (Eds.), *Nutraceuticals and Innovative Food Products for Healthy Living and Preventive Care* (pp. 304–322). IGI Global. doi:10.4018/978-1-5225-2970-5.ch014

Mishra, P., Gupta, V. C., & Dham, S. O. (2019). Demystifying the Indian Way of Doing Business. In Breaking Down Language and Cultural Barriers Through Contemporary Global Marketing Strategies (pp. 65-78). IGI Global. doi:10.4018/978-1-5225-6980-0.ch004

Misra, R., Singh, S., & Mahajan, R. (2020). An Analysis on Consumer Preference of Ayurvedic Products in Indian Market. *International Journal of Asian Business and Information Management, 11*(4), 1–15. doi:10.4018/IJABIM.2020100101

Mohanty, C., Satpathy, S. S., & Mohanty, S. (2021). An Eco-Friendly Approach for the Eradication of Heavy Metal Contaminants by Nano-Bioremediation. In S. Dey & B. Acharya (Eds.), *Recent Advancements in Bioremediation of Metal Contaminants* (pp. 220–236). IGI Global. doi:10.4018/978-1-7998-4888-2.ch012

Mthembu, N. (2021). Survival of Indigenous African Healing Systems in Post-Apartheid South Africa. In R. Nicholson (Ed.), *Natural Healing as Conflict Resolution* (pp. 1–20). IGI Global. doi:10.4018/978-1-7998-3665-0.ch001

Mubarak, S. M. (2019). Nanotechnology for Air Remediation. In R. Nazir (Ed.), *Nanotechnology Applications in Environmental Engineering* (pp. 121–142). IGI Global. doi:10.4018/978-1-5225-5745-6.ch006

Mundra, S., & Singh, P. (2020). Herbal Bioactives. In A. Singh, P. Singh, & N. Bithel (Eds.), *Advanced Pharmacological Uses of Medicinal Plants and Natural Products* (pp. 200–215). IGI Global. doi:10.4018/978-1-7998-2094-9.ch010

Mustafa, K., Kanwal, J., Khakwani, S., & Musaddiq, S. (2020). Cognitive Enhancement and Neuroprotective Abilities of Plants. In A. Singh, P. Singh, & N. Bithel (Eds.), *Advanced Pharmacological Uses of Medicinal Plants and Natural Products* (pp. 128–152). IGI Global. doi:10.4018/978-1-7998-2094-9.ch007

Mustafa, K., Kanwal, J., Musaddiq, S., & Khakwani, S. (2020). Ethnobotanical and Pharmacological Importance of the Herbal Plants With Anti-Hair Fall and Hair Growth Activities. In A. Singh, P. Singh, & N. Bithel (Eds.), *Advanced Pharmacological Uses of Medicinal Plants and Natural Products* (pp. 49–66). IGI Global. doi:10.4018/978-1-7998-2094-9.ch003

N., K., & P., V. (2021). Applications of Nano Materials in Cold Storage. In *Applications of Nanomaterials in Agriculture, Food Science, and Medicine* (pp. 252-269). IGI Global. http://doi:10.4018/978-1-7998-5563-7.ch014

Nag, A. (2020). Phytochemistry and Antigenotoxic Properties of Six Ethnobotanically Important Members From the Family Zingiberaceae. In Ethnomedicinal Plant Use and Practice in Traditional Medicine (pp. 131-153). IGI Global. doi:10.4018/978-1-7998-1320-0.ch007

Nayak, A., Saini, V. K., & Bhushan, B. (2021). Nanomaterials for Energy Harvesting and Storage. In M. A. Bhat, I. A. Wani, & S. Ashraf (Eds.), *Applications of Nanomaterials in Agriculture, Food Science, and Medicine* (pp. 188–203). IGI Global. doi:10.4018/978-1-7998-5563-7.ch011

Nazir, R. (2019). Introduction to Environmental Nanotechnology. In R. Nazir (Ed.), *Nanotechnology Applications in Environmental Engineering* (pp. 1–27). IGI Global. doi:10.4018/978-1-5225-5745-6.ch001

Nişancı, E., Yıldırım, D. Ç., Çevik, N. K., & Sırım, V. (2018). Indonesia. In S. Ozdemir, S. Erdogan, & A. Gedikli (Eds.), *Handbook of Research on Sociopolitical Factors Impacting Economic Growth in Islamic Nations* (pp. 332–357). IGI Global. doi:10.4018/978-1-5225-2939-2.ch017

Noor, N., Shah, A., Gani, A., Ashraf, Z. U., & Masoodi, F. A. (2021). Nanomaterials in Food Packaging. In M. A. Bhat, I. A. Wani, & S. Ashraf (Eds.), *Applications of Nanomaterials in Agriculture, Food Science, and Medicine* (pp. 270–287). IGI Global. doi:10.4018/978-1-7998-5563-7.ch015

Nosheen, S. (2019). Nanomembrane Applications in Environmental Engineering. In R. Nazir (Ed.), *Nanotechnology Applications in Environmental Engineering* (pp. 103–120). IGI Global. doi:10.4018/978-1-5225-5745-6.ch005

Nyika, J. M. (2021). Nanotechnology and Its Applications in Environmental Remediation. In M. A. Bhat, I. A. Wani, & S. Ashraf (Eds.), *Applications of Nanomaterials in Agriculture, Food Science, and Medicine* (pp. 29–48). IGI Global. doi:10.4018/978-1-7998-5563-7.ch002

Ogwu, M. C. (2019). Understanding the Composition of Food Waste. In Global Initiatives for Waste Reduction and Cutting Food Loss (pp. 212-236). IGI Global. doi:10.4018/978-1-5225-7706-5.ch011

Paholpak, P., Obler, L., & Lavretsky, H. (2018). Yoga for Seniors. In S. Telles & N. Singh (Eds.), *Research-Based Perspectives on the Psychophysiology of Yoga* (pp. 151–174). IGI Global. doi:10.4018/978-1-5225-2788-6.ch009

Paixão, S. M., Silva, T. P., Arez, B. F., & Alves, L. (2020). Advances in the Reduction of the Costs Inherent to Fossil Fuel Biodesulfurization Towards Its Potential Industrial Applications. In T. A. Saleh (Ed.), *Nanocomposites for the Desulfurization of Fuels* (pp. 235–283). IGI Global. doi:10.4018/978-1-7998-2146-5.ch007

Pandey, A., Tiwari, S. P., Dutta, V., & Kumar, V. (2018). Photon Upconversion in Lanthanide-Activated Inorganic Luminescent Materials. In R. Tiwari, V. Dubey, & S. J. Dhoble (Eds.), *Emerging Synthesis Techniques for Luminescent Materials* (pp. 86–116). IGI Global. doi:10.4018/978-1-5225-5170-6.ch004

Pandey, S., & Alam, A. (2020). Bryo-Pharmaceuticals. In A. Singh, P. Singh, & N. Bithel (Eds.), *Advanced Pharmacological Uses of Medicinal Plants and Natural Products* (pp. 269–284). IGI Global. doi:10.4018/978-1-7998-2094-9.ch014

Patel, A., Tiwari, S., Pandey, N., Gupta, D., & Prasad, S. M. (2020). Role of Spices Beyond a Flavouring Agent. In N. Mishra (Ed.), *Ethnopharmacological Investigation of Indian Spices* (pp. 5–35). IGI Global. doi:10.4018/978-1-7998-2524-1.ch002

Patra, B., Deep, S. K., & Pradhan, S. N. (2021). Different Bioremediation Techniques for Management of Waste Water. In S. Dey & B. Acharya (Eds.), *Recent Advancements in Bioremediation of Metal Contaminants* (pp. 1–18). IGI Global. doi:10.4018/978-1-7998-4888-2.ch001

Philip, J. M., Aravind, U. K., & Aravindakumar, C. T. (2018). Use of Antibiotics in Animals and Its Possible Impacts in the Environment. In D. Bogueva, D. Marinova, & T. Raphaely (Eds.), *Handbook of Research on Social Marketing and Its Influence on Animal Origin Food Product Consumption* (pp. 77–91). IGI Global. doi:10.4018/978-1-5225-4757-0.ch005

Polat, V., & Morkan, B. (2020). Green but How Green? In V. Naidoo & R. Verma (Eds.), *Green Marketing as a Positive Driver Toward Business Sustainability* (pp. 47–68). IGI Global. doi:10.4018/978-1-5225-9558-8.ch003

Pramodh, S. (2021). Male Infertility Management With Alternative Medicine. In A. Hussain & S. Behl (Eds.), *Treating Endocrine and Metabolic Disorders With Herbal Medicines* (pp. 164–186). IGI Global. doi:10.4018/978-1-7998-4808-0.ch007

Priya, E., & Subramanian, S. (2018). Automated Method of Analysing Sputum Smear Tuberculosis Images Using Multifractal Approach. In M. H. Kolekar & V. Kumar (Eds.), *Biomedical Signal and Image Processing in Patient Care* (pp. 184–215). IGI Global. doi:10.4018/978-1-5225-2829-6.ch010

Psaltis, A., & Mourlas, C. (2018). Realtime BioSensing System Assessing Subconscious Responses of Engagement. In R. Zheng (Ed.), *Digital Technologies and Instructional Design for Personalized Learning* (pp. 307–333). IGI Global. doi:10.4018/978-1-5225-3940-7.ch015

Radisavljević, S., & Petrović, B. (2020). Biogenic Synthesis of Gold Nanoparticles and Their Antimicrobial Activities. In A. Singh, P. Singh, & N. Bithel (Eds.), *Advanced Pharmacological Uses of Medicinal Plants and Natural Products* (pp. 216–233). IGI Global. doi:10.4018/978-1-7998-2094-9.ch011

Rahmatullah, M., Jannat, K., Reeck, G. R., Jahan, R., Rahman, T., Shova, N. A., & Islam, M. (2020). Ethnic Use, Phytochemistry, and Pharmacology of Cyperus rotundus. In Ethnomedicinal Plant Use and Practice in Traditional Medicine (pp. 82-104). IGI Global. http://doi:10.4018/978-1-7998-1320-0.ch005

Ramadhas, G., Sankar, A. S., & Sugathan, N. (2018). The Scientific Communication Process in Homoeopathic Toxicology. In J. J. Jeyasekar & P. Saravanan (Eds.), *Innovations in Measuring and Evaluating Scientific Information* (pp. 163–179). IGI Global. doi:10.4018/978-1-5225-3457-0.ch011

Ramalingam, K., & Amaechi, B. T. (2019). Phytochemicals. In K. Kamalanand, B. Thayumanavan, & P. Jawahar (Eds.), *Computational Techniques for Dental Image Analysis* (pp. 238–275). IGI Global. doi:10.4018/978-1-5225-6243-6.ch011

Ramírez-Rosas, M. B., Perales-Torres, A. L., & Santiago-Adame, R. (2020). The Therapeutic Potential of Ethnobotanical Plants in the Treatment of Different Diseases. In Ethnomedicinal Plant Use and Practice in Traditional Medicine (pp. 105-130). IGI Global. doi:10.4018/978-1-7998-1320-0.ch006

Rani, M., Chongtham, R., & Singh, A. (2020). Anti-Ulcer Activities of Medicinal Plants and Natural Products. In A. Singh, P. Singh, & N. Bithel (Eds.), *Advanced Pharmacological Uses of Medicinal Plants and Natural Products* (pp. 114–127). IGI Global. doi:10.4018/978-1-7998-2094-9.ch006

Rao, S. K. (2021). An Insight on Polycystic Ovary Syndrome (PCOS) and Use of Herbal Medicines as Alternative Treatment. In A. Hussain & S. Behl (Eds.), *Treating Endocrine and Metabolic Disorders With Herbal Medicines* (pp. 125–163). IGI Global. doi:10.4018/978-1-7998-4808-0.ch006

Rather, G. A., Chakravorty, A., Bhat, B. A., Malik, I. M., Mir, F. H., Sana, S. S., Raghavan, V., Nanda, A., & Choudhury, M. (2021). Routes of Synthesis and Characterizations of Nanoparticles. In M. A. Bhat, I. A. Wani, & S. Ashraf (Eds.), *Applications of Nanomaterials in Agriculture, Food Science, and Medicine* (pp. 288–309). IGI Global. doi:10.4018/978-1-7998-5563-7.ch016

Roos, G., Cheshire, A., Nayar, S., Clarke, S., & Zhang, W. (2019). *Macroalgal Value Chain*. IGI Global. doi:10.4018/978-1-5225-5577-3.ch002

Roy, P. P., Singh, J., & Ray, S. (2018). Exploring QSAR of Some Antitubercular Agents. *International Journal of Quantitative Structure-Property Relationships, 3*(1), 25–42. doi:10.4018/IJQSPR.2018010102

Ryumkina, I., & Ryumkin, S. (2020). World Market of Organic Food Products. In V. Erokhin & T. Gao (Eds.), *Handbook of Research on Globalized Agricultural Trade and New Challenges for Food Security* (pp. 63–86). IGI Global. doi:10.4018/978-1-7998-1042-1.ch004

Sadegh, H., Ali, G. A., Nia, H. J., & Mahmoodi, Z. (2019). Nanomaterial Surface Modifications for Enhancement of the Pollutant Adsorption From Wastewater. In R. Nazir (Ed.), *Nanotechnology Applications in Environmental Engineering* (pp. 143–170). IGI Global. doi:10.4018/978-1-5225-5745-6.ch007

Sadiku, E. R., Agboola, O., Mochane, M. J., Fasiku, V. O., Owonubi, S. J., Ibrahim, I. D., Abbavaram, B. R., Kupolati, W. K., Jayaramudu, T., Uwa, C. A., Daramola, O. O., Iheaturu, N. C., Nkuna, C., Sanni, S. E., Biotidara, O. F., Eze, A. A., Varaprasad, K., Adeyeye, O. A., Selatile, K. M., & Ndamase, A. S. (2019). The Use of Polymer Nanocomposites in the Aerospace and the Military/Defence Industries. In N. Ramdani (Ed.), *Polymer Nanocomposites for Advanced Engineering and Military Applications* (pp. 316–349). IGI Global. doi:10.4018/978-1-5225-7838-3.ch011

Sakthivel, S., Srivastava, M., Barman, M. K., Ravichandiran, N., Martins, S., Singh, M., Chand, K., Sonkar, S. C., & Bhukya, P. (2021). Cancer Stem Cells and Advanced Novel Technologies in Oncotherapy. In S. Kumar, M. A. Rizvi, & S. Verma (Eds.), *Handbook of Research on Advancements in Cancer Therapeutics* (pp. 486–513). IGI Global. doi:10.4018/978-1-7998-6530-8.ch017

Sansi, M. S., Iram, D., Narayan, K. S., Kumar, S., Prakash, O., & Misra, D. (2020). Antidiabetic Activity (Anti-Hyperglycemic Activity, Anti-Hyperlipidemic Activity)/ Agents From Medicinal Plants. In A. Singh, P. Singh, & N. Bithel (Eds.), *Advanced Pharmacological Uses of Medicinal Plants and Natural Products* (pp. 25–48). IGI Global. doi:10.4018/978-1-7998-2094-9.ch002

Sarilo, C. (2021). Healing Conflict With Grigri. In R. Nicholson (Ed.), *Natural Healing as Conflict Resolution* (pp. 21–41). IGI Global. doi:10.4018/978-1-7998-3665-0.ch002

Satpathy, R. (2020). Therapeutic Importance and Application of Boswellic Acid From the Plant Boswellia serrata. In A. Singh, P. Singh, & N. Bithel (Eds.), *Advanced Pharmacological Uses of Medicinal Plants and Natural Products* (pp. 302–315). IGI Global. doi:10.4018/978-1-7998-2094-9.ch016

Satpathy, R. (2021). Application of Some Medicinal Plants and Their Constituents in the Treatment of Diabetes Mellitus. In A. Hussain & S. Behl (Eds.), *Treating Endocrine and Metabolic Disorders With Herbal Medicines* (pp. 32–47). IGI Global. doi:10.4018/978-1-7998-4808-0.ch002

Sen, K. C. (2020). The Role of Crowdsourcing in the Healthcare Industry. In N. Wickramasinghe (Ed.), *Handbook of Research on Optimizing Healthcare Management Techniques* (pp. 191–201). IGI Global. doi:10.4018/978-1-7998-1371-2.ch013

Sertyesilisik, B. (2018). Global Sustainability Leadership. In A. H. Campbell (Ed.), *Global Leadership Initiatives for Conflict Resolution and Peacebuilding* (pp. 48–66). IGI Global. doi:10.4018/978-1-5225-4993-2.ch003

Shahrin, R., & Jamil, R. B. (2018). Consumers Consciousness Towards Environmental Aesthetics in Using Nutricosmetics Products. In F. Quoquab, R. Thurasamy, & J. Mohammad (Eds.), *Driving Green Consumerism Through Strategic Sustainability Marketing* (pp. 23–36). IGI Global. doi:10.4018/978-1-5225-2912-5.ch002

Shaikh, V. A., & Boubekri, N. (2020). Using Vegetable-Oil-Based Sustainable Metal Working Fluids to Promote Green Manufacturing. *International Journal of Manufacturing, Materials, and Mechanical Engineering, 10*(1), 1–19. doi:10.4018/IJMMME.2020010101

Shamim, A., Ahmad, S., Khitab, A., Anwar, W., Khushnood, R. A., & Usman, M. (2018). Applications of Nano Technology in Civil Engineering. *International Journal of Strategic Engineering, 1*(1), 48–64. doi:10.4018/IJoSE.2018010104

Sherwani, S. I., & Bates, B. R. (2021). Role of Wearable Technology and Fitness Apps in Obesity and Diabetes. In D. Sen & R. Ahmed (Eds.), *Privacy Concerns Surrounding Personal Information Sharing on Health and Fitness Mobile Apps* (pp. 31–59). IGI Global. doi:10.4018/978-1-7998-3487-8.ch002

Sim, M. S., You, K. Y., Esa, F., & Chan, Y. L. (2021). Nanostructured Electromagnetic Metamaterials for Sensing Applications. In M. A. Bhat, I. A. Wani, & S. Ashraf (Eds.), *Applications of Nanomaterials in Agriculture, Food Science, and Medicine* (pp. 141–164). IGI Global. doi:10.4018/978-1-7998-5563-7.ch009

Sima, V., & Gheorghe, I. G. (2018). Trends in the Evolution of Organic Food Market in Romania. *International Journal of Sustainable Economies Management, 7*(2), 45–55. doi:10.4018/IJSEM.2018040105

Simonovic, Z., & Ćurčić, N. V. (2020). Problems of Marketing Rural Tourism in Serbia. In A. Jean Vasile, J. Subic, A. Grubor, & D. Privitera (Eds.), *Handbook of Research on Agricultural Policy, Rural Development, and Entrepreneurship in Contemporary Economies* (pp. 296–313). IGI Global. doi:10.4018/978-1-5225-9837-4.ch015

Singh, A. (2020). Pharmacological Properties of Curcumin. In A. Singh, P. Singh, & N. Bithel (Eds.), *Advanced Pharmacological Uses of Medicinal Plants and Natural Products* (pp. 235–248). IGI Global. doi:10.4018/978-1-7998-2094-9.ch012

Singh, P., Mishra, N., & Gupta, E. (2020). Phytochemistry and Ethanopharmacology of Illicium verum (Staranise). In N. Mishra (Ed.), *Ethnopharmacological Investigation of Indian Spices* (pp. 93–105). IGI Global. doi:10.4018/978-1-7998-2524-1.ch007

Singh, R., & Dutta, S. (2020). Current Approaches of Nanotechnology for Potential Drinking Water Purification. In G. Duca & A. Vaseashta (Eds.), *Handbook of Research on Emerging Developments and Environmental Impacts of Ecological Chemistry* (pp. 307–324). IGI Global. doi:10.4018/978-1-7998-1241-8.ch014

Singh, S., Semwal, B. C., & Murti, Y. (2021). Role of Herbal Supplements in the Treatment of Obesity and Diabetes. In A. Hussain & S. Behl (Eds.), *Treating Endocrine and Metabolic Disorders With Herbal Medicines* (pp. 74–103). IGI Global. doi:10.4018/978-1-7998-4808-0.ch004

Sinha, S. (2018). Trends and Challenges in Enzymatic Bioengineering of Natural Products to Industrially Valuable Products. In S. L. Bharati & P. K. Chaurasia (Eds.), *Research Advancements in Pharmaceutical, Nutritional, and Industrial Enzymology* (pp. 351–374). IGI Global. doi:10.4018/978-1-5225-5237-6.ch016

Sivaprakasam, P., Anandasadagopan, S., Alagumuthu, T., & Pandurangan, A. K. (2020). Current Update on Natural Agents Against Triple Negative Breast Cancer. In A. Singh, P. Singh, & N. Bithel (Eds.), *Advanced Pharmacological Uses of Medicinal Plants and Natural Products* (pp. 91–113). IGI Global. doi:10.4018/978-1-7998-2094-9.ch005

Sizochenko, N., Kar, S., Syzochenko, M., & Leszczynski, J. (2019). Predicting Thermal Conductivity Enhancement of Al2O3/Water and CuO/Water Nanofluids Using Quantitative Structure-Property Relationship Approach. *International Journal of Quantitative Structure-Property Relationships, 4*(1), 18–27. doi:10.4018/IJQSPR.2019010102

Sofi, A. H., Akhoon, S. A., Mir, J. F., & Rather, M. U. (2021). Magnesium Oxide (MgO). In M. A. Bhat, I. A. Wani, & S. Ashraf (Eds.), *Applications of Nanomaterials in Agriculture, Food Science, and Medicine* (pp. 98–105). IGI Global. doi:10.4018/978-1-7998-5563-7.ch005

Som, S., Dutta, S., Kumar, V., & Swart, H. C. (2018). Swift Heavy Ion Synthesis and Modifications of Nanophosphors for Dosimetric Application. In R. Tiwari, V. Dubey, & S. J. Dhoble (Eds.), *Emerging Synthesis Techniques for Luminescent Materials* (pp. 1–25). IGI Global. doi:10.4018/978-1-5225-5170-6.ch001

Soriano, M. L. (2018). Moving into Nanotechnology Roles to Mimic and Boost Enzyme Activity. In S. L. Bharati & P. K. Chaurasia (Eds.), *Research Advancements in Pharmaceutical, Nutritional, and Industrial Enzymology* (pp. 421–440). IGI Global. doi:10.4018/978-1-5225-5237-6.ch019

Srivani, M., Mala, T., & Murugappan, A. (2020). A Survey on Precision Treatment for Humans Using Cognitive Machine Learning Techniques. In A. Solanki, S. Kumar, & A. Nayyar (Eds.), *Handbook of Research on Emerging Trends and Applications of Machine Learning* (pp. 79–106). IGI Global. doi:10.4018/978-1-5225-9643-1.ch005

Szewczuk, N. A., Duchowicz, P. R., & Pomilio, A. B. (2020). QSAR Analysis for the Inhibition of the Mutagenic Activity by Anthocyanin Derivatives. *International Journal of Quantitative Structure-Property Relationships, 5*(4), 69–82. doi:10.4018/IJQSPR.20201001.oa1

Tahir, L. (2019). Green Approaches to Environmental Sustainability. In R. Nazir (Ed.), *Nanotechnology Applications in Environmental Engineering* (pp. 81–101). IGI Global. doi:10.4018/978-1-5225-5745-6.ch004

Taiwo, V. (2020). Indigenous Communication and Traditional Health Practices Among the Yoruba During Pregnancy, Child Delivery, and Baby Caring. In K. O. Oyesomi & A. Salawu (Eds.), *Emerging Trends in Indigenous Language Media, Communication, Gender, and Health* (pp. 45–59). IGI Global. doi:10.4018/978-1-7998-2091-8.ch003

Tekiner, İ. H., Al-Baarri, A. N., & Bogueva, D. (2018). Genius, Creativity and (Not) Eating Meat. In D. Bogueva, D. Marinova, & T. Raphaely (Eds.), *Handbook of Research on Social Marketing and Its Influence on Animal Origin Food Product Consumption* (pp. 187–202). IGI Global. doi:10.4018/978-1-5225-4757-0.ch013

Thanh Tung, B. (2021). Natural Product Compounds for Breast Cancer Treatment. In S. Kumar, M. A. Rizvi, & S. Verma (Eds.), *Handbook of Research on Advancements in Cancer Therapeutics* (pp. 606–626). IGI Global. doi:10.4018/978-1-7998-6530-8.ch023

Torrens, F., & Castellano, G. (2018). QSPR Prediction of Retention Times of Methylxanthines and Cotinine by Bioplastic Evolution. *International Journal of Quantitative Structure-Property Relationships*, *3*(1), 74–87. doi:10.4018/IJQSPR.2018010104

Tripathi, S., Sanjeevi, R., Anuradha, J., Chauhan, D. S., & Rathoure, A. K. (2018). Nano-Bioremediation. In A. K. Rathoure (Ed.), *Biostimulation Remediation Technologies for Groundwater Contaminants* (pp. 202–219). IGI Global. doi:10.4018/978-1-5225-4162-2.ch012

Türkoğlu, E. A. (2021). Molecularly-Imprinted Nanomaterial-Based Surface Plasmon Resonance Biosensors in Molecular Diagnosis. In M. A. Bhat, I. A. Wani, & S. Ashraf (Eds.), *Applications of Nanomaterials in Agriculture, Food Science, and Medicine* (pp. 1–28). IGI Global. doi:10.4018/978-1-7998-5563-7.ch001

Ugemuge, N. S., Gayner, C., Natrajan, V., & Dhoble, S. J. (2018). Perspectives of Hydrothermal Synthesis of Fluorides for Luminescence Applications. In R. Tiwari, V. Dubey, & S. J. Dhoble (Eds.), *Emerging Synthesis Techniques for Luminescent Materials* (pp. 277–303). IGI Global. doi:10.4018/978-1-5225-5170-6.ch008

Vargas-Bernal, R. (2021). Advances in Electromagnetic Environmental Shielding for Aeronautics and Space Applications. In C. D. Nikolopoulos (Ed.), *Recent Trends on Electromagnetic Environmental Effects for Aeronautics and Space Applications* (pp. 80–96). IGI Global. doi:10.4018/978-1-7998-4879-0.ch003

Vargas-Bernal, R. (2021). The Role of Two-Dimensional Materials in Electromagnetic Interference Shielding. In Encyclopedia of Information Science and Technology, Fifth Edition (pp. 1254-1270). IGI Global. doi:10.4018/978-1-7998-3479-3.ch086

Wani, Z. A., & Pant, S. (2020). Plant Species Used by Locals as Ethno-Medicine in Gulmarg Wildlife Sanctuary, Kashmir Himalaya (India). In Ethnomedicinal Plant Use and Practice in Traditional Medicine (pp. 285-300). IGI Global. http://doi:10.4018/978-1-7998-1320-0.ch015

Yadav, V. K., Choudhary, N., Khan, S. H., Malik, P., Inwati, G. K., Suriyaprabha, R., & Ravi, R. K. (2020). Synthesis and Characterisation of Nano-Biosorbents and Their Applications for Waste Water Treatment. In G. Duca & A. Vaseashta (Eds.), *Handbook of Research on Emerging Developments and Environmental Impacts of Ecological Chemistry* (pp. 252–290). IGI Global. doi:10.4018/978-1-7998-1241-8.ch012

Yaseen, T., & Yaseen, A. (2019). Wastewater Treatment in Removal of Heavy Metals. In R. Nazir (Ed.), *Nanotechnology Applications in Environmental Engineering* (pp. 220–240). IGI Global. doi:10.4018/978-1-5225-5745-6.ch010

Yasin, D., Zafaryab, M., Fakhri, K. U., Husain, S., Afzal, B., Sami, N., Hemlata, H., Rizvi, M. M., & Fatma, T. (2021). Apoptotic Pathway. In S. Kumar, M. A. Rizvi, & S. Verma (Eds.), *Handbook of Research on Advancements in Cancer Therapeutics* (pp. 290–311). IGI Global. doi:10.4018/978-1-7998-6530-8.ch009

Yue, T., & Zou, Y. (2019). Online Teaching System of Sports Training Based on Mobile Multimedia Communication Platform. *International Journal of Mobile Computing and Multimedia Communications*, *10*(1), 32–48. doi:10.4018/IJMCMC.2019010103

# About the Contributors

**Ankur Kaul** is a pharmacist by profession with a blend of expertise in basic as well as radio-pharmaceutical research. She has acquired multi-disciplinary qualification with her Bachelor's degree in Pharmacy from Mumbai University, Master's Degree in Bioinformatics and Ph.D. (Life Science). After a brief stint of working at pharmaceutical companies like Novartis India ltd. and FDC India ltd, she joined INMAS, DRDO in year 2006. She is AERB-certified Radiation Safety Officer (RSO) for biomedical research and is faculty- member for Diploma in Radiation Medicine course (affiliated to University of Delhi) for PG MBBS students. She is also nominated as General Secretary for INMAS-Institutional Human Ethics Committee. Her research interests are radiopharmaceuticals-development for diagnostic application which includes radiolabeling of drugs and nano-particles and their preclinical evaluation. She has successfully developed and translated a 'diagnostic kit formulation' from mouse to man for application in early detection of infection. Her research contribution is recognized over 30 international peer-reviewed journals and is life-member of professional bodies like Indian Pharmaceutical Association (IPA) and Society of Nuclear Medicine, India (SNMI) and Delhi Pharmacy Council (DPC) and Indian Society of Nanomedicine (ISNM).

* * *

**Geeta Aggarwal** is M. Pharm., Ph.D. in Pharmaceutics with experience of more than 15 years in academics, administrative functions and research in areas of controlled and novel drug delivery systems. Presently she is working as Assistant Registrar (Academics) in Delhi Pharmaceutical Sciences and Research University. Along with teaching and research, she is Secretary of DPSRU Incubation centre (DIIF), Vice-President SPER- Central Women Forum, Vice President APTI Delhi State Branch and Head-Training & Placement, DPSRU. She has worked as Associate Professor and Departmental head (2010-2016), Convenor Academic Planning and Monitoring Committee (2013-2016), Research Coordinator (2011-2016), Coordinator Finishing School (2014-2016), Placement Coordinator (2008-2016) for University

School of Pharmaceutical Sciences Dr. Aggarwal was awarded for her outstanding services by DPSRU in 2017 & 2018 and as best faculty by Rayat Bahra University. She has been awarded for various paper presentations in international and national conferences. She has supervised 38 M. Pharm. students and 2 Ph.D. students. Presently she is supervising 5 Ph.D. and M. Pharm. students. Dr. Aggarwal has more than 100 publications in National and International Journals of repute with H-index of 19, i10-index of 32 and more than 1000 citations. She is author of various books and book chapters on Physical Pharmacy, Industrial training and Polymer gels. She is reviewer of various journals of repute. She has attended various conferences/ seminars/workshops/FDPs as organizer/co-ordinator/resource person/participant. Dr. Aggarwal has filed international and national patents for her research work and in the process of technology transfer. She is working on 6 Research projects by Industries. Dr. Aggarwal and her team has been selected for funding of 2 start-ups. She has keen interest in Entrepreneurship Development activities and research related to targeted drug delivery particularly to colon and lungs, nanovesicular drug delivery for topical and Transdermal application, value added natural polymers and bioavailability enhancement of poorly soluble drugs.

**Umesh Gupta** is currently working as Assistant Professor at Department of Pharmacy, Central University of Rajasthan, India, did Ph.D. in Pharmaceutical Sciences from Dr. H. S. Gour University, India under the mentorship of Prof. NK Jain. He has recently awarded "First Chancellor's Award" for Best Faculty at Central university of Rajasthan, Ajmer, India. He was also awarded "DAAD Research Stays for Academics and Scientists" at Leibniz-institut fur Polymer forschung Dresden, Germany. He has the past-experience of working as Research Scientist at Ranbaxy Research Laboratories, India and Post-Doctoral Research Associate at South Dakota State University, USA. He has more than ten years of research and academic experience. The area of research is mainly focused on dendrimer mediated drug delivery, solubilization and targeting. He has more than 59 publications, one edited book and ten book chapters to his credit with total citations of more than 3329 (h index of 26 as per Google Scholar).

**Mümtaz Güran** works as an Associate Professor for the Department of Medical Microbiology of the Eastern Mediterranean University, Faculty of Medicine. He is also a co-director of the COVID-19 Molecular Genetics Research Laboratory in Famagusta, Cyprus. He serves as an Associate Editor for "Cyprus Journal of Medical Sciences". He has published more than 20 SCI indexed publications. Main research interests; AST, Bacterial resistance, Antimicrobial development, Infectious disease epidemiology, Malassezia pathogenesis, Covid-19

**Radhika R. Jaswal** completed her Bachelor's, Master's and Doctoral studies from the Department of Biophysics, Panjab University, Chandigarh, India. During her Ph.D (2017), she computationally studied and designed a novel antibiofouling coating through the N-methylation approach. She is currently teaching in her parent department and is actively involved in the in-silico studies of proteins using quantum mechanics and molecular mechanics based softwares. Her main area of interest is studying the structural preferences of unusual amino acids and exploring their potential in therapeutics.

**Shubhi Joshi** received her BSc from Panjab University (India) in 2013. Then she moved to Guru Nanak Dev University (India), where she successfully completed her MSc in 2015. She is currently working towards a Ph.D. from Panjab University (India). Her research interests include interaction and characterization studies of novel peptide bioconjugates and their application as a biosensor.

**Tanzeer Kaur** is presently working in the department of Biophysics, Panjab University, Chandigarh. She has done her Ph.D in 2009 in biotechnology, where she isolated and characterized novel proteins from plant having potential to treat kidney stone diseases and worked on an Indo-European postdoctoral fellowship at NIPER, SAS nagar to study development of novel drugs for the treatment of tuberculosis. She is actively engaged in research and handling research grants from DST, SERB and ICMR. So far she has published 30 research articles in international journals and received DST PURSE grant award in 2018 for her publications.

**Anil K. Mishra** was born in 1963 in India. He obtained his M.S. degree (Chemistry) in 1984 from Gorakhpur University, Gorakhpur and his Ph.D. (Chemistry) in 1988 from Banaras Hindu University, Varanasi, India. In 1989, he moved to Université de Bourgogne, France where he worked with Prof. Guilard on macrocycle chemistry. In 1992, he joined Prof. Meares at the University of California, Davis,USA. He also served as a Scientist at INSERM, Nantes, France in the team of Prof. Chatal. In 1997, he was appointed as a Senior Scientist at INMAS, Delhi, India. His research group focuses on the development of targeted radiopharmaceuticals using different classes of molecules and applying novel chemistries and approaches.

**Manju Nagpal** is working as Asso. Prof. and coordinator handling M.Pharm program. 65 publications and 4 patents to her credit. Working in novel drug delivery systems such as binary dispersions, FDT, superporous hydrogels, interpenetrating networks, nanosponges for topical delivery, modified gums for solubility enhancement, etc.

**Simran Preet** was born in October 1980 in Chandigarh, India. She completed her BSc. Hons. (School), MSc. Hons. (School) and her doctorate in Microbiology from Department of Microbiology, Panjab University, Chandigarh, India. She is currently working as an Assistant Professor in department of Biophysics, Panjab University, Chandigarh, India since last eight years. She was awarded with Bhajan Kaur Arora medal (young scientist) in 2017 by Panjab University and is currently guiding four Ph. D students and has two ongoing ambitious research projects funded by DST-SERB in her lab.

**Sarita Rani** is currently working as Ph. D scholar at Department of Pharmacy, Central University of Rajasthan, India. She has completed her masters degree in Pharmaceutical Sciences from Central University of Rajasthan, India under the mentorship of Dr. Umesh Gupta. The area of her research is mainly focused on polymeric drug delivery and targeting through nano-carriers. She has contributed 9 publications, including research and review articles in reputed journals and one book chapter to her name.

**Avneet Saini** attended the Department of Biophysics, Panjab University, Chandigarh, India, where she gained a BSc, MSc and PhD— all in Biophysics. She received university medals for coming first throughout her bachelors and masters courses. Her PhD, completed in 2011, was entitled "Peptoids as Antimicrobial and Collagen Mimetics: A Computational Study", and involved conformational analysis of peptoids using quantum mechanics and molecular dynamics software. She started her teaching career in 2007 as an Assistant Professor at Goswami Ganesh Dutta Sanatan Dharma College, Chandigarh. In 2010 she rejoined Panjab University as Assistant Professor, where she is currently involved in teaching and research. In her post-doctoral research in London, she analyzed and characterized peptides derived from the family of intermediate filament proteins that have an ability to bind unpolymerized tubulin. This binding alters tubulin polymerization making it possible to disrupt intracellular transport and/or cell mobility and proliferation. She has three PhD students enrolled under her. In Dec 2017, she was awarded a research grant worth 60,00,000 INR by the Defense Research and Development Organization, India. Research Interests: • Drug Design and Peptide Design: Design and Structural Characterization using Computational Biophysics, Molecular Biophysics and Biophysical Chemistry Techniques. • Nanoparticle formulation, functionalization of peptides with different nanoparticles, conjugation studies involving peptides with graphene and its derivatives.

**Pranav J. Shah** is currently a Professor of Pharmaceutics and Pharmaceutical Technology at Maliba Pharmacy College, Bardoli, Gujarat, India. He has around 17 years of experience which includes amalgamation of industry, teaching and research experience. He has completed his B. Pharm, M. Pharm and Ph. D from The Maharaja Sayajirao University of Baroda, Vadodara, Gujarat. He had received UGC scholarships during his M. Pharm and Ph.D. He has worked for 6 years in F&D of various reputed organisations such as Macleods, Glenmark and Ipca. He was actively involved in formulation development of Para-III and Para-IV drugs. He has over 46 refereed publications, 1 book and 6 book chapters. His areas of research include development of nanocarriers and solubility/bioavailability enhancement. He has been awarded International Travel Grant by CSIR in 2017 and ICMR in 2015. He is also a recipient of Research Promotion Scheme by Uka Tarsadia University in the years 2019 and 2017 and Minor Research Project by Gujarat Council on Science and Technology in 2015. He is a life member of APTI and IPA and presently the Executive Committee member of CRS- IC. He is a member of Board of Studies of Faculty of Pharmacy and Academic Council of Uka Tarsadia University. He is a referee for M. Pharm & Ph. D at several universities and is an invited reviewer for many journals of repute. Presently, he is also the Coordinator of Training and Placement Cell of Uka Tarsadia University. He has successfully organised several national and international conferences, workshops and seminars at his institute.

**Pratibha Sharma** is a PhD student in the Department of Biophysics at Panjab University, Chandigarh. Her doctoral research focuses on designing a novel peptide analog with enhanced antimicrobial activity and low cytotoxicity based on the physiochemical properties like hydrophobicity, cationicity and amphipathicity and further in-silico and in-vitro structural characterization of the peptide analogs and their biological evaluation. She earned a master's degree in Biochemistry (Hons.) in the Department of Biochemistry at Panjab University, Chandigarh with a focus on acute lung injury. Her research is funded by UGC-BSR (University Grants Commission – Basic Scientific Research).

**Satyavir Singh** is a highly experienced faculty member working as Head of the Department, Department of Soil Science, Institute of Agricultural Sciences, Bundelkhand University Jhansi. He is actively engaged in the agricultural research work at Agricultural Farm of University Campus to carry out the DBT Sponsored research projects on traditional and herbal medicines to study their phytochemical potentials, growth, yield and crop production. He has guided 4 PhD Students and 12 M.Sc students under his supervision.

**Pavitra Solanki**, Research Associate working at Department of Pharmaceutics, School of Pharmaceutical Education and Research, Jamia Hamdard, New Delhi. He is actively engaged in research work on phytomedicines and synthetic compounds to determine the potential health benefits to treat ailments. He has published many research papers in indexed journals and participated in many conferences in India as well.

**Yasmin Sultana** is working as Prof. and Head of the Dept. of Pharmaceutics, SPER, Jamia Hamdard, New Delhi. She has more than 200 publications and is actively involved in research work.

**Abhishek Thakur** graduated with Bachelor's degree in Pharmaceutical sciences in 2017 and he is pursuing his master's degree in pharmacy with specialization in pharmaceutics. Currently, he is undergoing internship programme of six months at INMAS, DRDO, Delhi. His keen interest is in research and development, and he is working on formulation-development and evaluation of topical drug delivery systems.

# Index

## E



## I



## K



## M



## N



## P



## R



## S

# T



# W

www.ingramcontent.com/pod-product-compliance
Lightning Source LLC
LaVergne TN
LVHW060003031026
849161LV00025B/1351